ELEMENTS OF FUNCTIONAL ANALYSIS

L. A. Liusternik and V. J. Sobolev

ELEMENTS OF FUNCTIONAL ANALYSIS

Translated from the Russian by

ANTHONY E. LABARRE, JR., *University of Idaho*
HERBERT IZBICKI, *University of Vienna*
H. WARD CROWLEY, *University of Idaho*

FREDERICK UNGAR PUBLISHING COMPANY
NEW YORK

Second Printing, 1965

Printed in the United States of America

Library of Congress Catalog Card No. 60-53137

FROM THE PREFACE TO THE RUSSIAN EDITION

During the last decades, the methods of functional analysis were used a great deal in different branches of pure and applied mathematics—especially in works of Soviet mathematicians.[1] The fundamentals of functional analysis became an indispensable part of the mathematical training at the universities. As a first collective presentation, there was published a report by L. A. Liusternik in the first edition of the periodical "Advances of Mathematical Sciences" (in Russian). It was an introduction to a series of articles which were printed in the "Advances." Therefore it made no claim to completeness, not even for the parts of functional analysis dealt with therein. Based on this book, V. J. Sobolev built up lectures on functional analysis which he delivered several times at the State University of Voronezh.

These lectures deal with a series of questions which were not considered in the article mentioned above as, for instance, the principle of contraction mapping, the general form of linear functionals, spaces with bases, the theory of completely continuous operators, the spectral decomposition of bounded self adjoint operators, and so on. These lectures form the foundation for the present book. The last paragraph of the article, which deals with calculus of variations from the functional-analytic point of view, was expanded by L. A. Liusternik in the last chapter of the present book.

The first chapter is devoted to metric spaces. They are considered from another point of view, as, e.g., in the book "Introduction to the General Theory of Sets and Functions" (in Russian) by Aleksandrov or in Hausdorff's "Theory of Sets." In a certain sense, the first chapter is a completion of these books.

The two following chapters are devoted to normed linear spaces and linear operators in them.

The short fourth chapter serves as an introduction to the theory of completely continuous operators in spaces with basis. By this, all spaces important to functional analysis are covered.

The fifth chapter is an introduction to the spectral theory of self adjoint operators. This theory is fundamental in modern analysis and much has been written on it. Some of it is mentioned in the bibliography at the end of the book. Therefore, we restrict ourselves to an introduction and consider only the bounded self adjoint operators because it is possible here to become acquainted with the methods of spectral theory.

The last chapter is concerned with operations for abstract functions. Mainly, differentiation and its applications are considered.

The authors have taken care to illustrate the theory by applications to different

[1] And others. (Trans.)

branches of mathematics—Theory of Differential and Integral Equations, Methods of Approximation in Analysis, Theory of Almost Periodic Functions, Calculus of Variations, etc.

Several examples occupy whole paragraphs of the book. They are included to make clear the relations between Functional Analysis and other branches of mathematics.

The authors express their thanks to U. B. Germeier, A. I. Plessner and M. R. Shura-Bura for valuable remarks, and also B. G. Ashkenuz and D. A. Paikov for doing a large amount of work on the correction of proof sheets.

TABLE OF CONTENTS

Chapter III. LINEAR FUNCTIONALS

Chapter IV. COMPLETELY CONTINUOUS OPERATORS

Chapter V. ELEMENTS OF SPECTRAL THEORY OF SELF ADJOINT OPERATORS ON HILBERT SPACES

Chapter VI. SEVERAL TOPICS IN NON-LINEAR FUNCTIONAL ANALYSIS

APPENDIX

INTRODUCTION

Generalization of Basic Concepts from Analysis, Geometry and Algebra

At the beginning of this century, there was developed a new analytical discipline, so-called *Functional Analysis*.

The fundamental concepts and methods of functional analysis arose gradually in the oldest parts of mathematical analysis: in the Calculus of Variations [27], in the Theory of Differential Equations, in the Theory of Approximations of Functions, in practical analysis, and especially in the theory of integral equations [33]. The development of functional analysis made great progress, especially during the last decades. An important role in this development was played by Soviet, Polish, and Hungarian mathematicians.

It is the essence of functional analysis that the concepts and methods of elementary analysis and related branches of algebra and geometry are extended to more general objects. To a large degree, geometric and algebraic ideas are utilized. Such a generalization makes it possible to approach, from a unified point of view, questions which earlier were considered in an isolated manner in special branches of analysis. Furthermore, connections between quite different mathematical theories can be established and this enables us to discover new mathematical results. In order to convince ourselves of the truth of this assertion, it suffices to point out certain existence theorems for differential and integral equations, etc., which were found by methods of functional analysis during the last years [25], [8].

These generalizations became possible because in the course of the development, the basic concepts of analysis appeared to be very general. These concepts and methods often found analogies in algebra and geometry. Thus the solution of different problems of algebra and analysis is found by successive approximations. Moreover, the definitions of a functional, of an extremum of a functional, and the conditions for the existence of an extremum in the calculus of variations, are entirely analogous to the definitions of a function (of one or more variables), of an extremum of a function, and the conditions for the existence of such an extremum in the differential calculus.

The analogies between ordinary linear differential equations and linear difference equations, on one hand, and systems of linear algebraic equations on the other hand, are well known. These analogies became even more apparent in the theory of integral equations, which historically was developed later.

Paralleling analysis, geometry was also generalized in the 19th century. An impetus was given by the discovery of non-Euclidean geometry by Lobachevskii. The creation of n-dimensional geometry permits geometrical interpretation of functions of several variables. Simultaneously new relations between analysis and geometry were discovered. This connection between geometry and analysis requires broadening of the geometrical concepts. Here are some examples. The set of all solutions of an ordinary homogeneous linear differential equation of the nth order is isomorphic to the n-dimensional vector space. The geometric analog to the set of all solutions of a linear homogeneous partial differential equation is the vector space of infinitely many dimensions. A remarkable example of far reaching analogy between analysis and geometry is given by the development of functions with respect to elements of an orthogonal system. These systems resemble orthogonal systems of vectors in the Euclidean space. This analogy is emphasized by the terminology. The decomposition of a function into a Fourier series corresponds to the decomposition of a vector into its components, the completeness relation to the theorem of Pythagoras, the theorem of Parseval-Steklov, etc. The geometric representation of an infinite orthogonal system of functions requires the Euclidean space of infinitely many dimensions.

Not only the generalization but also the geometrization of the basic concepts and methods of classical analysis is typical of functional analysis. Functions with certain properties are considered as elements of "function spaces." As we mentioned earlier, such considerations require extensions of the concepts—as, for instance, infinite dimensional space, metric space, etc. This leads then to the "abstract space" which embraces the geometric as well as the functional spaces.

The abstract space enables us to deal geometrically with questions of analysis. Such a geometric representation of analytic theories is widespread in mathematical literature, also in physics and mechanics. Many conjectures were suggested by analogies with n-dimensional geometry. The proofs of several theorems were obtained in a geometrical manner. Thus a new geometric method was achieved in analysis.

Simultaneously with the generalization of geometric concepts, the algebraic concepts were generalized in the 19th century.

Algebraic operations on numbers were carried over to objects of a more general nature (matrices, operators, etc.). Concepts such as group, ring, field, linear manifold, etc. were developed.

In connection with the algebraic concepts in analysis, algebraic mappings involving a limit process began to be considered. Thus a special discipline was created during the last century: topological algebra (Haar, Kolmogorov, Pontriagin, and others), which investigates from a unified point of view algebraic objects with a limit defined for them (topological groups, rings, fields, etc.).

Concepts of analysis originate from algebraic operations by passage to the limit. Elementary chapters of algebra play the same role for ordinary analysis as the corresponding generalizations for functional analysis.

Thus, linear algebra corresponds to the theory of linear operators, to which a large part of this book is devoted.

An essential process of analysis—the approximation of nonlinear objects by linear ones—is also carried over to functional analysis (see Ch. VI). The transition from "polynomials on rings" (rings of matrices, operators, etc.) to arbitrary functions of these arguments corresponds to the transition from polynomials with numerical arguments to arbitrary functions. Important disciplines are based on this, as, for instance, calculus of matrices, calculus of operators, and the spectral theory of linear operators (see Ch. V).

During the early development of functional analysis, generalized linear algebra played a dominating role. Lately, the methods and concepts from the different branches of algebraic topology [6] were introduced in functional analysis due to the work of Soviet mathematicians (I. M. Gel'fand, D. A. Karkov, M. A. Neumark, and others). These methods enable us to solve many difficult problems of analysis.

You will find a geometric and algebraic introduction to functional analysis in §§ 2, 12, 13, 15, 17.

CHAPTER I

METRIC SPACES

§1. FUNCTIONS (OPERATORS). THE LIMIT PROCESS

ONE of the fundamental ideas of mathematical analysis is the notion of a function. We define a function in the following way: Let X and Y be sets of real numbers. If to every number x from X there is associated a unique number y from Y, then we say that a *single-valued function* $y=f(x)$ whose *range* lies in the set Y is defined on the set X. The set X is called the *domain of definition* of the function.

As we easily see, it is not essential to the concept of a functional dependence that X and Y be sets of real numbers. If we understand by X and Y arbitrary sets, we arrive at a more general concept of a function. There are examples from different branches of analysis.

Examples. 1. Let $y=f(x_1, x_2, \ldots, x_n)$ be a real function of n real variables. Then X is a set of n-tuples of real numbers, Y is a set of real numbers.

2. Let $\vec{Y}=f(x)$ be a vector function which associates with every real number x an n-dimensional vector. Here X is the set of real numbers, Y a set of n-dimensional vectors.

3. In the calculus of variations we consider the functionals

$$I(\gamma)=\int_a^b F(x,y,y')\,dx,$$

where γ is a curve given by an equation $y=f(x)$, which (i) belongs to the class C_1 of those functions which have a continuous derivative and (ii) passes through the two given points $A(a,y_a)$ and $B(b,y_b)$. In this case, X is the set of those curves with the given properties, and Y is a set of real numbers.

4. In the theory of integral equations we consider expressions of the form

$$y(s)=\int_a^b K(s,t)\,x(t)\,dt.$$

We assume that the kernel $K(s,t)$ is defined and continuous on the square $a\leqq s,t\leqq b$. We can consider the above equation as a prescription by which there is associated with every function $x(s)$ continuous on the interval $[a,b]$ another function continuous on the same interval. Here X and Y are sets of continuous functions.

We now introduce a general definition of a function.

Let there be given two arbitrary sets X and Y and a rule by which there is associated with every $x\in X$ a uniquely determined element $y\in Y$. Then we say that there is determined an *abstract function* (or an *operator*) $y=f(x)$ or $y=fx$, which is defined on the set X and whose range lies in the set Y.[1] If, in particular, the values of the operator are real numbers, then the operator is called a *functional*.[2]

---

[1] We agree to say of a property that it is fulfilled "on a set" or "in a set," if it holds for all elements of this set or perhaps not for all elements of the set, respectively.

[2] An operator defined on X with range in Y is also called a *mapping* of the set X into the set Y.

As to the properties of such general operators almost nothing can be said. Let us now introduce some further hypotheses.

THE LIMIT PROCESS AND ABSTRACT SPACES. Together with the function concept there appears another important concept in analysis, namely that of a limit and in connection with it, the idea of continuity. A set in which there is defined a limit of a sequence is called an *abstract space*. Abstract spaces, whose elements are functions or sequences of numbers, are called *function spaces*. The study of certain classes of operators which are defined in function spaces is the essential content of functional analysis.

§2. METRIC AND TOPOLOGICAL SPACES

IN analysis there are several concepts of convergence. In certain cases we introduce for one and the same sequence, in connection with different problems, different concepts of limit. Above all, we meet with the concept of a limit if we are dealing with sequences of real numbers. This concept admits of immediate generalization to sequences of complex numbers and to sequences of n-dimensional vectors. For a sequence of functions we know of many kinds of convergence: ordinary (not uniform), uniform, almost everywhere, and so forth. These concepts are well known. Furthermore, P. L. Chebyshev has considered, in connection with the interpolation process in the method of least squares, convergence in the mean with weight $\rho(t)$ of a sequence of functions [52]. A sequence of functions $x_n(t)$ defined on an interval $[a, b]$ *converges in the mean with weight* $\rho(t)$ to a function $x(t)$ defined on the same interval if

$$\int_a^b [x_n(t) - x(t)]^2 \rho(t)dt \to 0$$

as $n \to \infty$. Here $\rho(t)$ is an arbitrary positive function on the interval. The Hungarian mathematician F. Riesz introduced in his research on systems of integrable functions the concept of *convergence in the mean of the pth power*. Here p is a real number which satisfies the condition $p \geqq 1$ [48].

All these concepts of convergence of a sequence of elements x_n (which can be numbers, vectors, or functions) to an element x mean an 'approaching' of x_n to x, i.e., a 'decreasing to zero of the distance' between these elements with increasing n. According to our definition of the distance between the elements x_n and x, we obtain different definitions of the limit. It is therefore necessary to give a general definition of the distance between elements which includes the special cases considered, and then to introduce, with the help of this distance defined in the set, the idea of a limiting process. This set then becomes an abstract space.

DEFINITION. A set X is called a *metric space*, if to every pair of elements x, y of this set there corresponds a non-negative real number $\rho(x, y)$ for which the following conditions hold:

1) $\rho(x, y) = 0$ if and only if $x = y$ (identity).
2) $\rho(x, y) = \rho(y, x)$ (symmetry).
3) $\rho(x, y) + \rho(y, z) \geqq \rho(x, z)$ (triangle inequality).

The number $\rho(x, y)$ is called the distance between the elements x and y, and the above three conditions are called the *metric axioms*. Obviously, these metric axioms express the fundamental properties of the distance between the points of three-dimensional Euclidean space [51].

LIMIT. An element x of a metric space M is called *limit* of a sequence of elements $x_1, x_2, \ldots, x_n, \ldots$ of M, if $\rho(x_n, x) \to 0$ as $n \to \infty$. In this case we write $x_n \to x$ or $\lim\limits_n x_n = x$. The elements of a metric space are also called points. Finally we note that every set Y, which lies in a metric space X, and for which we take the same definition for the distance of two elements, is also a metric space.

For convergent sequences of points of a metric space we have several general theorems.

THEOREM 1. *If a sequence $\{x_n\}$ of points of a metric space X converges to a point $x \in X$, then every subsequence $\{x_{n_k}\}$ of the sequence $\{x_n\}$ converges to the same point.*

The proof is almost trivial. For, if $\rho(x_n, x) < \varepsilon$ for $n \geqq n_0(\varepsilon)$, then $\rho(x_{n_k}, x) < \varepsilon$ for $n_k \geqq n_0(\varepsilon)$.

THEOREM 2. *A sequence $\{x_n\}$ of points of a metric space can converge to at most one limit.*

Let $x_n \to x$ and $x_n \to y$. Then

$$\rho(x, y) \leqq \rho(x, x_n) + \rho(x_n, y) < \varepsilon$$

for sufficiently large n. Since x and y are fixed and ε is an arbitrary positive number, the inequality holds only if $\rho(x, y) = 0$, i.e., $x = y$.

THEOREM 3. *If a sequence $\{x_n\}$ of points of X converges to a point $x \in X$, then it is bounded in the following sense: for every fixed point θ of the space, the set of numbers $\rho(x_n, \theta)$ is bounded.*

According to the triangle inequality, we have for every n, $\rho(x_n, \theta) \leqq \rho(x_n, x) + \rho(x, \theta) \leqq a + \rho(x, \theta) = K$, since $\{\rho(x_n, x)\}$ as a convergent number sequence is bounded, and hence the numbers $\rho(x_n, x)$ are not greater than a constant a.

We call the set of all points x of the space X which satisfy the inequality $\rho(x, a) < r$, $(\rho(x, a) \leqq r)$ a *sphere* (*closed sphere*), with center a and radius r. We designate this sphere by $S(a, r)$ $(\bar{S}(a, r))$.

Further, by a *neighborhood* of a point x we understand any sphere with center x. We easily see that x is the limit of a sequence $\{x_n\}$ if and only if every neighborhood of the point x contains all points of the sequence beginning with a certain index n. A set is called *bounded* if it lies in a sphere.

CLOSURE. Many of the concepts which we have met in the theory of linear sets of points can be introduced in general metric spaces. If a set $M \subset X$ is given, then we call a $\in X$ a *limit point* of this set if every neighborhood of the point a contains at least one point of the set $M - a$, i.e., if $S(a, r) \cdot (M - a) \neq 0$ for every r. The set obtained by adding to M all of its limit points is called the *closure* of M, and is designated by $\bar{M}$.

We easily recognize the following fundamental properties of the closure process on sets of points of a metric space:

1) $\overline{M+N}=\bar{M}+\bar{N}$.
2) $M\subset\bar{M}$.
3) $\bar{\bar{M}}=\bar{M}$.
4) $\bar{O}=O$.

Some of these properties are evident.

A set M is called *closed* if $M=\bar{M}$. A set is *open* if its complement $X-M$ is closed. A set M is *dense* in an open set G if $G\subset\bar{M}$. In particular M is called *everywhere dense in a space* X or simply *everywhere dense* if $\bar{M}=X$. Finally M is *nowhere dense in a space* X, if every sphere of this space contains a sphere which is free of points of the set M.

TOPOLOGICAL SPACE. By defining the concept of a limit with the help of the idea of distance, we arrived at the notion of a metric space. We obtain a much more inclusive class of abstract spaces if we introduce axiomatically the notion of a limit. We shall see that it is much more convenient not to introduce the limit of a sequence axiomatically, but the concept of closure of a set. From the latter we secure the concept of a limit and thus we arrive at the concept of a topological space [1].

According to P. S. Aleksandrov and V. Sierpinski we call a set X a *topological space* if in this set certain subsets, so-called *closed sets*, are designated. We require that these sets fulfill the following topological axioms:

1) *The intersection of arbitrarily many and the union of finitely many closed sets is a closed set.*

2) *The whole space X as well as the empty set are closed.*

The intersection of all the closed sets which contain a given set M is called the *closure* of M in the space X and we designate this by $\bar{M}$. We can show that the closure operation on sets of a topological space satisfies the conditions 1) - 4). which we listed above. The points of the set $\bar{M}$ are called *contact points* of M.

We can give another definition of a contact point. We call a set *open* if its complement with respect to the whole space X is a closed set. As we easily see, the following hold:

1) The union of arbitrarily many and the intersection of finitely many open sets is an open set.

2) The whole space and the empty set are open.

We define a *neighborhood* of a point $x\in X$ to be an arbitrary open set containing this point. We may then define a contact point of a set M to be a point such that every neighborhood of that point contains at least one point of M. We can easily prove the equivalence of these two definitions.

A special case of a contact point is a limit point of a set. A point x is called a *limit point* of a set M if in every neighborhood of the point x infinitely many points of the set M lie. Now let $x_1, x_2, \ldots, x_n, \ldots$ be a sequence of points of a topological space X. A point x is called a *limit* of this sequence if in every neighborhood of the point x all the points of the sequence $\{x_n\}$ lie from a certain index on. Thus we arrive at the definition of limit of a sequence of points of a topological space which is formally the same as the definition of limit of a sequence of points in a metric space. The limit of a sequence of points of an arbitrary topological space is, however, not uniquely determined. There exist sequences which converge to several limit points as the following example shows.

Let X be an arbitrary countable set. We assume that the whole set X, every finite subset of X, and the empty set are closed. Evidently the axioms are fulfilled and X is a topological space. Further we easily see that $\bar{M} = X$ in case M is infinite and $\bar{M} = M$ in case M is finite. Neighborhoods of a point $x \in X$ are evidently the whole set X and the subsets containing x which we obtain from X if we omit a finite number of points.

Now let $\{x_n\}$ be a sequence of the space X whose terms are all distinct. Obviously, every open set contains all points of the sequence from a certain index on. This means, however, that every point of the space X is a limit of the sequence considered.

In order to determine *uniquely* the limits of sequences in a topological space, we add one axiom, the so-called *second separation axiom.*

For every two points x and y of the space X there exist two disjoint *neighborhoods of these points.*

In a space with the second separation axiom a sequence can converge to at most one limit. For, if $x_n \to x$ and if y is different from x, then a neighborhood of y which is disjoint from a certain neighborhood of x can contain only a finite number of points of the given sequence.

Let X be a topological space defined by means of giving its closed sets. A space X is called *metrizable* if we can introduce a metric in this space, in such a way that the collection of all closed sets in the sense of the introduced metric coincides with the collection of the originally given closed sets. The conditions under which a given topological space is metrizable were discovered by the Soviet mathematician P. S. Uryson. They are the content of the theorem on metrization.

In order to formulate this theorem, we must introduce several definitions. We call every open set containing a given closed set a *neighborhood of the closed set.* A topological space X is called *regular* if every point $x \in X$ and every subset $F \subset X$ not containing x have disjoint neighborhoods. Finally we call every system of open sets a *basis* of the space X if every open $G \subset X$ can be represented as the union of sets of the system.

URYSON'S THEOREM ON METRIZATION
A topological space with a countable basis is metrizable if and only if it is regular.

§3. EXAMPLES OF METRIC SPACES

THE REAL NUMBER LINE. Let $X = R$, where R is the set of all real numbers (the real number line). If $x \in R$, $y \in R$ we put $\rho(x, y) = |x - y|$. The metric axioms hold. Convergence in X is the usual convergence of sequences of real numbers.

EUCLIDEAN SPACE. Let X be an n-dimensional arithmetic space, that is the set of all n-tuples of real numbers. If $x = (\xi_1, \ldots, \xi_n)$ and $y = (\eta_1, \ldots \eta_n)$ are elements of X, we set

$$\rho(x, y) = \sqrt{\sum_{i=1}^{n} (\xi_i - \eta_i)^2}.$$

Clearly the axioms for a metric are fulfilled.

Let $x_k = (\xi_1^{(k)}, \xi_2^{(k)}, \ldots, \xi_n^{(k)})$, $k = 1, 2, \ldots$ and $\rho(x, x_k) \to 0$, i.e., let $\sqrt{\sum_{i=1}^{n} (\xi_i^{(k)} - \xi_i)^2} \to 0$ as $k \to \infty$. This is equivalent to the condition $\xi_i^{(k)} \to \xi_i$, $i = 1, 2, \ldots, n$ as $k \to \infty$. Therefore convergence in this space is coordinate-wise convergence.

A space X with this metric is called n-dimensional *Euclidean space*. We designate it by E_n.

THE SPACE OF CONTINUOUS FUNCTIONS WITH CHEBYSHEV METRIC. Let X be the set of all continuous functions defined on the interval $[0, 1]$.[3]

We introduce a metric in X by setting $\rho(x, y) = \max_{0 \leqq t \leqq 1} |x(t) - y(t)|$. We see that the metric axioms are fulfilled, for $\rho(x, y) \geqq 0$ and $\rho(x, y) = 0$ if and only if $x(t) = y(t)$. Also we note that $\rho(x, y) = \rho(y, x)$. Furthermore for every $t \in [0, 1]$ we have:

$$\begin{aligned} |x(t) - z(t)| &= |[x(t) - y(t)] + [y(t) - z(t)]| \\ &\leqq |x(t) - y(t)| + |y(t) - z(t)| \\ &\leqq \max_t |x(t) - y(t)| + \max_t |y(t) - z(t)| \\ &= \rho(x, y) + \rho(y, z). \end{aligned}$$

Therefore $\rho(x, z) = \max_t |x(t) - z(t)| \leqq \rho(x, y) + \rho(y, z)$.

The set of continuous functions defined on the interval $[0, 1]$ with the above metric is called the *space of continuous functions*. We denote it by $C[0, 1]$. We shall also say the *space of continuous functions with Chebyshev metric*.

We now consider convergence in the space $C[0, 1]$. Let $\{x_n(t)\}$ be a given sequence of elements of $C[0, 1]$ which converges to $x(t)$ (i.e. $\rho(x_n, x) \to 0$ as $n \to \infty$). This means that $\max_t |x(t) - x_n(t)| \to 0$ as $n \to \infty$, i.e., for every $\varepsilon > 0$ there exists a natural number $n_0 = n_0(\varepsilon)$ such that $\max |x(t) - x_n(t)| < \varepsilon$ for all $n \geqq n_0(\varepsilon)$. Consequently, $|x(t) - x_n(t)| < \varepsilon$ for all $n \geqq n_0(\varepsilon)$ and for all $t \in [0, 1]$. This means, however, that the sequence $\{x_n(t)\}$ converges *uniformly* to $x(t)$. We easily see that the converse also holds: if a sequence $\{x_n(t)\}$ converges uniformly to $x(t)$, then $\rho(x_n, x) \to 0$. Therefore convergence in the space $C[0, 1]$ is uniform convergence in the interval $[0, 1]$.

THE SPACE OF BOUNDED MEASURABLE FUNCTIONS. Let X be the set of bounded measurable functions defined on the interval $[0, 1]$. We consider two functions to be equal if they coincide almost everywhere on $[0, 1]$.

Let $x(t) \in X$, $y(t) \in X$, and $E \subset [0, 1]$ be an arbitrary set of measure zero. We introduce the notation

$$\rho_E(x, y) = \sup_{t \in [0, 1] - E} |x(t) - y(t)|.$$

Then we put $\rho(x, y) = \inf_E \rho_E(x, y)$, where we take the greatest lower bound over all sets of measure zero which lie in $[0, 1]$. Therefore

[3] If the interval of the independent variable t is $[a, b]$, then we can transform it into the interval $[0, 1]$ by introducing a new independent variable $\tau = \dfrac{t - a}{b - a}$.

$$\rho(x,y) = \inf_E \sup_{t\in[0,1]-E} |x(t) - y(t)|.$$

We show that the metric axioms are fulfilled.

1) Evidently,

$$\rho_E(x,y) = \sup_{t\in[0,1]-E} |x(t) - y(t)| \geqq 0.$$

Therefore $\rho(x,y) = \inf_E \rho_E(x,y) \geqq 0$. If $\rho(x,y) = 0$, i.e., $\inf_E \rho_E(x,y) = 0$, there exists for every $\varepsilon_n > 0$ a set E_n of measure zero such that

$$\rho_{E_n}(x,y) = \sup_{t\in[0,1]-E_n} |x(t) - y(t)| < \varepsilon_n,$$

therefore

$$|x(t) - y(t)| < \varepsilon_n,$$

if $t \varepsilon [0,1] - E_n$. Let the sequence $\{\varepsilon_n\}$ converge to zero as $n \to \infty$ and let E_n be the set corresponding to ε_n. Then

$$|x(t) - y(t)| < \varepsilon_n$$

holds on the set

$$\prod_{n=1}^{\infty} \{[0,1] - E_n\} = [0,1] - \sum_{n=1}^{\infty} E_n$$

for all n, i.e.,

$$x(t) - y(t) = 0.$$

Because $\sum_{n=1}^{\infty} E_n$ is of measure zero, it follows that $x(t) = y(t)$ almost everywhere on $[0,1]$, q.e.d.

2) $\rho(x,y) = \rho(y,x)$ is evident.

3) Let $x(t)$, $y(t)$ and $z(t)$ be functions belonging to X.

For every fixed $t \in [0,1] - E$ with E a set of measure zero,

$$|x(t) - z(t)| \leqq |x(t) - y(t)| + |y(t) - z(t)|$$
$$\leqq \sup_{t\in[0,1]-E} |x(t) - y(t)| + \sup_{t\in[0,1]-E} |y(t) - z(t)| = \rho_E(x,y) + \rho_E(y,z).$$

It follows that

$$\rho_E(x,z) = \sup_{t\in[0,1]-E} |x(t) - z(t)| \leqq \rho_E(x,y) + \rho_E(y,z).$$

Therefore $\rho(x,z) \leqq \rho_E(x,y) + \rho_E(y,z)$. Since $\rho(x,y) = \inf_E \rho_E(x,y)$ and $\rho(y,z) = \inf_E \rho_E(y,z)$, for every $\varepsilon_0 > 0$ there exist sets E_0 and E_0' of measure zero such that

$$\rho_{E_0}(x,y) \leqq \rho(x,y) + \frac{\varepsilon_0}{2}, \quad \rho_{E_0'}(y,z) \leqq \rho(y,z) + \frac{\varepsilon_0}{2}.$$

Consequently

$$\rho_{E_0}(x,y) + \rho_{E_0'}(y,z) \leqq \rho(x,y) + \rho(y,z) + \varepsilon_0.$$

On the other hand,

$$\rho_{E_0+E_0'}(x,y) = \sup_{t\in[0,1]-E_0-E_0'} |x(t) - y(t)| \leqq \sup_{t\in[0,1]-E_0} |x(t) - y(t)| = \rho_{E_0}(x,y)$$

and analogously

$$\rho_{E_0'} + E_0'(y,z) \leqq \rho_{E_0'}(y,z).$$

If we put $E = E_0 + E_0'$, we obtain

$$\rho(x,z) \leqq \rho_E(x,y) + \rho_E(y,z) \leqq \rho_{E_0}(x,y) + \rho_{E_0'}(y,z) \leqq \rho(x,y) + \rho(y,z) + \varepsilon_0.$$

Since ε_0 is arbitrarily small,

$$\rho(x,z) \leqq \rho(x,y) + \rho(y,z),$$

q.e.d.

The space obtained is designated by M(0,1).

We investigate convergence in this space. Let $x(t) \in M(0,1), x_n(t) \in M(0,1) (n = 1,$

2,...) and $\rho(x, x_n) \to 0$ as $n \to \infty$. This means that for a given $\varepsilon_k > 0$ and $n \geqq n_0(\varepsilon_k)$,

$$\rho(x_n, x) = \inf_{E} \sup_{t \in [0,1] - E} |x_n(t) - x(t)| < \varepsilon_k.$$

Then there exists a set E_k of measure zero such that if $n \geqq n_0(\varepsilon_k)$, then

$$\sup_{t \in [0,1] - E_k} |x_n(t) - x(t)| < \varepsilon_k.$$

Therefore $|x_n(t) - x(t)| < \varepsilon_k$ for $n \geqq n_0(\varepsilon_k)$ and for every $t \in [0,] - E_k$.

We now take a null sequence $\{\varepsilon_m\}$ (i.e., $\varepsilon_m \to 0$ as $m \to \infty$) and the corresponding sets E_m. Let $\varepsilon > 0$ be arbitrary. Then $|x_n(t) - x(t)| < \varepsilon_k < \varepsilon$ for $n \geqq n_0(\varepsilon) = n_0(\varepsilon_k)$ and for all $t \in [0,1] - \sum_{m=1}^{\infty} E_m$.

Therefore $x_n(t) \to x(t)$ almost everywhere on $[0, 1]$, i.e., $x_n(t)$ converges almost everywhere uniformly on $[0, 1]$.

Conversely, let $\{x_n(t)\}$ converge uniformly almost everywhere to $x(t)$. Then for every $\varepsilon > 0$ there exists an index $n_0(\varepsilon)$ and a set E_ε of measure zero such that $|x_n(t) - x(t)| < \varepsilon$ for n $\geqq n_0(\varepsilon)$ and every $t \in [0, 1] - E_\varepsilon$. Then, however, for $n \geqq n_0(\varepsilon)$,

$$\sup_{t \in [0,1] - E_\varepsilon} |x_n(t) - x(t)| \leqq \varepsilon,$$

It follows that

$$\inf_{E} \sup_{t \in [0,1] - E} |x_n(t) - x(t)| \leqq \varepsilon$$

for $n \leqq n_0(\varepsilon)$, i.e., $\rho(x_n, x) \to 0$ as $n \to \infty$.

Therefore convergence in the space $M(0, 1)$ is *almost everywhere uniform convergence.*

THE SPACE OF BOUNDED SEQUENCES OF NUMBERS. Let X be the space of bounded sequences of numbers

$$x = \{\xi_1, \xi_2, \ldots, \xi_n, \ldots\} = \{\xi_i\}.$$

This means that for every x there exists a constant K_x such that $|\xi_i| \leqq K_x$ for all i.

If $x = \{\xi_i\}$ and $y = \{\eta_i\}$ belong to X, we introduce the distance

$$\rho(x, y) = \sup_i |\xi_i - \eta_i|.$$

We assert that the metric axioms are fulfilled.

1. $\rho(x, y) \geqq 0$; $\rho(x, y) = 0$ only if $x = y$ and
2. $\rho(x, y) = \rho(y, x)$ are evident.
3. Let $z = \{\zeta_i\}, z \in X$. Then

$$|\zeta_i - \xi_i| \leqq |\zeta_i - \eta_i| + |\eta_i - \xi_i|$$
$$\leqq \sup_i |\zeta_i - \eta_i| + \sup_i |\eta_i - \xi_i| = \rho(z, y) + \rho(y, x).$$

Therefore

$$\rho(x, z) = \sup_i |\xi_i - \zeta_i| \leqq \rho(z, y) + \rho(y, x).$$

The space so obtained is the *space m of bounded sequences of numbers.*

Let x_n and x be elements of m, $x_n = \{\xi_i^{(n)}\}$, $x = \{\xi_i\}$ and $\rho(x_n, x) \to 0$ as $n \to \infty$. This means that there exists for every $\varepsilon > 0$ an index $n_0(\varepsilon)$ such that

$$\rho(x_n, x) = \sup |\xi_i^{(n)} - \xi_i| < \varepsilon \text{ for } n \geqq n_0(\varepsilon).$$

Therefore, for $n \geqq n_0(\varepsilon)$ and every i,

$$|\xi_i^{(n)} - \xi_i| < \varepsilon.$$

We easily see that conversely: if $|\xi_i^{(n)} - \xi_i| < \varepsilon$ for $n \geqq n_0(\varepsilon)$ and all i then

$\rho(x_n, x) \to 0$ as $n \to \infty$. Consequently, convergence in the space m is *uniform* (*index*) *coordinate-wise convergence*.

THE SPACE OF CONVERGENT SEQUENCES OF NUMBERS. Let X be the set of convergent sequences of numbers

$$x = \{\xi_1, \xi_2, \ldots, \xi_n, \ldots\},$$

where

$$\lim_{i \to \infty} \xi_i = \xi$$

exists. If

$$x = \{\xi_i\}, y = \{\eta_i\}, x \in X, y \in X,$$

we put

$$\rho(x, y) = \sup_i |\xi_i - \eta_i|.$$

The space so obtained we designate by c.

Obviously $c \subset m$. It follows that the metric axioms are fulfilled and that convergence in c is again uniform (index) coordinate-wise convergence.

There follows an example of a topological metric space.

THE SPACE OF ALL SEQUENCES OF NUMBERS. Let X be the set of all sequences of real numbers. We introduce a notion of a limit in this set, by saying that $x_n = \{\xi_i^{(n)}\}$ converges to $x = \{\xi_i\}$ if $\xi_i^{(n)} \to \xi_i$ as $n \to \infty$ for all $i = 1, 2, \ldots$ (in general non-uniformly with respect to i). We obtain a certain abstract space, which we designate by s.

We now show that the space s is metrizable.

Let $x = \{\xi_i\}, x \in s$ and $y = \{\eta_i\}, y \in s$. We put

$$\rho(x, y) = \sum_{i=1}^{\infty} \frac{1}{2^i} \frac{|\xi_i - \eta_i|}{1 + |\xi_i - \eta_i|}$$

and prove that the metric axioms are fulfilled by this distance.[4] That

1) $\rho(x, y) \geqq 0$ and $\rho(x, y) = 0$ only for $x = y$ and
2) $\rho(x, y) = \rho(y, x)$

hold, we see immediately.

3) The triangle inequality follows from the relation

$$\frac{|a + b|}{1 + |a + b|} \leqq \frac{|a|}{1 + |a|} + \frac{|b|}{1 + |b|},$$

which we prove as follows.

Let a and b have the same signs. We can assume that $a > 0$ and $b > 0$, and so

$$\frac{|a + b|}{1 + |a + b|} = \frac{a + b}{1 + a + b} = \frac{a}{1 + a + b} + \frac{b}{1 + a + b}$$
$$< \frac{a}{1 + a} + \frac{b}{1 + b} = \frac{|a|}{1 + |a|} + \frac{|b|}{1 + |b|}.$$

If a and b have different signs, we can assume that $|a| > |b|$. Then $|a + b| \leqq |a|$.

We consider the function $f(x) = \dfrac{x}{1 + x}$. $f'(x) = \dfrac{1}{(1 + x)^2}$, $f'(x) > 0$, therefore $f(x)$ is a monotonic increasing function.

[4] This notion of distance was first introduced by M. Fréchet [Trans.].

Therefore

$$\frac{|a+b|}{1+|a+b|} \leqq \frac{|a|}{1+|a|} < \frac{|a|}{1+|a|} + \frac{|b|}{1+|b|} .$$

Let us return to the triangle inequality and we will find that

$$\rho(x,z) = \sum_{i=1}^{\infty} \frac{1}{2^i} \frac{|\xi_i - \zeta_i|}{1+|\xi_i - \zeta_i|} = \sum_{i=1}^{\infty} \frac{1}{2^i} \frac{|\xi_i - \eta_i + \eta_i - \zeta_i|}{1+|\xi_i - \eta_i + \eta_i - \zeta_i|}$$

$$\leqq \sum_{i=1}^{\infty} \frac{1}{2^i} \frac{|\xi_i - \eta_i|}{1+|\xi_i - \eta_i|} + \sum_{i=1}^{\infty} \frac{1}{2^i} \frac{|\eta_i - \S_i|}{1+|\eta_i - \zeta_i|} = \rho(x,y) + \rho(y,z),$$

q.e.d.

Convergence as determined by this metric is coordinate-wise (in general non-uniformly with respect to the index), for let $x_n = \{\xi_i^{(n)}\}$, $x = \{\xi_i\}$ and $\rho(x_n, x) \to 0$; this last means that

$$\sum_{i=1}^{\infty} \frac{1}{2^i} \frac{|\xi_i^{(n)} - \xi_i|}{1+|\xi_i^{(n)} - \xi_i|} < \varepsilon$$

holds for $n \geqq n_0(\varepsilon)$. Moreover

$$\frac{1}{2^i} \frac{|\xi_i^{(n)} - \xi_i|}{1+|\xi_i^{(n)} - \xi_i|} < \varepsilon$$

for $n \geqq n_0(\varepsilon)$ and since ε was arbitrary and i was fixed, $|\xi_i^{(n)} - \xi_i| \to 0$ as $n \to \infty$. Conversely, let $|\xi_i^{(n)} - \xi_i| \to 0$ as $n \to \infty$ for every i. We choose an arbitrary $\varepsilon > 0$ and determine an m such that

$$\sum_{i=m+1}^{\infty} \frac{1}{2^i} < \frac{\varepsilon}{2}$$

holds. Then

$$\rho(x_n, x) = \sum_{i=1}^{m} \frac{1}{2^i} \frac{|\xi_i^{(n)} - \xi^i|}{1+|\xi_i^{(n)} - \xi_i|} + \sum_{i=m+1}^{\infty} \frac{1}{2^i} \frac{|\xi_i^{(n)} - \xi_i|}{1+|\xi_i^{(n)} - \xi_i|}$$

$$< \sum_{i=1}^{m} \frac{1}{2_i} \frac{|\xi_i^{(n)} - \xi_i|}{1+|\xi_i^{(n)} - \xi_i|} + \frac{\varepsilon}{2} .$$

Since the number of terms in the first sum is finite and fixed, we can choose an $n_0(\varepsilon)$ such that

$$\sum_{i=1}^{m} \frac{1}{2^i} \frac{|\xi_i^{(n)} - \xi_i|}{1+|\xi_i^{(n)} - \xi_i|} < \frac{\varepsilon}{2}$$

for $n \geqq n_0(\varepsilon)$. For this n,

$$\rho(x_n, x) < \varepsilon,$$

q.e.d.

From what has been proved, it follows that convergence in the sense of the Fréchet metric coincides with that of the space s.

THE SPACE OF CONVERGENCE IN MEASURE. Let X be the set of all measurable functions $x(t)$ defined on the interval $[0, 1]$.

Two functions which coincide almost everywhere are considered to be equal.

For every function $x(t)$, we construct a set E_x of real numbers ε such that the set of points t of the interval $[0, 1]$ for which $|x(t)| > \varepsilon$ has measure not greater than ε.

The set E_x is not empty; for instance, $\varepsilon = 1$ belongs to this set.

With every $\varepsilon \in E_x, E_x$ contains all $\varepsilon' > \varepsilon$. We define $\varepsilon_0(x) = \inf_{\varepsilon \in E_x} E_x$ and set $\rho(x, y) = \varepsilon_0(x - y)$.

We now show that the metric axioms are fulfilled.

1. $\rho(x, y) \geqq 0$ is obvious.

If $\rho(x, y) = 0$, this means that for every positive ε, $|x(t) - y(t)| > \varepsilon$ only on a set of measure $\leqq \varepsilon$. Let $a > 0$ be given. If M_a is the set of points t for which $|x(t) - y(t)| > a$ and if $\varepsilon < a$, then $M_a \subset M_\varepsilon$ and measure $M_\varepsilon \leqq \varepsilon$. Since ε was arbitrary, measure $M_a = 0$. Therefore, almost everywhere $|x(t) - y(t)| \leqq a$. Since a was given arbitrarily too, it follows that $x(t) = y(t)$ almost everywhere. Conversely, if $x(t) = y(t)$ almost everywhere, then $\rho(x, y) = 0$.

2. $\rho(x, y) = \rho(y, x)$ is clear.

3. Let x, y and z belong to X, and $\rho(x, y) = a$, $\rho(y, z) = b$. For every δ the set M_a of the points t for which $|x(t) - y(t)| > a + \delta$, has a measure not greater than $a + \delta$. Likewise for the set M_b of points t for which $|y(t) - z(t)| > b + \delta$, measure $M_b \leqq b + \delta$. Therefore

$$|x(t) - y(t)| \leqq a + \delta \tag{*}$$

on the set $N_a = [0, 1] - M_a$ with measure $N_a > 1 - (a + \delta)$ and analogously

$$|y(t) - z(t)| \leqq b + \delta \tag{**}$$

on the set $N_b = [0, 1] - M_b$ with measure $N_b > 1 - (b + \delta)$. We put $N = N_a N_b = [0, 1] - M_a - M_b$. Then the inequalities (*) and (**) are fulfilled on N, and consequently on N,

$$|x(t) - z(t)| \leqq |x(t) - y(t)| + |y(t) - z(t)| \leqq a + b + 2\delta. \tag{***}$$

Furthermore measure $N = 1 -$ measure $(M_a + M_b) \geqq 1 -$ measure $M_a -$ measure $M_b \geqq 1 - (a + b + 2\delta)$.

The inequality (***) is in every case fulfilled on N, but it can also hold at other points. We designate by $\tilde{N}$ the set of all points $t \in [0, 1]$ for which the inequality (***) holds. Evidently $\tilde{N} \supset N$, i.e., measure $\tilde{N} \geqq$ measure $N \geqq 1 - (a + b + 2\delta)$. Let $M = [0, 1] - \tilde{N}$. Then

$$\text{measure } M \leqq a + b + 2\delta$$

and

$$|x(t) - z(t)| > a + b + 2\delta \text{ on } M.$$

Therefore the number $a + b + 2\delta \in E_{x-z}$ and consequently

$$\rho(x, z) = \varepsilon_0(x - z) = \inf E_{x-z} \leqq a + b + 2\delta = \rho(x, y) + \rho(y, z) + 2\delta.$$

Since δ was arbitrary, we obtain for $\delta \to 0$

$$\rho(x, z) \leqq \rho(x, y) + \rho(y, z).$$

The space thus obtained is designated by $S(0, 1)$. Convergence in $S(0, 1)$ is called *convergence in measure*.

THE SPACE OF THE FUNCTIONS WITH INTEGRABLE pTH POWER. Let X be the set of all functions $x(t)$ which belong[5] to $L_p(0, 1)$. Two functions which differ only on a set of measure zero are again considered as equal.

---

[5] viz Appendix I.

If $x(t) \in L_p(0,1)$ and $y(t) \in L_p(0,1)$, we put

$$\rho(x,y) = \left(\int_0^1 | x(t) - y(t) |^p dt \right)^{\frac{1}{p}} .$$

We easily prove that the metric axioms hold. It is immediate that $\rho(x,y) \geqq 0$ and $\rho(x,y) = 0$ only if $x = y$, and that $\rho(x,y) = \rho(y,x)$. The verification of the triangle axiom follows from the Minkowski inequality for integrals.[6]

The space $L_2(0,1)$ is called the *Hilbert function-space.*

Let $x_n(t) \in L_p(0,1)$, $n = 1,2,3,\ldots$ and let $\{x_n(t)\}$ converge to $x(t)$, $x(t) \in L_p(0,1)$, i.e.,

$$\int_0^1 | x_n(t) - x(t) |^p \, dt \to 0$$

as $n \to \infty$. Then we say that the sequence $\{x_n(t)\}$ of functions converges in the mean to the pth power to the function $x(t)$. For $p = 2$ we ordinarily speak simply of convergence in the mean.

THE SPACE l_p OF NUMBER SEQUENCES. Let X be the set of the real number sequences $\{\xi_1, \xi_2, \ldots, \xi_n, \ldots\}$. Let
$x = \{\xi_1, \xi_2, \ldots, \xi_n, \ldots\}$ and $y = \{\eta_1, \eta_2, \ldots \eta_n, \ldots\}$; then we define the distance by

$$\rho(x,y) = \left(\sum_{i=1}^{\infty} | \xi_i - \eta_i |^p \right)^{\frac{1}{p}} .$$

We see at once that the metric axioms hold.
Evidently $\rho(x,y) \geqq 0$, $\rho(x,y) = 0$ only if $x = y$, and $\rho(x,y) = \rho(y,x)$. The triangle axiom follows from the Minkowski inequality for sums.[7]

The space l_2 is called the *Hilbert coordinate-space.*

We can show[8] that the convergence of a sequence $\{x_n\}$, $x_n = \{\xi_1^{(n)}, \xi_2^{(n)}, \ldots, \xi_i^{(n)}\}$ to an element $x = \{\xi_1, \xi_2, \ldots\}$ in the space l_p means that

1) $\xi_i^{(n)} \to \xi_i$ as $n \to \infty$ for all i and
2) For every $\varepsilon > 0$ there exists an index $N_0(\varepsilon)$ such that

$$\left(\sum_{i=N+1}^{\infty} | \xi_i^{(n)} |^p \right)^{\frac{1}{p}} < \varepsilon$$

for $N \geqq N_0(\varepsilon)$ and all n.

Let X be a space of n-tuples of reals (see p. 9), $x = \{\xi_1, \xi_2, \ldots, \xi_n\}$ and $y = \{\eta_1, \eta_2, \ldots \eta_n\}$. We put

$$\rho(x,y) = \left(\sum_{i=1}^{n} |\xi_i - \eta_i|^p \right)^{\frac{1}{p}} .$$

We name the space so obtained $l_p^{(n)}$. In particular $l_2^{(n)}$ is the n-dimensional Euclidean space.

We can assume[9] that $l_p^{(n)} \subset l_p$, if we regard every element $\{\xi_1, \xi_2, \ldots \xi_n\} \in l_p^{(n)}$ as

---

[6] viz. Appendix I.
[7] viz. Appendix I.
[8] Compare with the criterion for compactness in the space l_p, p. 45.
[9] See p. 21.

an element $\{\xi_1, \xi_2, \ldots, \xi_n, 0, 0, \ldots\} \in l_p$. Convergence in the space $l_p^{(n)}$ is coordinatewise convergence.

THE COMPLEX SPACES, C, L_p, c, l_p. Besides the spaces C, L_p, c, l_p we can consider the corresponding complex spaces C, L_p, c, l_p. Elements of the complex spaces C and L_p are complex-valued functions of a real variable. Elements of the complex spaces c and l_p are sequences of complex numbers.

All definitions given above for real spaces can be framed immediately in the corresponding complex spaces.

There follows an example of a non-metrizable space.

THE SPACE OF POINTWISE CONVERGENCE. We consider the collection $F(0, 1)$ of all real functions $x(t)$, which are defined on the interval $[0,1]$. We define in $F(0,1)$ a topology by stipulating that a function $x(t)$ belongs to the closure $\bar{A}$ of a set A, $A \subset F(0,1)$, if for every $\varepsilon > 0$ and every finite system of points $t_1, \ldots, t_n, t_i \in [0,1]$ there is a function $y(t) \in A$ such that $|x(t_i) - y(t_i)| < \varepsilon$ for $i = 1, 2, \ldots, n$. We easily convince ourselves that the closure process defined in this way makes $F(0, 1)$ a topological space. Here convergence in $F(0,1)$ means convergence of a function at every point of the interval $[0,1]$.

The space $F(0,1)$ is not metrizable. For, $\bar{M} = F(0,1)$ for the set M of all functions which are continuous on $[0,1]$. If $F(0,1)$ were metrizable, then $\bar{M}$ would coincide with the set of limits of sequences of M, i.e., with the set M_1 of functions of the first category of Baire, which is not the same as $F(0,1)$.

Some function spaces whose elements are functions of several variables were considered by S. L. Sobolev [25].

§4. COMPLETE SPACES. EXAMPLES

DEFINITIONS. A sequence $\{x_n\}$ of elements of a metric space X is called a *Cauchy sequence* or *fundamental sequence*, if for every $\varepsilon > 0$ there is an index $n_0(\varepsilon)$ such that $\rho(x_m, x_n) < \varepsilon$ for $m \geqq n_0(\varepsilon)$ and $n \geqq n_0(\varepsilon)$.

THEOREM. *If a sequence $\{x_n\}$ converges to a limit x, then it is a fundamental sequence.*

Let $x = \lim_n x_n$. Then there exists for every $\varepsilon > 0$ an index $n_0(\varepsilon)$ such that $\rho(x_n, x) < \frac{\varepsilon}{2}$ for $n \geqq n_0$. Consequently

$$\rho(x_m, x_n) \leqq \rho(x_m, x) + \rho(x_n, x) < \varepsilon$$

for $n \geqq n_0$ and $m \geqq n_0$, q.e.d.

The converse is false for an arbitrary metric space since there exist metric spaces in which there are fundamental sequences which do not converge to a limit in the space.

Examples. 1. Let X be the set of rational numbers. Distance is defined by

$$\rho(r_1, r_2) = |r_1 - r_2|$$

and X is a metric space. The sequence $\{r_n\} = \left\{\frac{1}{2^n}\right\}$ is a fundamental sequence and converges to the limit 0. Now take $r_n = \left(1 + \frac{1}{n}\right)^n$. This sequence is a fundamental sequence but it has no limit in the space X.

2. Let X be the space of the polynomials $P(t), 0 \leqq t \leqq 1$ with the Tchebycheff metric, i.e. for $P(t)$ and $Q(t) \in X$,

$$\rho(P, Q) = \max_{0 \leqq t \leqq 1} |P(t) - Q(t)|.$$

Let $\{P_n(t)\}$ be a sequence of polynomials converging to a continuous function which is not a polynomial. The sequence $\{P_n(t)\}$ is obviously a fundamental sequence which, however, has no limit in the space X.

If every fundamental sequence of a metric space X has a limit in X, then X is called *complete.*

We remark that a *closed subset of a complete space is also a complete space.* In the following we will prove that certain metric spaces are complete.

COMPLETENESS OF E_n. For E_n, the Euclidean n-dimensional space, completeness follows from the Cauchy criterion for sequences of points.

COMPLETENESS OF $C[0, 1]$. Let a sequence $\{x_i(t)\}$ be given with $x_i(t) \in C[0, 1]$, $i = 1, 2, \ldots$ and let

$$\rho(x_m, x_n) \to 0$$

as $m, n \to \infty$. This means that the sequence $\{x_i(t)\}$ converges uniformly in the sense of Cauchy on $[0, 1]$. Let $x(t)$ be the limit of the sequence $\{x_i(t)\}$ of functions. Since $x(t)$ is the limit of a uniformly convergent sequence of continuous functions, it is also continuous on $[0, 1]$. Therefore $x(t) \in C[0, 1]$ and $\rho(x_n, x) \to 0$.

Thus we see the following: If we choose an arbitrary fundamental sequence $\{x_n(t)\}$ from $C[0, 1]$, then we can find an element $x(t)$ of the same space to which this sequence converges. This means, however, that the space $C[0, 1]$ is complete.

THE COMPLETENESS OF m. Let $\{x_n\}$ be a fundamental sequence of elements of m, and let $x_n = \{\xi_i^{(n)}\}$. Since $x_n \in m$, $|\xi_i^{(n)}| \leqq K_n$ for $i = 1, 2 \ldots$. Moreover, because $\{x_n\}$ is fundamental, for every $\varepsilon > 0$ there exists an index n_0 such that $\rho_n(x_n, x_k) < \varepsilon$ whenever $n \geqq n_0$ and $k \geqq n_0$, or, what is the same

$$\sup_i |\xi_i^{(n)} - \xi_i^{(k)}| < \varepsilon \text{ for } n, k \geqq n_0.$$

It follows that

$$|\xi_i^{(n)} - \xi_i^{(k)}| < \varepsilon \tag{1}$$

for $n \geqq n_0, k \geqq n_0$ and uniformly for i.

Let i be fixed. Then, according to (1), the sequence $\{\xi_i^{(1)}, \xi_i^{(2)}, \ldots, \xi_i^{(n)}, \ldots\}$ of numbers satisfies the Cauchy condition for existence of a limit and consequently converges to a number ξ_i. We thus obtain a sequence $\{\xi_1, \xi_2, \ldots, \xi_n, \ldots\}$ of numbers.

We consider the inequality (1) and let k increase indefinitely. In the limit we obtain

$$|\xi_i^{(n)} - \xi_i| \leqq \varepsilon \tag{2}$$

for $n \geqq n_0$ and all i.

It follows that $\quad |\xi_i| \leqq |\xi_i^{(n)} - \xi_i| + |\xi_i^{(n)}| \leqq \varepsilon + K_n,$

for all i. This means, however, that $\{\xi_i\}$ is a bounded sequence of numbers, i.e., $x = \{\xi_i\} \in m$. From (2) we obtain $\sup_i |\xi_i^{(n)} - \xi_i| \leqq \varepsilon$ for $n \geqq n_0$, i.e., $\rho(_n, x) \leqq \varepsilon$ for $n \geqq n_0$.

Since $\varepsilon > 0$ was arbitrary it follows that $x_n \to x$ as $n \to \infty$. We therefore have proved the completeness of the space m.

THE COMPLETENESS OF c. Let $\{x_n\}$ be a fundamental sequence of elements of the space c. Let $x_n = \{\xi_i^{(n)}\}$. This means that for every $\varepsilon > 0$ there exists an n_0 such that $\rho(x_m, x_n) = \sup_i |\xi_i^{(m)} - \xi_i^{(n)}| < \varepsilon$ for $n \geqq n_0$. Consequently

$$|\xi_i^{(n)} - \xi_i^{(m)}| < \varepsilon \tag{3}$$

for $n, m \geqq n_0$ and for all i.

We fix i and consider the sequence $\{\xi_i^{(1)}, \xi_i^{(2)}, \ldots\}$. According to (3) this sequence converges by the Cauchy criterion to the limit ξ_i. We obtain a sequence $\{\xi_1, \xi_2 \ldots\}$ of numbers. Let $m \to \infty$ in the inequality (3); then

$$|\xi_i^{(n)} - \xi_i| \leqq \varepsilon \tag{4}$$

for $n, m \geqq n_0$ and for all i. Since $x_n \in c$, there exists $\lim_i \xi_i^{(n)} = \xi^{(n)}$. It follows that $|\xi^{(n)} - \xi_i^{(n)}| < \varepsilon$ for $i \geqq N(\varepsilon, n)$. We thus obtain, for $n \geqq n_0$ and $m, k \geqq N(n, \varepsilon)$:

$$|\xi_m - \xi_k| \leqq |\xi_m - \xi_m^{(n)}| + |\xi_m^{(n)} - \xi^{(n)}| + |\xi^{(n)} - \xi_k^{(n)}| + |\xi_k^{(n)} - \xi_k| < \varepsilon + \varepsilon + \varepsilon + \varepsilon = 4\varepsilon.$$

Therefore the sequence $\{\xi_i\}$ satisfies the condition of the Cauchy criterion. Hence $\lim_i \{\xi_i\} = \xi$ and consequently $x = \{\xi_i\} \in c$.

Next, we obtain from (4)

$$\rho(x_n, x) = \sup_i |\xi_i^{(n)} - \xi_i| \to 0$$

for $n \to \infty$, i.e., $x_n \to x$ for $n \to \infty$ and the completeness of the space c is proved.

THE COMPLETENESS OF $M(0, 1)$. Let $\{x_n\}$ be a fundamental sequence of elements of the space $M(0, 1)$. For every $\varepsilon > 0$ with $n, m \geqq n_0(\varepsilon)$,

$$\inf_{\substack{E \subset [0,1] \\ \text{measure } E = 0}} \sup_{t \in [0,1] - E} |x_n(t) - x_m(t)| < \varepsilon$$

holds. Consequently there exists a set E_{n_0} of measure zero such that

$$\sup_{t \in [0,1] - E_{n_0}} |x_n(t) - x_m(t)| < \varepsilon \text{ for } m, n > n_0$$

i.e., $|x_n(t) - x_m(t)| < \varepsilon$ for n, $m \geqq n_0$ almost everywhere on $[0, 1]$. From this we easily see (cf. p. 10) that a sequence $\{x_n(t)\}$ of bounded measurable functions is uniformly almost everywhere on [0,] a fundamental sequence. Therefore, there exists a bounded measurable function $x(t)$ which is almost everywhere on $[0, 1]$ the limit of this sequence. It is readily seen that $\rho(x_n, x) \to 0$ and consequently the completeness of the space $M(0, 1)$ is proved.

THE COMPLETENESS OF $L_p(0, 1)$. We select an arbitrary fundamental sequence $\{x_n(t)\}$ of functions of $L_p(0, 1)$. This means that for every $\varepsilon_k > 0$ there exists an index $N = N(\varepsilon_k)$ such that

$$\left(\int_0^1 |x_n(t) - x_m(t)|^p dt\right)^{\frac{1}{p}} < \varepsilon_k \text{ for } n,\ m \geqq N(\varepsilon_k)$$

i.e.,

$$\int_0^1 |x_n(t) - x_m(t)|^p dt < \varepsilon_k^p \text{ for } n,\ m \geqq N(\varepsilon_k).$$

We put
$$\varepsilon_k^p = \frac{1}{2^{kp+k}}.$$
Then, for $n, m \geqq N(\varepsilon_k)$
$$|x_n(t) - x_m(t)| > \frac{1}{2^k}$$
only on a set A_k with measure less than $\frac{1}{2^k}$. In particular, for an increasing sequence of natural numbers n_k, for which $n_k \geqq N(\varepsilon_k)$,
$$|x_{n_{k+1}}(t) - x_{n_k}(t)| > \frac{1}{2^k}$$
only on a set A_k with measure less than $\frac{1}{2^k}$.

We designate $x_{n_k}(t)$ by $z_k(t)$ and construct the set
$$B_k = \prod_{i=k}^{\infty}\{[0, 1] - A_i\} = [0, 1] - \sum_{i=k}^{\infty} A_i.$$
Then measure $B_k \geqq 1 - \sum_{i=k}^{\infty}$ measure $A_i \geqq 1 - \frac{1}{2^{k-1}}$

and $|z_{n+1}(t) - z_n(t)| \leqq \frac{1}{2^n}$ for $n \geqq k$ on B_k. This means that $\{z_n(t)\}$ is a uniform fundamental sequence on B_k. It is clear that $B_k \subset B_{k+1}$. Let $B_0 = \lim_k B_k = \sum_{k=1}^{\infty} B_k$, then measure $B_0 = 1$.

Since every point of the set B_0 belongs to a certain set B_k, $\{z_n(t)\}$ is a fundamental sequence on B_0 and converges therefore to a limit function $z(t)$, where the convergence is uniform on every set B_k. Now B_0 differs from $[0, 1]$ only by a set of measure zero, therefore $z(t)$ is defined almost everywhere on $[0, 1]$. $\{z_n(t)\}$, being a subsequence of a fundamental sequence (in the mean), has the same properties, therefore
$$\int_0^1 |z_n(t) - z_m(t)|^p \, dt < \varepsilon$$
for $n, m \geqq n_0(\varepsilon)$. Since the integrand is non-negative,
$$\int_{B_k} |z_n(t) - z_m(t)|^p \, dt < \varepsilon.$$
The sequence $z_n(t)$ is uniformly convergent on B_k, therefore we can take the limit $m \to \infty$ under the integral sign. We obtain
$$\int_{B_k} |z_n(t) - z(t)|^p \, dt \leqq \varepsilon \text{ for } n \geqq n_0(\varepsilon).$$
For $k \to \infty$, we obtain
$$\int_{B_0} |z_n(t) - z(t)|^p dt \leqq \varepsilon \text{ if } n \geqq n_0(\varepsilon)$$

or
$$\int_0^1 |z_n(t) - z(t)|^p dt \leqq \varepsilon \text{ for } n \geqq n_0(\varepsilon). \tag{5}$$

It follows that $z(t) - z_n(t) \in L_p(0, 1)$. Then, however, $z(t) = [z(t) - z_n(t)] + z_n(t) \in L_p(0, 1)$. (See Appendix I.)

From (5), it follows that
$$\rho(z_n, z) \leqq \varepsilon^{\frac{1}{p}}$$
for $n \geqq n_0(\varepsilon)$, i.e., $\rho(z_n, z) \to 0$ as $n \to \infty$. Furthermore, since $\rho(x_n, z) \leqq \rho(x_n, x_{n_k}) + \rho(z_k, z) \to 0$ as $n \to \infty$ and $k \to \infty$, the completeness of the space $L_p(0, 1)$ is proved.

The completeness of $L_2(0, 1)$ was for the first time shown by F. Riesz [48].

The Completeness of l_p. We consider a fundamental sequence $\{x_n\}$, $x_n = \{\xi_i^{(n)}\} \in l_p$.
$$\rho(x_n, x_m) = \left(\sum_{i=1}^{\infty} | \xi_i^{(n)} - \xi_i^{(m)} |^p \right)^{\frac{1}{p}} < \varepsilon \text{ for } n, m \geqq n_0. \tag{6}$$
It follows that
$$| \xi_i^{(n)} - \xi_i^{(m)} | < \varepsilon \tag{7}$$
for $n, m \geqq n_0$ and for all i. We choose a fixed i. From (7) we conclude that the sequence $\xi_i^{(1)}, \xi_i^{(2)}, \ldots$ converges to a limit ξ_i:
$$\lim_n \xi_i^{(n)} = \xi_i. \tag{8}$$

We put $x = \{\xi_i\}$. From the inequality (6), we have
$$\sum_{i=1}^{k} | \xi_i^{(n)} - \xi_i^{(m)} |^p < \varepsilon^p$$
for every k.

As $m \to \infty$,
$$\sum_{i=1}^{k} | \xi_i^{(n)} - \xi_i |^p \leqq \varepsilon^p \text{ for } n \geqq n_0.$$
If we let k increase without bound, we find that
$$\sum_{i=1}^{\infty} | \xi_i^{(n)} - \xi_i |^p \leqq \varepsilon^p \text{ for } n \geqq n_0.$$
Hence: $x - x_n \in l_p$. Therefore $x = (x - x_n) + x_n \in l_p$. Finally $\rho(x_n, x) \to 0$ for $n \to \infty$ and the completeness of the space l_p is proved.

The completeness of $S(0, 1)$ and s can also be shown. We shall not dwell on this.

§5. COMPLETION OF METRIC SPACES

It is known that the completeness of the real number line plays an important role in analysis. It is possible to construct analysis rigorously only from the totality of real numbers. The completeness of metric spaces plays a similar role in Functional Analysis. We therefore seek a process of completing spaces which are not complete. Here we proceed in a manner analogously to the passing from the set of rational numbers to the set of all real numbers. We first introduce an idea which will be useful to us also in connection with other considerations.

Let two metric spaces X and Y be given. Let the distance between two elements x_1 and x_2 of the space X be $\rho(x_1, x_2)$ and the distance between two elements y_1 and y_2 of the space Y be $\rho(y_1, y_2)$.

If there exists a one-to-one correspondence between the elements of the spaces X and Y such that the distance between two elements of one of the spaces is equal to the distance between the corresponding elements of the other space, then X and Y are called *isometric*.

For questions which depend only on the distance of elements, e.g., convergence, completeness, etc., isometric spaces are to be considered as being equal.

Besides isometric spaces X and Y we can also consider isometric sets which are contained in these spaces. If we are concerned with questions dealing entirely with the metric, we can replace any set in a metric space with an isometric one.

If X_0 is an incomplete metric space, there exists in this space at least one fundamental sequence which contains no limit in X_0.

To every incomplete metric space X_0 there exists to within isometry exactly one complete metric X with a subspace X' which is everywhere dense in X and is isometric to X_0.

X is called the *completion of the space* X_0.

We consider all fundamental sequences

$$\{x_n\},\ \{y_n\},\ \{z_n\}, \ldots$$

which are formed from elements of X_0. Two arbitrary sequences $\{x_n\}$ and $\{x_n'\}$ which are fundamental and for which $\rho(x_n, x'_n) \to 0$ as $n \to \infty$ we place in the same class $\tilde{x}$. We consider this class as an element of a new space X. Let $\tilde{x}$ and $\tilde{y}$ be two elements of X. From the classes $\tilde{x}$ and $\tilde{y}$ we choose arbitrary sequences $\{x_n\}$ and $\{y_n\}$.

We show that $\lim\limits_{n\to\infty} \rho(x_n, y_n)$ exists. We have

$$\rho(x_n, y_n) \leqq \rho(x_n, x_m) + \rho(x_m, y_m) + \rho(y_m, y_n).$$

It follows that $\rho(x_n, y_n) - \rho(x_m, y_m) \leqq \rho(x_n, x_m) + \rho(y_m, y_n).$ (1)

If the indices n and m are permuted, we find

$$\rho(x_m, y_m) - \rho(x_n, y_n) \leqq \rho(x_n, x_m) + \rho(y_n, y_m). \tag{2}$$

From (1) and (2) it follows that

$$|\,\rho(x_m, y_m) - \rho(x_n, y_n)\,| \leqq \rho(x_n, x_m) + \rho(y_n, y_m).$$

The right side of this inequality converges to 0 as $n, m \to \infty$. Therefore, the number sequence $\{\rho(x_n, y_n\}$ is convergent according to the Cauchy criterion, i.e., there exists $\lim\limits_{n\to\infty} \rho(x_n, y_n)$.

We now define in X by means of

$$\rho(\tilde{x}, \tilde{y}) = \lim_{n\to\infty} \rho(x_n, y_n)$$

a distance and show that the so-defined distance is independent of the choices of the sequences $\{x_n\}$ and $\{y_n\}$ from the corresponding classes.

Let $\{x_n'\}$ and $\{y_n'\}$ be two other sequences from the classes $\tilde{x}$ and $\tilde{y}$. Then

$$\rho(x_n, y_n) \leqq \rho(x_n, x_n') + \rho(x_n', y_n') + \rho(y_n', y_n).$$

It follows that

$$\lim_{n\to\infty} \rho(x_n, y_n) \leqq \lim_{n\to\infty} \rho(x_n', y_n').$$

Similarly, we find that conversely

$$\lim_{n\to\infty} \rho(x_n', y_n') \leqq \lim_{n\to\infty} \rho(x_n, y_n).$$

Consequently
$$\lim_{n\to\infty} \rho(x_n, y_n) = \lim_{n\to\infty} \rho(x_n', y_n').$$

The metric axioms hold:

1. Since $\rho(x_n, y_n) \geqq 0$, so is also $\rho(\tilde{x}, \tilde{y}) = \lim_{n\to\infty} \rho(x_n, y_n) \geqq 0$. Moreover, the equation $\rho(\tilde{x}, \tilde{y}) = \lim_{n\to\infty} \rho(x_n, y_n) = 0$ means, according to the hypothesis, that the sequences $\{x_n\}$ and $\{y_n\}$ belong to one class.

Since $\{x_n\}$ was an arbitrary sequence of the class $\tilde{x}$ and $\{y_n\}$ was an arbitrary sequence of the class $\tilde{y}, \tilde{x} = \tilde{y}$.

2. That $\rho(\tilde{x}, \tilde{y}) = \rho(\tilde{y}, \tilde{x})$ is obvious.

3. If $\{x_n\} \in \tilde{x}, \{y_n\} \in \tilde{y}, \{z_n\} \in \tilde{z}$, then evidently

$$\rho(\tilde{x}, \tilde{z}) = \lim_n \rho(x_n, z_n) \leqq \lim_n \rho(x_n, y_n) + \lim_n \rho(y_n, z_n) = \rho(\tilde{x}, \tilde{y}) + \rho(\tilde{y}, \tilde{z}).$$

We show that X is a complete space. To do this, we choose an arbitrary fundamental sequence $\{\tilde{x}_1, \tilde{x}_2, \ldots, \tilde{x}_n, \ldots\}$ of elements of the space X, i.e., let

$$\rho(\tilde{x}_n, \tilde{x}_m) \to 0 \text{ as } n, m \to \infty.$$

From every class $\tilde{x}_n$ we select an arbitrary sequence

$$\{x_1^{(n)}, x_2^{(n)}, \ldots, x_k^{(n)}, \ldots\}.$$

Since this is a fundamental sequence, we can find a k_n such that

$$\rho(x_m^{(n)}, x_{k_n}^{(n)}) \leqq \frac{1}{n}$$

holds for $m > k_n$.

We consider now the sequence

$$\{x_{k_1}^{(1)}, x_{k_2}^{(2)}, \ldots, x_{k_n}^{(n)}, \ldots\}$$

and show that it is a fundamental sequence.

We call sequences of the form $\{x, x, \ldots, x, \ldots\}$ *constant*. They are obviously fundamental sequences and consequently each is contained in a certain class and is therefore associated with an element of X. Obviously, there can belong to one and the same class at most one constant sequence. Let

$$\{x, x, \ldots, x, \ldots\} \in \tilde{x},$$
$$\{y, y, \ldots, y, \ldots\} \in \tilde{y}.$$

Then $\rho(x, y) = \rho(\tilde{x}, \tilde{y})$.

We construct constant sequences

$$\{x_{k_n}^{(n)}, x_{k_n}^{(n)}, \ldots, x_{k_n}^{(n)}, \ldots\}$$

and designate the classes to which these sequences belong by $\tilde{x}_{k_n}^{(n)}$. We have

$$\rho(\tilde{x}_n, \tilde{x}_{k_n}^{(n)}) = \lim_{k\to\infty} \rho(x_k^{(n)}, x_{k_n}^{(n)}) \leqq \frac{1}{n} \tag{3}$$

and furthermore,

$$\rho(x_{k_n}^{(n)}, x_{k_m}^{(m)}) = \rho(\tilde{x}_{k_n}^{(n)}, \tilde{x}_{k_m}^{(m)}) \leqq \rho(\tilde{x}_{k_n}^{(n)} \tilde{x}_n) + \rho(\tilde{x}_n, \tilde{x}_m) + \rho(\tilde{x}_m, \tilde{x}_{k_m}^{(m)})$$
$$\leqq \rho(\tilde{x}_n, \tilde{x}_m) + \frac{1}{n} + \frac{1}{m}.$$

Let $\varepsilon > 0$ be a given number. Select n_0 such that

$$1)\ \frac{1}{n} < \frac{\varepsilon}{3},\ 2)\ \frac{1}{m} < \frac{\varepsilon}{3} \text{ and } 3)\ \rho(\tilde{x}_n, \tilde{x}_m) < \frac{\varepsilon}{3}$$

for $n, m \geqq n_0$. Then

$$\rho(x_{k_n}^{(n)}, x_{k_m}^{(m)}) < \varepsilon$$

for $n, m \geqq n_0$.

This means that $\{x_{k_1}^{(1)}, x_{k_2}^{(2)}, \ldots, x_{k_n}^{(n)}, \ldots\}$ is a fundamental sequence and consequently belongs to a class of X which we designate by $\tilde{x}$.

We show that

$$\tilde{x} = \lim_{n\to\infty} \tilde{x}_n.$$

With this, the proof of the completeness of X will be finished.

According to (3),

$$\rho(\tilde{x}, \tilde{x}_n) \leqq \rho(\tilde{x}, \tilde{x}_{k_n}^{(n)}) + \rho(\tilde{x}_{k_n}^{(n)}, \tilde{x}_n) \leqq \rho(\tilde{x}, \tilde{x}_{k_n}^{(n)}) + \frac{1}{n}.$$

In addition,

$$\rho(\tilde{x}, \tilde{x}_{k_n}^{(n)}) = \lim_{p\to\infty} \rho(x_{k_p}^{(p)}, x_{k_n}^{(n)}) < \frac{\varepsilon}{2}$$

for sufficiently large n since $\{x_{k_n}^{(n)}\}$ is a fundamental sequence. If we choose n such that $\frac{1}{n} < \frac{\varepsilon}{2}$, then we obtain $\rho(\tilde{x}, \tilde{x}_n) < \varepsilon$. Consequently $\tilde{x} = \lim \tilde{x}_n$, q.e.d.

We finally show that X_0 is isometric to a certain subspace X' of X which is everywhere dense in X. Let X' contain all classes $\tilde{x}$, among whose elements there are constant sequences $\{x, x, \ldots\}$. There is a one-to-one correspondence between the classes $\tilde{x} \in X'$ and the elements x of the constant sequence. In addition, if $\{x\} \in \tilde{x}$ and $\{y\} \in \tilde{y}$, then $\rho(\tilde{x}, \tilde{y}) = \rho(x, y)$. Therefore the one-to-one correspondence established between the elements of X_0 and the corresponding elements of X' is an isometry.

It is easily seen that X' is everywhere dense in X, i.e., for every $\varepsilon > 0$ and for every element $\tilde{x} \in X$ there is an element $\tilde{x}_\varepsilon \in X'$ such that $\rho(\tilde{x}, \tilde{x}_\varepsilon) \leqq \varepsilon$. Let x be a class and let the Cauchy sequence $\{x_1, x_2, \ldots, x_n, \ldots\}$ belong to $\tilde{x}$. We choose an n such that $\rho(x_m, x_n) < \varepsilon$ for $m > n$. Then we construct the constant sequence $\{x_n, x_n, \ldots x_n, \ldots\}$ and designate by $\tilde{x}_\varepsilon$ the class containing this sequence. Obviously, $x_\varepsilon \in X'$. Furthermore $\rho(\tilde{x}, \tilde{x}_\varepsilon) = \lim_m \rho(x_m, x_n) \leqq \varepsilon$, and we are through.

It remains to show that the completion of a space X_0 is uniquely determined to within isometry. That is, there exists—to within isometry—at most one complete space X which contains an everywhere dense subset isometric to X_0. For, if Y is another complete space in which X_0 is everywhere dense, then every point $\tilde{y} \in Y$ is determined by a certain sequence $\{x_1, x_2, \ldots x_n, \ldots\}$ with $x_i \in X_0$. On the other hand, since this is a fundamental sequence, it defines an element $\tilde{x} \in X$. We associate this element $\tilde{x}$ with the element $\tilde{y}$. Conversely, let an element $\tilde{x} \in X$ be given and let $\{x_1', x_2', \ldots, x_n', \ldots\}$ be a certain fundamental sequence from the class $\tilde{x}'$. Since X_0 lies in the complete space Y, this fundamental sequence defines a certain element $\tilde{y}' \in Y$. We associate this element with $\tilde{x}'$. We thus obtain a correspondence between the elements of the spaces X and Y which is obviously one-to-one. Since, moreover

$$\rho(\tilde{x}, \tilde{x}') = \lim_{n \to \infty} \rho(x_n, y_n) = \rho(\tilde{y}, \tilde{y}'),^{10}$$

this correspondence is isometric, and this completes the proof.

Examples. 1. We consider the space l_p' consisting of all ordered systems $\{\xi_1, \xi_2, \ldots, \xi_{k_1}\}$, where k_1 is a natural number and ξ_i are arbitrary real numbers. If

$$x = \{\xi_1, \xi_2, \ldots, \xi_{k_1}\},\ y = \{\eta_i, \eta_2, \ldots, \eta_{k_2}\}$$

and $k_2 \geqq k_1$, we introduce a metric by

$$\rho(x, y) = \left(\sum_{i=1}^{k_1} |\xi_i - \eta_i|^p + \sum_{i=k_1+1}^{k_2} |\eta_i|^p\right)^{\frac{1}{p}}.$$

l_p' is an incomplete space, since, for instance, even though the sequence

$$x_1 = \{1\},\ x_2 = \left\{1, \frac{1}{2}\right\},\ x_3 = \left\{1, \frac{1}{2}, \frac{1}{2^2}\right\}, \ldots,\ x_n = \left\{1, \frac{1}{2}, \frac{1}{2^2}, \ldots, \frac{1}{2^{n-1}}\right\}, \ldots$$

is fundamental, (i.e.,

$$\rho(x_n, x_m) = \left(\sum_{i=m}^{n-1} \frac{1}{2^{ip}}\right)^{\frac{1}{p}} \to 0 \text{ as } m, n \to \infty,\ m < n),$$

it has no limit in l_p'.

We designate the completion of the space l_p' by X. Since l_p' obviously is everywhere dense in the complete space l_p, it follows that $X = l_p$. Therefore the completion of the space l_p' leads to l_p (more precisely: to a space isometric to l_p).

2. Let $C_0[0, 1]$ be the space of all polynomials defined on the interval $[0, 1]$. In this space let the Chebyshev metric

$$\rho(P, Q) = \max_{0 \leqq t \leqq 1} |P(t) - Q(t)|$$

be introduced. $C_0[0, 1]$ obviously is not complete. Since $C_0[0, 1]$ lies everywhere dense in the complete space $C[0, 1]$, the completion of $C_0[0, 1]$ leads to $C[0, 1]$.

3. Let $L_p'(0, 1)$ be the set of all continuous functions defined on the interval $[0, 1]$ with the metric

$$\rho(x, y) = \left(\int_0^1 |x(t) - y(t)|^p \, dt\right)^{\frac{1}{p}}.$$

$L_p'(0, 1)$ is an incomplete space. For, a sequence of continuous functions which converges in the mean (pth power) to a discontinuous function, is a fundamental sequence in $L_1'(0, 1)$ but has no limit in this space. When we complete $L_p'(0, 1)$, we obtain $L_p(0, 1)$.

4. The set of solutions of a linear partial differential equation is, in general, not complete when considered as a metric space.

---

[10] One easily verifies: if, in a metric space $x_n \to x$ and $y_n \to y$, then also $\rho(x_n, y_n, \to \rho(x, y)$.

If we complete this space, we arrive at the *generalized solutions*, in the sense of S. L. Sobolev, of the given equation, which are the elements of the complete space.

For instance, the totality of solutions of the wave equation

$$\frac{\partial^2 u}{\partial x^2} = \frac{1}{a^2}\frac{\partial^2 u}{\partial t^2}$$

has the form:

$$u = f(x + at) + \varphi(x - at). \tag{*}$$

Here f and φ are twice differentiable functions of one variable. If we complete the set of solutions (*) by taking limits in the sense of uniform convergence, we obtain the set of generalized solutions, which again have the form (*). Here the functions f and φ are arbitrary continuous functions. These generalized solutions have physical meaning in many cases [25].

§6. TWO THEOREMS ON COMPLETE SPACES

THEOREM 1. *In a complete metric space X, let a nested sequence of closed spheres be given (i.e., every succeeding sphere is contained in all preceding spheres). If the sequence of radii of these spheres converges to* 0, *there exists exactly one point which belongs to all spheres.*

Let the spheres considered be

$$\bar{S}_1(a_1, \varepsilon_1), \bar{S}_2(a_2, \varepsilon_2), \ldots, \bar{S}_n(a_n, \varepsilon_n), \ldots .$$

According to the hypothesis $\bar{S}_1 \supset \bar{S}_2 \supset \ldots \supset \bar{S}_n \supset \ldots$.
We consider the sequence of centers of these spheres

$$a_1, a_2, \ldots a_n, \ldots, a_{n+p}, \ldots .$$

Since $\bar{S}_{n+p} \subset \bar{S}_n$, $a_{n+p} \in \bar{S}_n(a_n, \varepsilon_n)$. Therefore

$$\rho(a_{n+p}, a_n) \leqq \varepsilon_n.$$

Consequently

$$\rho(a_{n+p}, a_n) \to 0$$

as $n \to \infty$, i.e., the sequence of centers of the spheres is fundamental.

Since the space X is complete, this sequence converges to a limit $a \in X$. We select an arbitrary sphere $\bar{S}_k$ (k fixed). Then the points $a_k, a_{k+1}, \ldots, a_{k+n}, \ldots$ lie in this sphere. Since the sphere $\bar{S}_k$ is closed, the limit a of the sequence $a_k, a_{k+1}, \ldots, a_{k+n}, \ldots$ also belongs to $\bar{S}_k$, therefore

$$a = \lim_{n \to \infty} a_n$$

belongs to all the spheres.

Let us assume that there exists a point b different from a, which belongs to all spheres, such that

$$\rho(a, b) = \delta > 0.$$

Since the points a and b are in $\bar{S}_n$, $n = 1, 2, \ldots$, it follows that

$$\delta = \rho(a, b) \leqq \rho(a, a_n) + \rho(a_n, b) \leqq 2\varepsilon_n.$$

This is impossible, however, since $\varepsilon_n \to 0$, q.e.d.

A set M is called *of first category* if it can be represented as the union of at most countably many nowhere dense sets. A set which is not of first category is called *of second category*.

For example, the set of rational points of a straight line is of first category, the set of all irrational points is of second category.

THEOREM 2. *A complete space is a set of second category.*

We assume the converse: Let the complete space X be representable in the form $X = \sum_{n=1}^{\infty} M_n$, where $M_n, n = 1, 2, \ldots$ are nowhere dense sets. We choose a sphere $\bar{S}(a, 1)$ with center an arbitrary point a and radius 1. Since M_1 is nowhere dense, there exists a sphere $\bar{S}(a_1, r_1)$ interior to $\bar{S}(a, 1)$ with radius $r_1 < \frac{1}{2}$ which contains no point of the set M_1. Since the set M_2 is nowhere dense, there exists a sphere $\bar{S}(a_2, r_2)$ interior to $\bar{S}(a_1, r_1)$ with radius $r_2 < \frac{1}{2^2}$ which contains no point of the set M_2, etc.

We obtain a sequence of closed spheres

$$\bar{S}(a_1, r_1), \bar{S}(a_2, r_2), \ldots, \bar{S}(a_n, r_n), \ldots,$$

where every one lies in the preceding one and whose radii converge to 0. $\bar{S}(a_n, r_n)$ contains no points of the sets $M_1, M_2, \ldots, M_n$. According to the preceding theorem, there exists a point $a_0 \in X$ belonging to all spheres. On the other hand, this point a_0 belongs to none of the sets M_n, hence

$$a_0 \notin X = \sum_{n=1}^{\infty} M_n.$$

We therefore obtain a contradiction, and the theorem is proved.

§7. THE PRINCIPLE OF CONTRACTION MAPPING

The iteration method is well-known. We often use it in existence proofs for solutions of algebraic, differential, integral and other equations. With its help we can also find approximating solutions. This iteration method belongs to the general setting of Functional Analysis and leads to the principle of the contraction mapping, which was formulated by the Polish mathematician S. Banach.

THEOREM. *Let there be given in a complete metric space X an operator A which transforms the elements of the space X into elements of the same space. Suppose further that*

$$\rho(A(x), A(y)) \leqq \alpha\rho(x, y), \tag{1}$$

where $\alpha < 1$. Then there exists exactly one point x_0 such that $A(x_0) = x_0$.

The point x_0 is called *fixed point* of the operator A.

We select an arbitrary but fixed element $x \in X$, put

$$x_1 = A(x), x_2 = A(x_1), x_3 = A(x_2), \ldots, x_n = A(x_{n-1}), \ldots$$

and show that the sequence $\{x_n\}$ is fundamental. We remark that

$$\rho(x_1, x_2) = \rho(A(x), A(x_1)) \leqq \alpha\rho(x, x_1) = \alpha\rho(x, A(x)),$$

$$\rho(x_2, x_3) = \rho(A(x_1), A(x_2)) \leqq \alpha\rho(x_1, x_2) \leqq \alpha^2\rho(x, A(x)),$$

$$\cdots\cdots\cdots\cdots$$

$$\rho(x_n, x_{n+1}) \leqq \alpha^n\rho(x, A(x)),$$

$$\cdots\cdots\cdots\cdots$$

Furthermore

$$\rho(x_n, x_{n+p}) \leqq \rho(x_n, x_{n+1}) + \ldots + \rho(x_{n+p-1}, x_{n+p}) \\ \leqq (\alpha^n + \alpha^{n+1} + \ldots + \alpha^{n+p-1})\rho(x, A(x)) \\ = \frac{\alpha^n - \alpha^{n+p}}{1-\alpha}\rho(x, A(x)). \tag{2}$$

According to the hypothesis $\alpha < 1$, we obtain $\rho(x_n, x_{n+p}) < \frac{\alpha^n}{1-\alpha}\rho(x, A(x))$. From this it follows that $\rho(x_n, x_{n+p}) \to 0$ as $n \to \infty$. Therefore the sequence $\{x_n\}$ is fundamental. Since the space X is complete, there exists an element $x_0 \in X$ which is the limit of this sequence:

$$x_0 = \lim_{n\to\infty} x_n.$$

We show that $A(x_0) = x_0$. Now

$$\rho(x_n, A(x_0)) \leqq \rho(x_0, x_n) + \rho(x_n, A(x_0)) = \rho(x_0, x_n) + \rho(A(x_{n-1}), A(x_0)) \\ \leqq \rho(x_0, x_n) + \alpha\rho(x_0, x_{n-1}).$$

If ε is arbitrarily given and n is sufficiently large, then

$$\rho(x_0, x_{n-1}) < \frac{\varepsilon}{2\alpha} \text{ and } \rho(x_0, x_n) < \frac{\varepsilon}{2}.$$

Consequently $$\rho(x_0, A(x_0)) < \varepsilon.$$

But $\varepsilon > 0$ was arbitrary, hence this inequality holds if and only if $\rho(x_0, A(x_0)) = 0$, i.e., if $A(x_0) = x_0$, q.e.d.

We assume that there are two elements $x_0 \in X, y_0 \in X$ for which

$$A(x_0) = x_0 \text{ and } A(y_0) = y_0.$$

Then $$\rho(x_0, y_0) = \rho(A(x_0), A(y_0)) \leqq \alpha\rho(x_0, y_0).$$

If we assume $\rho(x_0, y_0) > 0$, then it follows from the preceding that $1 \leqq \alpha$; this is, however, a contradiction to the hypothesis.

If we take the limit as $p \to \infty$ in formula (2) we obtain an estimation for the error of the nth approximation:

$$\rho(x_n, x_0) \leqq \frac{\alpha^n}{1-\alpha}\rho(x, A(x)). \tag{3}$$

Remark 1. We can construct an approximating sequence $\{x_n\}$ which converges to the fixed point x_0, beginning with an arbitrary element $x \in X$. Only the speed with which the sequence $\{x_n\}$ tends to its limit depends on the choice of the element x.

Remark 2. Sometimes it is useful to consider a mapping with the property that (1) is not fulfilled for the whole space.

Let the inequality (1) hold for an r-neighborhood $S(x, r)$ of the point x. If $x_1 = A(x)$ and

$$\rho(x, x_1) < (1-\alpha)r,$$

then the sequence $x_n = A(x_{n-1}), n = 1, 2, \ldots$ converges to the unique (with respect to $S(x, r)$), solution x_0 of the equation

$$x = A(x).$$

To prove this, we show by induction that all x_n and x_0 lie in $S(x, r)$. Let the elements $x_k, k = 1, 2, \ldots, n-1$ be from $S(x, r)$. Then

$$\rho(x_2, x_1) \leqq a\rho(x_1, x) \leqq a(1-a)r,$$
$$\rho(x_3, x_2) \leqq a\rho(x_2, x_1) \leqq a^2(1-a)r,$$
$$\cdots\cdots\cdots\cdots$$
$$\rho(x_{n-1}, x_{n-2}) < a^{n-2}(1-a)r.$$

Therefore $\rho(x_n, x_{n-1}) \leqq \rho(A(x_{n-1}), A(x_{n-2})) \leqq a^{n-1}(1-a)r$

and $\rho(x_n, x) \leqq \rho(x_1, x) + \rho(x_2, x_1) + \ldots + \rho(x_n, x_{n-1})$

$$\leqq \sum_{k=1}^{n} a^{k-1}(1-a)r = (1-a^n)r < r.$$

Therefore x_n lies also in $S(x, r)$.

Hence the whole sequence $\{x_n\}$ is contained in a neighborhood in which (1) is fulfilled. The theorem is therefore applicable also in this case.

We offer several examples of the principle of the contraction mapping.

SOLUTION OF SYSTEMS OF LINEAR ALGEBRAIC EQUATIONS BY THE ITERATION METHOD.

We consider an n-dimensional arithmetic space. If $x = \{\xi_1, \xi_2, \ldots, \xi_n\}$ and $y = \{\eta_1, \eta_2, \ldots, \eta_n\}$, then we put $\rho(x, y) = \max_i |\xi_i - \eta_i|$. We can easily show that the metric space m_n thus defined is complete. In this space we consider the operator $y = A(x)$, which is determined by the equations

$$\eta_i = \sum_{j=1}^{n} a_{ij}\xi_j + b_i, \quad i = 1, 2, \ldots, n\ .$$

We have

$$\rho(y_1, y_2) = \rho(A(x_1), A(x_2)) = \max_i |\eta_i^{(1)} - \eta_i^{(2)}|$$
$$= \max_i \left| \sum_{j=1}^{n} a_{ij}(\xi_j^{(1)} - \xi_j^{(2)}) \right| \leqq \max_i \sum_{j=1}^{n} |a_{ij}|\,|\xi_j^{(1)} - \xi_j^{(2)}|$$
$$\leqq \max_j |\xi_j^{(1)} - \xi_j^{(2)}| \max_i \sum_{j=1}^{n} |a_{ij}| = \rho(x_1, x_2) \max_i \sum_{j=1}^{n} |a_{ij}|.$$

If we now assume
$$\sum_{j=1}^{n} |a_{ij}| \leqq a < 1 \tag{4}$$

for all i, then we convince ourselves of the applicability of the principle of the contraction mapping, and consequently the operator A has exactly one fixed point. We therefore obtain the

THEOREM. *If for a matrix (a_{ij}) $\sum_{j=1}^{n} |a_{ij}| \leqq a < 1$ holds for all i, then the system of equations*

$$\xi_i - \sum_{j=1}^{n} a_{ij}\xi_j = b_i, \quad i = 1, 2, \ldots, n$$

has exactly one solution $x_0 = (\xi_1^0, \xi_2^0, \ldots, \xi_n^0)$ for arbitrary $b_1, b_2, \ldots, b_n$.

We can find this solution by an iteration method beginning with an arbitrary vector $x = (\bar{\xi}_1, \bar{\xi}_2, \ldots, \bar{\xi}_n)$.

If

$$x_1 = A(x), x_2 = A(x_1), \ldots, x_p = A(x_{p-1}), \ldots \quad \text{and} \quad x_p = (\xi_1^{(p)}, \xi_2^{(p)}, \ldots, \xi_n^{(p)}),$$

then
$$x_0 = \lim_{n\to\infty} x_n,$$
i.e.,
$$\xi_i = \lim_{n\to\infty} \xi_i^{(n)}.$$
According to (3), we obtain
$$\max_i (|\xi_i^{(p)} - \xi_i|) = \rho(x_p, x_0) \leqq \frac{\alpha^p}{1-\alpha}\, \rho(x, A(x)).$$

The conditions (4) are sufficient for the convergence of the sequence of iterative approximations.

EXISTENCE AND UNIQUENESS OF THE SOLUTION OF AN INTEGRAL EQUATION.

THEOREM. *Let the function $K(s, t)$ be defined and measurable in $a \leqq s \leqq b$, $a \leqq t \leqq b$. Furthermore, let*
$$\int_a^b\int_a^b K^2(s,t)\,ds\,dt < \infty$$
and let $f(s) \in L_2(a, b)$. Then the integral equation
$$x(s) = f(s) + \lambda \int_a^b K(s,t)\,x(t)\,dt$$
has for every sufficiently small value of the parameter λ exactly one solution $x(s) \in L_2(a, b)$.

We consider the operator
$$A(x) = f(s) + \lambda \int_a^b K(s,t)\,x(t)\,dt$$
and show that it transforms every function $x(t) \in L_2(a, b)$ into a function which also belongs to this space. Since $f(s) \in L_2(a, b)$, it suffices to show that the operator
$$A_0(x) = \int_a^b K(s,t)\,x(t)\,dt$$
transforms every function $x(t) \in L_2(a, b)$ into a function belonging to $L_2(a, b)$.

Let
$$\int_a^b K(s,t)\,x(t)\,dt = y(s).$$
Then, according to the Buniakovskii [11] inequality
$$y^2(s) = \left(\int_a^b K(s,t)\,x(t)\,dt\right)^2 \leqq \int_a^b K^2(s,t)\,dt \int_a^b x^2(t)\,dt.$$

---

[11] Cauchy proved the analogous inequality for finite sums. Buniakovskii used the inequality for classical integrals. Schwarz proved the inequality for the Lebesgue integral. Hence the "C.B.S." inequality as suggested by E. Hewitt [Trans.].

Since
$$\int_a^b x^2(t)\,dt$$
is a constant and by hypothesis
$$\int_a^b K^2(s,t)\,dt$$
is an s-integrable function, then $y^2(s)$ is s-integrable also. Now
$$\int_a^b y^2(s)\,ds \leqq \int_a^b\int_a^b K^2(s,t)\,ds\,dt \int_a^b x^2(t)\,dt.$$
We estimate $\rho(A(x), A(z))$:
$$\rho(A(x),A(z)) = \left(\int_a^b \left[\lambda \int_a^b K(s,t)x(t)\,dt - \lambda \int_a^b K(s,t)z(t)\,dt\right]^2 ds\right)^{\frac{1}{2}}$$
$$= |\lambda| \left(\int_a^b \left[\int_a^b K(s,t)\,[x(t)-z(t)]\,dt\right]^2 ds\right)^{\frac{1}{2}}$$
$$\leqq |\lambda| \left(\int_a^b\int_a^b K^2(s,t)\,ds\,dt\right)^{\frac{1}{2}} \left(\int_a^b [x(t)-z(t)]^2\,dt\right)^{\frac{1}{2}}$$
$$= |\lambda| \left(\int_a^b\int_a^b K^2(s,t)\,dt\,ds\right)^{\frac{1}{2}} \rho(x,z).$$
If
$$|\lambda| < \left(\int_a^b\int_a^b K^2(s,t)\,ds\,dt\right)^{-\frac{1}{2}}, \tag{5}$$
then we can apply the principle of the contraction mapping. Hence existence and uniqueness of the solution are proved for such values λ which satisfy the inequality (5).

The proof of the following theorem, which is due to V. V. Nemytskii [15], leans on the principle of the contraction mapping.

THEOREM. *If a function $K(x, y, u)$ is continuous in a certain region G of the (x, y)-plane and if the inequality*
$$|K(x,y,u_1) - K(x,y,u_2)| \leqq C|u_1 - u_2|, \; C \textit{ constant},$$
is fulfilled for all $|u| \leqq L$ with L constant, then for sufficiently small λ the equation
$$u(x) = \lambda \int_G K(x,y,u(y))\,dy,$$
has exactly one solution $\bar{u}$ which satisfies the condition $|\bar{u}(x)| \leqq L$. If u_1 is an arbitrary continuous function which fulfills the inequality $|u_1(x)| \leqq L$, and if

$$u_n(x) = \lambda \int_G K(x, y, u_{n-1}(y))\,dy, n = 2, 3, \ldots,$$

then $\{u_n\}$ *converges uniformly to* $\bar{u}$.

With the help of the principle of the contraction mapping other results were secured too. This principle is the simplest case of a much more general topological theorem on fixed points. It was established under different circumstances by Brouwer, Birkhoff, Kellogg, P. S. Aleksandrov and V. V. Nemytskii. The most general formulation for metric spaces was given by the Polish mathematician J. Schauder, who lost his life in a tragic way to the Gestapo in the year 1943. Schauder also offered numerous examples. The theorem of Schauder was transferred to non-metric spaces by A. N. Tikhonov. A survey of these works can be found in the article by V. V. Nemytskii [15].

§8. SEPARABLE SPACE

A space X is called *separable*, if there exists in this space a countable everywhere dense set. In other words, if there exists in X a sequence

$$\{x_1, x_2, \ldots, x_n, \ldots\} \tag{1}$$

such that there is for every $x \in X$ a subsequence $\{x_{n_1}, x_{n_2}, \ldots, x_{n_k}, \ldots\}$ of (1) which converges to the element x.

If X is a metric space, we can formulate the definition of a separable space as follows:

In the space X there exists a sequence (1) such that for every $x \in X$ and every $\varepsilon > 0$ there is an element x_{n_0} of (1) with $\rho(x, x_{n_0}) < \varepsilon$.

THE SEPARABILITY OF THE N-DIMENSIONAL EUCLIDEAN SPACE E_n. The set E_n^0, which consists of all points of E_n with rational coordinates is countable and everywhere dense in E_n.

THE SEPARABILITY OF $C[0, 1]$. In the space $C[0, 1]$ we consider the set C_0 consisting of all polynomials with rational coefficients. C_0 is countable. We easily convince ourselves that C_0 is everywhere dense in $C[0, 1]$. If we take an arbitrary function $x(t) \in C[0, 1]$, then according to the theorem of Weierstrass, there exists a polynomial $P(t)$ such that

$$\max_{0 \leqq t \leqq 1} |x(t) - P(t)| < \frac{\varepsilon}{2}$$

where $\varepsilon > 0$ is given. On the other hand there is obviously another polynomial $P_0(t)$ with rational coefficients such that

$$\max_{0 \leqq t \leqq 1} |P(t) - P_0(t)| < \frac{\varepsilon}{2}.$$

Therefore $\rho(x, P_0) = \max\limits_{0 \leqq t \leqq 1} |x(t) - P_0(t)| < \varepsilon$, q.e.d.

THE SEPARABILITY OF l_p. Let E_0 be the set of all elements x of the form $\{r_1, r_2, \ldots, r_n, 0, 0, \ldots\}$, where r_i are rational numbers and n is an arbitrary natural number. E_0 is countable. We can easily show that E_0 is everywhere dense in l_p. We select an element $x = \{\xi_1, \xi_2 \ldots, \xi_n, \ldots\}$ of l_p and let $\varepsilon > 0$ be an arbitrarily given number.

We seek a natural number n such that

$$\sum_{k=n+1}^{\infty} |\xi_k|^p < \frac{\varepsilon^p}{2}$$

holds. We now take an element $x_0 = \{r_1, r_2, \ldots, r_n, 0, 0, \ldots\}$ of E_0 such that

$$\sum_{k=1}^{n} |\xi_k - r_k|^p < \frac{\varepsilon^p}{2}.$$

We then obtain

$$[\rho(x, x_0)]^p = \sum_{k=1}^{n} |\xi_k - r_k|^p + \sum_{k=n+1}^{\infty} |\xi_k|^p < \varepsilon^p$$

and from this $$\rho(x, x_0) < \varepsilon.$$
Hence the assertion is proved.

THE SEPARABILITY OF $L_p(0, 1)$. Every function $x(t)$ of the space $L_p(0, 1)$ is the limit in the mean of p-th power of a sequence of bounded measurable functions $x_n(t)$ which are defined by

$$x_N(t) = \begin{cases} x(t) \text{ if } |x(t)| \leqq N, \\ N, \quad \text{if } |x(t)| > N. \end{cases}$$

It readily follows from a theorem of Luzin (C-property) that every bounded measurable function is the limit in the mean of pth power of a sequence of continuous functions. Consequently, the set of all functions continuous on $[0, 1]$ is everywhere dense in $L_p(0, 1)$. On the other hand, the countable set of polynomials with rational coefficients is everywhere dense in $C[0, 1]$, and this in the sense of the metric of this space, therefore a fortiori in the sense of the metric of $L_p(0, 1)$. Therefore the considered set of polynomials is everywhere dense in $L_p(0, 1)$. This proves the separability of $L_p(0, 1)$.

THE SEPARABILITY OF s. Let E_0 be the set of the elements x of the form

$$\{r_1, r_2, \ldots, r_n, 0, 0, \ldots\},$$

where r_i are rational numbers and n is an arbitrary natural number. E_0 is countable. We show that from E_0 we can select a subsequence which converges to an arbitrarily chosen element

$$x = \{\xi_1, \xi_2, \ldots, \xi_n, \ldots\} \in s.$$

For every ξ_n we construct a sequence $\{r_n^{(k)}\}$, $k = 1, 2, 3, \ldots$, of rational numbers which converges to ξ_n as $k \to \infty$. We consider the sequence $\{x^{(k)}\}$ of elements of E_0 of the form

$$x^{(k)} = \{r_1^{(k)}, r_2^{(k)}, \ldots, r_k^{(k)}, 0, 0, \ldots\}.$$

We easily see that $x^{(k)} \to x$ as $k \to \infty$. To prove this, we have to show that the nth component of $x^{(k)}$ converges to the nth component of x as $k \to \infty$. This, however, we see immediately since

$$|\xi_n - r_n^{(k)}| < \varepsilon$$

if we only chose $k > n$ sufficiently large.

THE INSEPARABILITY OF m. We consider the set of all elements $x = \{\xi_i\}$ from m, where ξ_i is equal to 0 or 1. The set of these elements has the cardinality of the continuum. We choose two different elements $x = \{\xi_i\}$ and $y = \{\eta_i\}$ of this set. Then $\rho(x, y) = \sup_i |\xi_i - \eta_i| = 1$, and we have a continuum of elements such that the distance between any two of them is 1. From this, we easily obtain that m is inseparable. We assume that there exists a countable everywhere dense set E_0 in m. About every element of E_0 we construct a sphere of radius $\varepsilon = \frac{1}{3}$. Then all elements of the space E_0 lie in the interiors of these spheres. Because the cardinal number of the spheres is countable, in at least one of these spheres there must be two distinct elements x and y. Let the center of such a sphere be x_0. Then

$$1 = \rho(x, y) \leqq \rho(x, x_0) + \rho(x_0, y) \leqq \tfrac{1}{3} + \tfrac{1}{3} = \tfrac{2}{3},$$

which, however, is impossible. Consequently m is inseparable.

We can show, however, that the subspace c of m is separable.

INSEPARABILITY OF THE SPACE OF SQUARE-INTEGRABLE FUNCTIONS ON AN INFINITE INTERVAL.
We consider the set of all functions $x(t)$ which are defined on the real line, therefore for $-\infty < t < \infty$, and are measurable and such that the

$$\lim_{T\to\infty} \frac{1}{T} \int_{-T}^{T} x^2(t)\, dt$$

exists, where the integral is understood in the sense of Lebesgue. We can show that this set of functions becomes a complete metric space if we introduce a metric by the formula [30]

$$\rho(x, y) = \left(\lim_{T\to\infty} \frac{1}{T} \int_{-T}^{T} [x(t) - y(t)]^2\, dt \right)^{\frac{1}{2}}$$

We designate this space by $\tilde{L}_2$.

We easily prove that $\tilde{L}_2$ is inseparable. Consider a continuum of elements $\{\sin \alpha t\}$, where α is an arbitrary real number. For $\alpha \neq \beta$, we have

$$\rho^2(\sin \alpha t, \sin \beta t) = \lim_{T\to\infty} \frac{1}{T} \int_{-T}^{T} [\sin \alpha t - \sin \beta t]^2\, dt = 2.$$

Consequently, we obtain as in the above example an uncountable set of elements such that the distance between any two of them is $\sqrt{2}$! Similar to the earlier case it follows that the assumption of the separability of $\tilde{L}_2$ leads to a contradiction.

§9. COMPACTNESS IN METRIC SPACES

DEFINITION. More than one hundred years ago the Czech mathematician B. Bolzano proved the theorem that every bounded infinite set of points of the real number line has at least one limit point. He called attention to the importance of this theorem for a rigorous construction of analysis. The idea to select a convergent sequence from an arbitrary set which does not consist of points, but of functions and curves, was used in the existence proof for solutions of ordinary differential equations, in the calculus of variations, and so on. This leads to the general definition of compactness of a set lying in an abstract space.

A set M lying in an abstract (in particular, a metric) space X is called *B-compact* if every infinite subset of this set contains a convergent sequence. If these limits belong to M, then M is called *compact in itself*. When, however, they do belong to the space X and possibly not to the set M, then M is called *compact in X* or compact with respect to X. Necessary and sufficient for the compactness of M in itself is that M be closed and compact in X.

If, in particular, every infinite subset of the space X contains a sequence converging to an element of X, then the space X is called compact. Obviously, every compact space is complete.

Examples. 1. Let $X = [0, 1]$. Obviously X is a compact space according to the theorem of B. Bolzano.

2. $X = E_1$, the one-dimensional Euclidean space (the real number line), is not compact. The subset $M = \{0, 1, 2, \ldots, n, \ldots\}$ contains no convergent sequence. According to the theorem of Bolzano, however, every bounded set of elements of this space is compact.

3. Analogous to the above, we find that E_n, the n-dimensional Euclidean space, is not compact. Every bounded set of elements of this space, however, is compact.

4. $X = C[0, 1]$ is not compact. There exists in $C[0, 1]$ a bounded non-compact set (see remark p. 36).

5. The space $X = l_2$ is not compact. There exists in this space a bounded non-compact set. For example, the closed unit sphere $\bar{S}_1$, i.e., the set of all elements $x = \{\xi_1, \xi_2, \ldots, \xi_n, \ldots\}$ for which

$$\left(\sum_{i=1}^{\infty} \xi_i^2\right)^{\frac{1}{2}} \leqq 1$$

holds, is not compact. We consider the sequence of points of $\bar{S}_1$

$$x_1 = \{1, 0, 0, \ldots\},\ x_2 = \{0, 1, 0, \ldots\},\ x_3 = \{0, 0, 1, \ldots\}, \ldots$$

then $\rho(x_i, x_j) = \sqrt{2}$ for $i \neq j$. Therefore the sequence $\{x_n\}$ and every subsequence of it is divergent, which proves the non-compactness of $\bar{S}_1$. There exists, however, a compact set in l_2. A non-trivial example of such a set is the so-called "Hilbert cube U", which is determined by the totality of points $x = \{\xi_1, \xi_2, \ldots, \xi_n, \ldots\}$, whose coordinates satisfy the condition $|\xi_n| \leqq \frac{1}{n}$. The compactness of the set U follows from the general criteria for compactness, formulated on page 136. See also p. 137.

P. S. Uryson has proved that every *separable* metric space can be mapped in a one-to-one and continuous manner onto the Hilbert cube.

THE EXISTENCE OF AN EXTREMUM. The proofs of the theorems on continuous functions defined on a closed interval are based on compactness. Several of these theorems can be generalized to continuous functionals defined on compact sets of an arbitrary metric space. The following generalizations of the well-known theorems of Weierstrass hold.

THEOREM 1. *If the set M in a space X is compact in itself, and if $f(x)$ is a continuous functional defined on this set, then:*

1. *The functional $f(x)$ is bounded on M.*
2. *The functional $f(x)$ assumes on M its least upper and greatest lower bounds.*

Proof. 1. We show that the functional $f(x)$ is bounded above (analogously, we can show boundedness below). We assume the converse. Then there exists a sequence $\{x_n\}$ of points of M, such that $f(x_n) > n$. Since the set M is compact in itself, the sequence $\{x_n\}$ contains a subsequence $\{x_{n_k}\}$ which converges to a point $x_0 \in M$. On the one hand, $f(x_{n_k}) > n_k$ and consequently $f(x_{n_k}) \to \infty$ as $n_k \to \infty$. On the other hand, however,

$$f(x_{n_k}) \to f(x_0) \text{ as } n_k \to \infty,$$

since the functional is continuous everywhere on M; therefore, in particular, at the point x_0. We have arrived at a contradiction, and the boundedness of the functional $f(x)$ is proved.

2. Let $\beta = \sup_{x \in M} f(x)$. This means that $f(x) \leqq \beta$ for all $x \in M$, and that for every $\varepsilon > 0$, there exists a point $x' \in M$ such that

$$f(x') > \beta - \varepsilon.$$

Consequently we can choose a sequence $\{x_n\}$ such that

$$\beta - \frac{1}{n} < f(x_n) \leqq \beta. \tag{1}$$

Since M is compact in itself, the sequence $\{x_n\}$ contains a subsequence $\{x_{n_k}\}$ which converges to $x_0 \in M$. Then, according to (1),

$$\beta - \frac{1}{n_k} < f(x_{n_k}) \leqq \beta$$

and so

$$\lim_{n_k \to \infty} f(x_{n_k}) = \beta.$$

On the other hand,

$$\lim_{n_k \to \infty} f(x_{n_k}) = f(x_0),$$

because $f(x)$ is continuous on M and therefore also at the point x_0. Thus $f(x_0) = \beta$, q.e.d.

Analogously we can show that if $\alpha = \inf_{x \in M} f(x)$, then there exists a point $x_0' \in M$ such that $f(x_0') = \alpha$.

Remark. If a functional is defined on a set M which is not compact in itself, then $\sup_M f(x)$ and $\inf_M f(x)$ need not be assumed.

Let us consider, for instance, in $C[0, 1]$ the set M of all functions $x(t)$ for which $x(0) = 0, x(1) = 1$ and $\max_{0 \leqq x \leqq 1} |x(t)| \geqq 1$. The continuous functional

$$f(x) = \int_0^1 x^2(t)\, dt$$

defined on M does not assume its greatest lower bound on M.

For, if $x(t) = t^n$, then $f(x) = \dfrac{1}{2n+1}$. Therefore $\inf f(x) = 0$. Obviously, however, for every continuous curve $x(t)$ which passes through the points $(0, 0)$ and $(1, 1)$, $f(x) > 0$. From this it follows, in particular, that the set of curves considered is not compact in itself, although it is a bounded and closed set in $C[0, 1]$.

Therefore, compactness is an essential hypothesis for Theorem 1. The assumption of the existence of a maximum or a minimum of a functional on a set non-compact in itself can lead to a false conclusion, as the preceding example shows.

As a second example of this kind, we give a "pseudo-proof" of the fifth postulate of Euclid. As is well-known, this postulate is equivalent to the assumption that the sum of the angles of a triangle is equal to π. We can prove rigorously that it cannot be greater than π. Here we show that the sum of the angles of an arbitrary triangle is

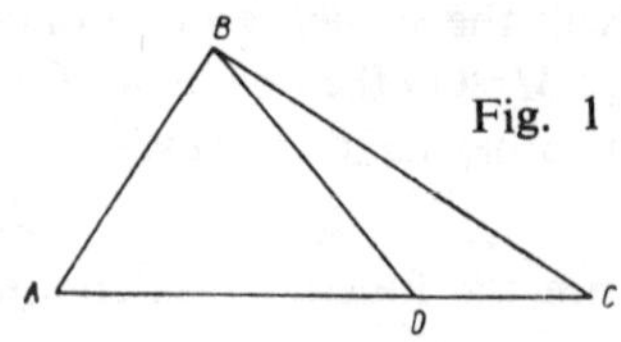

Fig. 1

equal to π. Let K be the least upper bound of the sum of the angles of the triangle, then $K \leqq \pi$, and suppose that there exists a triangle ABC (see Fig. 1) in which the maximum sum of angles is assumed. We join an arbitrary point D lying in the interior of the side AC by a segment BD with B. BD divides our triangle into two triangles, ADB and DCB, and the sum of the angles of each of these triangles is not greater than K. On the other hand, the sum of the angles in both triangles together is equal to $K + \pi$. Consequently, $K + \pi \leqq 2K$. Since, however, K is not greater than π, it follows that $K = \pi$. Therefore, there exists a triangle whose angle-sum is equal to π, and the fifth postulate of Euclid is proved.

The mistake was in the hypothesis that there exists a triangle whose angle-sum assumes the least upper bound. In the geometry of N. I. Lobachevskii the difference between π and the angle-sum of a triangle is proportional to the area of the triangle, and when the difference tends to 0, then the triangle degenerates to a point.

Theorem 1 admits of generalization to semi-continuous functionals. A functional $f(x)$ is called *lower* (*upper*) *semi-continuous* if $x_n \to x$ implies $\underline{\lim}_n f(x_n) \geqq f(x)$ ($\overline{\lim}_n f(x_n) \leqq f(x)$). For such functionals we have:

THEOREM 2. *A lower (upper) semi-continuous functional $f(x)$ defined on a set which is compact in itself, is bounded below (above) and assumes its greatest lower (least upper) bound.*

This theorem finds wide-spread application in the calculus of variations, because the most important classes of functionals considered there are semi-continuous.

CRITERIA FOR COMPACTNESS. We shall now prove a general criterion for compactness of a set lying in a metric space. In order to do this, we introduce the following definition:

A set E of a metric space X is called an ε-*net* for the set M of the same space, if for every point $x \in M$ there exists a point $x_\varepsilon \in E$ such that $\rho(x, x_\varepsilon) < \varepsilon$. (In particular, M can coincide with the whole space X.)

THEOREM 3 (HAUSDORFF). *In order that a set M in a metric space X be compact, it is necessary that for every $\varepsilon > 0$ there exists a finite ε-net for M. If the space X is complete, then this condition is also sufficient.*

Necessity. We assume that M is compact. Let x_1 be an arbitrary point of M. If $\rho(x, x_1) < \varepsilon$ for all $x \in M$, then a finite ε-net exists. If, however, this is not the case, then there exists a point $x_2 \in M$ such that $\rho(x_1, x_2) \geqq \varepsilon$. If for every point $x \in M$ either $\rho(x_1, x) < \varepsilon$ or $\rho(x_2, x) < \varepsilon$, then we have found a finite ε-net. If, however, this does not hold, then there exists a point x_3 such that

$$\rho(x_1, x_3) \geqq \varepsilon, \rho(x_2, x_3) \geqq \varepsilon.$$

Continuing in this way, we obtain points $x_1, x_2, \ldots, x_n$ for which

$$\rho(x_i, x_j) \geqq \varepsilon$$

if $i \neq j$. There exist two possibilities. Either the procedure ceases after the kth step, i.e., for every $x \in M$ one of the inequalities

$$\rho(x_i, x) < \varepsilon, i = 1, 2, \ldots, k,$$

holds and the $x_1, x_2, \ldots, x_k$ form a finite ε-net for M, or we can continue indefinitely the present procedure. The latter, however, cannot occur, since otherwise we would obtain an infinite sequence $\{x_n\}$ of points such that

$$\rho(x_i, x_j) \geqq \varepsilon$$

for $i \neq j$, and neither this sequence nor any of its subsequences would converge. This is a contradiction to the hypothesis that M is compact.

Sufficiency. We assume that the space X is complete and that to every $\varepsilon > 0$ there is a finite ε-net for M. We choose a null sequence $\{\varepsilon_n\}$. For every ε_n we construct a finite ε_n-net

$$[x_1^{(n)}, x_2^{(n)}, \ldots, x_{k_n}^{(n)}]$$

for the set M. Then we choose an arbitrary infinite subset $T \subset M$. Around every point of the net $x_1^{(1)}, x_2^{(1)}, \ldots, x_{k_1}^{(1)}$ belonging to ε_1, we put a closed sphere with a radius ε_1. Then every point of T is contained in one of these spheres. Since the number of the spheres is finite, there is at least one sphere containing an infinite set of points of T. We denote this subset of T by T_1. Around every point $x_1^{(2)}, x_2^{(2)}, \ldots, x_{k_2}^{(2)}$ of the ε_2-net we place a closed sphere with a radius ε_2. By the same reasoning as above, we obtain an infinite set $T_2 \subset T_1$, situated in one of the constructed spheres of radius ε_2. Continuing this procedure, we obtain a sequence of infinite subsets of $T: T_1 \supset T_2 \supset \ldots \supset T_n \ldots$, where the subset T_n is contained in a closed sphere of radius ε_n. Consequently, the distance between two arbitrary points of T_n does not exceed $2\varepsilon_n$.

Now we choose a point $a_1 \in T_1$, a point $a_2 \in T_2$, different from a_1, a point $a_3 \in T_3$, different from a_1 and a_2, and so on, and we obtain a sequence of points

$$T_\omega = \{a_1, a_2, \ldots, a_n, \ldots\}.$$

This sequence is fundamental. For, $a_n \in T_n$ and $a_{n+p} \in T_{n+p}$ for every natural number p. Hence

$$\rho(a_{n+p}, a_n) < 2\varepsilon_n \to 0$$

as $n \to \infty$. Because, by hypothesis the space X is complete, the sequence T_ω converges to a point $a \in X$. This proves the compactness of the set M.

COROLLARY 1. *A set M of a complete metric space X is compact if and only if for every $\varepsilon > 0$ there exists a compact ε-net for M.*

Let E be a compact $\frac{\varepsilon}{2}$-net for the set M. Applying the preceding theorem to E, we find that there exists a finite $\frac{\varepsilon}{2}$-net E_0 for E. Then E_0 is a finite ε-net for M. For, for every point $x \in M$ there exists a point $x_1 \in E$ such that

$$\rho(x, x_1) < \frac{\varepsilon}{2}.$$

Furthermore, for every point $x_1 \in E$ there exists a point $x_2 \in E_0$ such that

$$\rho(x_1, x_2) < \frac{\varepsilon}{2}.$$

Consequently, for every point $x \in M$, there exists a point x_2, such that

$$\rho(x, x_2) \leqq \rho(x, x_1) + \rho(x_1, x_2) < \frac{\varepsilon}{2} + \frac{\varepsilon}{2} = \varepsilon,$$

i.e., E_0 is a finite ε-net for M.

Since the space X is complete, we conclude by Theorem 3 that M is compact.

COROLLARY 2. *A compact space X is separable.*

We choose an arbitrary sequence $\{\varepsilon_n\}$ with $\lim_n \varepsilon_n = 0$ and construct, for every ε_n a finite ε_n-net. Let

$$T_n = [x_1^{(n)}, x_2^{(n)}, \ldots, x_{k_n}^{(n)}]$$

and $E = \sum_{n=1}^{\infty} T_n$. Obviously, E is a countable set which is everywhere dense in X.

COROLLARY 3. *A compact set M of a metric space is bounded.*

Let $T_1 = x_1, \ldots, x_n$ be a 1-net for M, a a fixed element of the space X and

$$d = \max_i \rho(a, x_i).$$

Evidently, we have for every point $x \in M$,

$$\rho(x, a) < 1 + d,$$

which completes the proof.

§10. CRITERIA FOR COMPACTNESS OF SETS.

CRITERION FOR COMPACTNESS IN $C[0,1]$. Functions of a set M are called *uniformly bounded* if there exists a constant K such that $|x(t)| \leqq K$ for all functions $x(t)$ belonging to M. They are called *equicontinuous* if for every $\varepsilon > 0$ and for all functions $x(t) \in M$, there exists a $\delta < 0$ depending only on ε such that $|t_1 - t_2| < \delta(\varepsilon)$ implies $|x(t_1) - x(t_2)| < \varepsilon$ for every t_1 and $t_2 \varepsilon [0, 1]$.

THEOREM 1 (ARZELÀ). *A set $M \subset C[0,1]$ is compact if and only if the functions $x(t) \in M$ are uniformly bounded and equicontinuous.*

Necessity. Let M be compact. The uniform boundedness of the functions $x(t) \in M$ follows from Corollary 3 to Theorem 3. We now prove the equicontinuity of the functions $x(t) \in M$. For a given $\varepsilon > 0$, we construct a finite $\frac{\varepsilon}{3}$-net, $[x_1(t), x_2(t), \ldots, x_k(t)]$ for the set M. Since the functions $x_i(t)$ are continuous on $[0, 1]$ they are also uniformly continuous on this interval.

For every function $x_i(t)$, we select a δ_i, such that

$$|x_i(t_1) - x_i(t_2)| < \frac{\varepsilon}{3}$$

holds for $|t_1 - t_2| < \delta_i$ and all t_1, $t_2 \varepsilon [0, 1]$. Let δ be the smallest number δ_i, $i = 1, 2, \ldots, k$. Now, if $|t_1 - t_2| < \delta$

$$|x(t_1) - x(t_2)| \leqq \max_{0 \leqq t_1 \leqq 1} |x(t_1) - x_i(t_1)| + |x_i(t_1) - x_i(t_2)|$$
$$+ \max_{0 \leqq t_2 \leqq 1} |x_i(t_2) - x(t_2)| < 2\rho(x, x_i) + \frac{\varepsilon}{3}$$

holds for every function $x(t) \in M$. If we select from the net that $x_i(t)$ for which

$$\rho(x_i, x) < \frac{\varepsilon}{3}$$

holds, then

$$|x(t_1) - x(t_2)| < \varepsilon.$$

Since $\varepsilon > 0$ was arbitrary and the estimation thus obtained does not depend on the location of the points $t_1, t_2, \in [0, 1]$ and on the choice of the function $x(t) \in M$, the equicontinuity of all functions belonging to M is proved.

Sufficiency. By virtue of the hypotheses of the theorem, we can give for every $\varepsilon > 0$, a $\delta > 0$ such that

$$|x(t_1) - x(t_2)| < \varepsilon$$

for $|t_1 - t_2| < \delta$, for every $t_1 \in [0, 1]$, $t_2 \in [0, 1]$ and for every function $x(t) \in M$. We select a natural number n such that $\frac{1}{n} < \delta$ and divide $[0, 1]$ into n equal parts

$$\left[\frac{i}{n}, \frac{i+1}{n}\right], \qquad i = 0, 1, \ldots n-1.$$

Then

$$|x(t_1) - x(t_2)| < \varepsilon \tag{1}$$

holds for every function $x(t) \in M$ and for all $t_1, t_2 \in [0, 1]$, for which $|t_1 - t_2| < \frac{1}{n} < \delta$; in particular for all t_1, t_2 from one and the same subinterval $\left[\frac{i}{n}, \frac{i+1}{n}\right]$.

We associate with every $x(t)$ a function $x_n(t)$ such that

1) $x_n\left(\frac{i}{n}\right) = x\left(\frac{i}{n}\right)$ for $i = 0, 1, 2, \ldots, n$, and

2) the functions $x_n(t)$ are linear on the intervals $\left[\frac{i}{n}, \frac{i+1}{n}\right]$.

Therefore $x_n(t)$ is a polygonal curve of n segments inscribed in the curve $x(t)$.

Let

$$x\left(\frac{i}{n}\right) \leqq x\left(\frac{i+1}{n}\right).$$

Then, according to the linearity of $x_n(t)$ on $\left[\frac{i}{n}, \frac{i+1}{n}\right]$,

$$x\left(\frac{i}{n}\right) \leqq x_n(t) \leqq x\left(\frac{i+1}{n}\right),$$

hence

$$x(t) - x\left(\frac{i+1}{n}\right) \leqq x(t) - x_n(t) \leqq x(t) - x\left(\frac{i}{n}\right).$$

If

$$x\left(\frac{i+1}{n}\right) \leqq x\left(\frac{i}{n}\right),$$

then we obtain

$$x(t) - x\left(\frac{i}{n}\right) \leqq x(t) - x_n(t) \leqq x(t) - x\left(\frac{i+1}{n}\right).$$

Consequently

$$|x(t) - x_n(t)| < \varepsilon$$

for all $t \in [0, 1]$, i.e., $\rho(x, x_n) < \varepsilon$.

Therefore the set E of functions $x_n(t)$ forms an ε-net for M.

Furthermore, according to the uniform boundedness of the set M,

$$|x_n(t)| \leqq |x(t)| + |x(t) - x_n(t)| < \varepsilon + K = K_1,$$

i.e., E is uniformly bounded.

E can be interpreted as a bounded set in $(n+1)$-dimensional Euclidean space (every polygonal curve $x_n(t)$ is determined by the $n+1$ ordinates of its vertices). Thus, E is a compact set.

Hence, for every $\varepsilon > 0$, we can form a compact ε-net for M. Then, by virtue of the completeness of $C[0,1]$ and Corollary 1 to Theorem 3, M is a compact set.

CRITERIA FOR COMPACTNESS IN $L_p(0,1)$.

First we introduce the Steklov function.

Let $y(t) \in L_p(0,1)$. We put $y(t) = 0$ if t lies outside of $[0,1]$ and form the *Steklov function*

$$y_h(t) = \frac{1}{2h} \int_{t-h}^{t+h} y(s)\, ds.$$

The integral on the right side exists because $y(t) \in L_p(0,1)$ implies $y(t) \in L(0,1)$ and therefore $y(s) \in L(t-h, t+h)$. We now show that $y_h(t) \in L_p(0,1)$.

Applying the Hölder inequality, we obtain

$$|y_h(t)|^p = \left(\frac{1}{2h} \left| \int_{t-h}^{t+h} y(s)\, ds \right| \right)^p \leqq \frac{1}{(2h)^p} \left[\int_{t-h}^{t+h} ds \right]^{\frac{p}{q}} \left[\int_{t-h}^{t+h} |y(s)|^p\, ds \right]$$

$$= \frac{1}{(2h)^p} (2h)^{p\left(1-\frac{1}{p}\right)} \int_{t-h}^{t+h} |y(s)|^p\, ds = \frac{1}{2h} \int_{t-h}^{t+h} |y(s)|^p\, ds.$$

Hence

$$\int_0^1 |y_h(t)|^p\, dt \leqq \frac{1}{2h} \int_0^1 \left[\int_{t-h}^{t+h} |y(s)|^p\, ds \right] dt = \frac{1}{2h} \int_0^1 \left[\int_{-h}^{h} |y(t+\sigma)|^p\, d\sigma \right] dt$$

$$= \frac{1}{2h} \int_{-h}^{h} \left[\int_0^1 |y(t+\sigma)|^p\, dt \right] d\sigma. \tag{2}$$

Here we have put $s = t + \sigma$, $ds = d\sigma$. If s traverses the interval $[t-h, t+h]$, then σ varies in $[-h, h]$. Moreover

$$\frac{1}{2h} \int_{-h}^{h} \left[\int_0^1 |y(t+\sigma)|^p\, dt \right] d\sigma = \frac{1}{2h} \int_{-h}^{h} \left[\int_{\sigma}^{\sigma+1} |y(\tau)|^p\, d\tau \right] d\sigma$$

$$= \frac{1}{2h} \int_{-h}^{h} \left[\int_{\sigma}^{1} |y(\tau)|^p\, d\tau \right] d\sigma \leqq \frac{1}{2h} \int_{-h}^{h} \left[\int_0^1 |y(\tau)|^p\, d\tau \right] d\sigma = \int_0^1 |y(\tau)|^p\, d\tau, \tag{3}$$

if we substitute τ for $t + \sigma$ and the hypothesis is invoked that $y(t) = 0$ if $t \,\bar{\in}\, [0,1]$.

Therefore from (2) and (3), we obtain

$$\int_0^1 |y_h(t)|^p\, dt \leqq \int_0^1 |y(t)|^p\, dt, \tag{4}$$

and from this $y_h(t) \in L_p(0,1)$.

Putting $y(t) = x(t) - z(t)$ in the inequality (4) and using $y_h(t) = x_h(t) - z_h(t)$, we arrive at

$$\int_0^1 |x_h(t) - z_h(t)|^p dt \leqq \int_0^1 |x(t) - z(t)|^p dt.$$

If we raise both sides to the $\frac{1}{p}$ power, there results

$$\rho(x_h, z_h) \leqq \rho(x, z). \tag{5}$$

THEOREM 2. (KOLMOGOROV). *A set X of the space $L_p(0, 1)$ is compact if and only if*

1) *there exists a constant K such that*

$$\int_0^1 |x(t)|^p dt \leqq K^p$$

for every function $x(t) \in X$ and

2) *for every $\varepsilon > 0$, there exists a $\delta > 0$ such that for every $x(t) \in X$ and $h < \delta$, $\rho(x, x_h) < \varepsilon$.*

Necessity. Condition 1 follows from Corollary 3 to Theorem 3, §9. Let us consider 2. Because the set X is compact, there exists for every $\varepsilon > 0$ a finite $\frac{\varepsilon}{3}$-net E for X. Because every function $f(t) \in L^p(0, 1)$ can be approximated arbitrarily closely by a continuous function $\phi(t)$ in the sense of the metric of this space, we can assume that E consists of continuous functions

$$x_1(t), x_2(t), \ldots, x_n(t).$$

Moreover,

$$y_h(t) = \frac{1}{2h} \int_{t-h}^{t+h} y(s) ds = y(t + \theta h) \to y(t) \text{ as } h \to 0$$

uniformly on $[0, 1]$ if $y(s)$ is continuous. Therefore

$$y_h(t) \to y(t)$$

in the mean as $h \to 0$ and hence $\rho(y, y_h) \to 0$ as $h \to 0$, if $y(t)$ is continuous.

Let us now choose an $x_i(t) \in E$. Since $x_i(t)$ is continuous there exists, for every $\varepsilon > 0$, a $\delta_i > 0$ such that

$$\rho((x_i)_h, x_i) < \frac{\varepsilon}{3}$$

for all $h < \delta_i$. Let $\delta = \min \delta_i$. Then, for $h < \delta$,

$$\rho((x_i)_h, x_i) < \frac{\varepsilon}{3}$$

for all $i = 1, 2, \ldots, n$.

Now, let $x(t)$ be an arbitrary function of X. We seek an $x_i(t) \in E$ such that

$$\rho(x, x_i) < \frac{\varepsilon}{3}.$$

Invoking the inequality (5), we discover that

$$\rho(x_h, x) \leqq \rho(x_h, (x_i)_h) + \rho((x_i)_h, x_i) + \rho(x_i, x)$$

$$\leqq 2\rho(x_i x) + \rho((x_i)_h, x_i) < 2\frac{\varepsilon}{3} + \frac{\varepsilon}{3} = \varepsilon$$

for $h < \delta$.

The necessity for the second condition is therefore proved.

Sufficiency. First we show that under the hypotheses of the theorem, the function $x_h(t)$ corresponding to the functions $x(t)$ of the set X are equicontinuous for every fixed h.

We have

$$|x_h(t_1) - x_h(t)| = \frac{1}{2h}\left|\int_{t_1-h}^{t_1+h} x(\tau)\,d\tau - \int_{t-h}^{t+h} x(\tau)\,dr\right|$$

$$= \frac{1}{2h}\left|\int_{t+h}^{t_1+h} x(\tau)\,d\tau - \int_{t-h}^{t_1-h} x(\tau)\,d\tau\right| \leqq \frac{1}{2h}\left(\left|\int_{t+h}^{t_1+h} x(\tau)\,d\tau\right| + \left|\int_{t-h}^{t_1-h} x(\tau)d\tau\right|\right)$$

$$\leqq \frac{1}{2h}2|t_1 - t|^{\frac{1}{q}}\left(\int_0^1 |x(\tau)|^p\,d\tau\right)^{\frac{1}{p}} \leqq \frac{K}{h}|t_1 - t|^{\frac{p-1}{p}}, \tag{6}$$

and from this, the equicontinuity of the functions $x_h(t)$ for fixed h.

We now show that the set of the functions $x_h(t)$ with fixed h is uniformly bounded. From (6)

$$|x_h(t_1) - x_h(t)| \leqq Kh^{-\frac{1}{p}},$$

follows for $|t_1 - t| < h$, i.e., the variation of every function $x_h(t)$ does not exceed $Kh^{-\frac{1}{p}}$ on an interval of length h. If n is the smallest integer for which $nh \geqq 1$, then the variation of $x_h(t)$ on $[0, 1]$ does not exceed $nKh^{-\frac{1}{p}}$. On the other hand,

$$\int_0^1 |x_h(t)|^p\,dt \leqq \int_0^1 |x(t)|^p\,dt \leqq K^p,$$

and this implies that there must exist on $[0, 1]$ for every function $x_h(t)$ at least one point t_0 such that

$$|x_h(t_0)| \leqq K.$$

Since, as we have proved, the variation of each function $x_h(t)$ on $[0, 1]$ is not greater than $nKh^{-\frac{1}{p}}$, then the inequality

$$|x_h(t)| \leqq |x_h(t_0)| + nKh^{-\frac{1}{p}} \leqq K + nKh^{-\frac{1}{p}} = K_1$$

is fulfilled for every $x_h(t)$. The uniform boundedness of the $x_h(t)$ for a fixed h is therefore proved.

According to the theorem of Arzelà the set of functions $x_h(t)$ is compact for every fixed h in the sense of uniform conevrgence, a fortiori in the sense of convergence in the mean.

We select an arbitrary $\varepsilon > 0$. By 2), there exists an h such that for all $x \in X$,

$$\rho(x, x_h) < \varepsilon.$$

This means that the functions $x_h(t)$ form an ε-net for the set X. Since, however, this ε-net is compact, and $L_p(0, 1)$ is complete, X is also compact and the theorem is proved.

We owe to Riesz the following theorem.

Let M be an arbitrary set of functions of $L_p(0, 1)$. If

1) *there exists a constant K such that*

$$\int_0^1 |x(t)|^p \, dt \leqq K^p$$

for all $x(t) \in M$ and

2) *for every $\varepsilon > 0$, there exists a $\sigma > 0$ such that for all $x(t) \in M$ and h with $|h| < \sigma$*

$$\int_0^1 |x(t+h) - x(t)|^p \, dt < \varepsilon^p,$$

then the set M is compact in the sense of the metric of the space $L_p(0, 1)$.

For proof of this theorem, see Nemytskii [15].

CRITERION FOR COMPACTNESS IN THE SPACE Q. AS AN EXAMPLE, PROOF OF AN EXISTENCE THEOREM IN THE CALCULUS OF VARIATIONS. We consider the set of all curves q determined by

$$x = x(t), y = y(t), z = z(t) \quad (0 \leqq t \leqq 1). \tag{7}$$

Let $x(t), y(t), z(t)$ be continuous functions of the parameter t. If q and q_1 are two curves given in the form (7), we let curve-points associated with the same value of the parameter correspond to each other. Let ε be the maximum of the distance between two corresponding points of the two curves. The number ε depends on the choice of the parametric representation of the two curves. As the distance $\rho(q, q_1)$ we take the g.l.b. of the numbers ε over all possible parametric representations.

We easily conclude that the so-defined distance between two curves satisfies all metric axioms. We designate the space thus obtained by Q. Q plays an important role in the calculus of variations. We can show that Q is complete.

THEOREM 3 (HILBERT). *The set Q_1 of all rectifiable curves ($Q_1 \subset Q$) which lie in a finite part of the space and have uniformly bounded lengths, is compact.*

Proof. Let the lengths of the curves $q \in Q_1$ not exceed l. We divide every curve $q \in Q_1$ into n arcs of equal lengths, connect the partitioning points by segments and obtain a polygonal curve q_n. At least one of the n arcs of the curve q and at least one segment of the polygonal curve q_n is not greater than $\frac{l}{n}$. The distance between the points of such an arc and the points of the corresponding segment $\leqq \frac{2l}{n}$. For q and q_n, we introduce a parametric representation such that we obtain a sub-arc of the curve q, the associated segment of the polygonal curve q_n, correspondingly, as t traverses the interval $\left(\frac{i}{n}, \frac{i+1}{n}\right)$, $i = 0, 1, 2, \ldots, n-1$. The distance between two points of q and q_n which correspond to the same value of the parameter is not greater than $\frac{2l}{n}$, i.e.,

$$\rho(q, q_n) \leqq \frac{2l}{n}.$$

Therefore, the set $Q_1^{(n)}$ of the polygonal curves q_n forms a $\frac{2l}{n}$-net for Q_1. Every polygonal curve is determined by the $3(n+1)$ coordinates of its $n+1$ vertices By hypothesis, these coordinates are uniformly bounded. Hence $Q_1^{(n)}$ is compact. Then, however, according to Corollary 1 to Theorem 3, §9, Q is compact too.

From Theorem 3 which we just proved there follows a series of theorems on existence of geodesic curves.

The length of a curve is a lower semi-continuous functional in the space Q (see p. 37).

Let S be a closed surface. We consider the set of rectifiable curves which lie on the surface and pass through two points A and B. Let l be the length of one of these curves.

We use the notation $\{q\}$ for the totality of all curves of our set whose lengths $l(q) \leqq l$. Let $l_0 = \inf_{\{q\}} l(q)$. The set $\{q\}$ is not empty and is, according to Theorem 3, compact. Furthermore, we can show that $\{q\}$ is closed (in Q). Finally $l(q)$ is a lower semi-continuous functional on $\{q\}$. According to the theorem on the minimum (Theorem 2, §9) $\{q\}$ contains a curve q_0 such that $l(q_0) = l_0$. In the calculus of variations, this curve q_0 is called a geodesic curve.

CRITERION FOR COMPACTNESS IN l_p. A set $M \subset l_p$ is compact if and only if M is bounded and if, for every $\varepsilon > 0$, there exists an index n_0, depending only on ε such that $\sum_{i=n+1}^{\infty} |\xi_i|^p < \varepsilon^p$ for $n \geqq n_0$ and for all $x = \{\xi_1, \xi_2, \ldots, \xi_n, \ldots\}$. The proof follows from the general theorem which will be deduced in §28 (see p. 136).

§11. ALMOST PERIODIC FUNCTIONS [12]

We consider the set of all functions $x(t)$ defined and bounded on the real number line $-\infty < t < \infty$. If $x(t)$ and $y(t)$ are two such functions, we define their distance by

$$\rho(x, y) = \sup_{-\infty < t < \infty} |x(t) - y(t)|.$$

As we easily see, all axioms of a metric are fulfilled here. The space thus obtained is denoted by M_∞. We can show that M_∞ is complete.

According to Bohr [31], [38] a function $x(t), -\infty < t < \infty$, is called *almost periodic* if it is continuous on the real number line and if there exists for every $\varepsilon > 0$ a length $l(\varepsilon)$ such that every interval $x < t < x + l$ contains at least one number $\tau = \tau(\varepsilon)$ for which the inequality

$$|x(t+\tau) - x(t)| \leqq \varepsilon \text{ on } -\infty < t < \infty$$

---

[12] For a generalization see J. von Neumann, Almost Periodic Functions in Groups I. *Trans. Amer. Math. Soc.* v. 36, pp. 445-492, 1932. [Trans.]

is fulfilled. Such a number $\tau(\varepsilon)$ is called an ε-period of the function $x(t)$. We can show that every almost periodic function is bounded on the real number line and therefore belongs to the space M_∞.

THEOREM. *A set X of almost periodic functions is compact in the sense of the metric of M_∞ if and only if*

1. *the functions of the set X are uniformly bounded and equicontinuous, and*
2. *for every $\eta > 0$, there exists an $l = l(\eta)$ such that every interval $(\alpha, \alpha + l)$ of the real number line contains a number ρ which is an η-period for all functions of the set X.*

Necessity. The proof of 1 is analogous to the proof of the corresponding assertion in the theorem of Arzelà. We consider Condition 2.

Since X is compact, for every $\eta > 0$, there exists a finite $\frac{\eta}{3}$-net for the set X. Let it consist of the elements $x_1(t), x_2(t), \ldots, x_n(t)$ (that we can take for the $\frac{\eta}{3}$-net points which belong to X can be seen from the proof of the necessity of the condition of Theorem 3, §9). As Bohr has shown, there exists a number $l > 0$ with the property that every interval $(\alpha, \alpha + l)$ contains a number s which is an $\frac{\eta}{3}$-period for all $x_i(t), i = 1, 2, \ldots, n$:

$$|x_i(t+s) - x_i(t)| \leqq \frac{\eta}{3}, i = 1, 2, \ldots, n, -\infty < t < \infty.$$

Since, on the other hand, $\{x_i(t)\}$ is an $\frac{\eta}{3}$-net for X, there exists for every function $x(t) \in X$ an $x_i(t)$ such that

$$|x(t) - x_i(t)| < \frac{\eta}{3} \quad \text{in} \quad -\infty < t < \infty.$$

From these inequalities, it follows that

$$\begin{aligned} |x(t+s) - x(t)| &\leqq |x(t+s) - x_i(t+s)| \\ &+ |x_i(t+s) - x_i(t)| + |x_i(t) - x(t)| < 3\frac{\eta}{3} = \eta \end{aligned}$$

for $-\infty < t < \infty$. Therefore s is an η-period for all $x(t) \in X$ and the necessity of 2 is proved.

Sufficiency. We assume that for a set X of almost periodic functions, 1 and 2 are fulfilled and choose $\eta > 0$. Let $l = l(\eta)$ be determined such that every interval of length l has an η-period for all $x(t) \in X$. We associate with every $x(t) \in X$ a function $\bar{x}(t)$ defined by

$$\bar{x}(t) = \begin{cases} x(t) & \text{if } -l \leqq t \leqq l, \\ x(t - r_n) & \text{if } \begin{cases} n\,l < t \leqq (n+1)\,l, & n = 1, 2, 3, \ldots, \\ n\,l \leqq t < (n+1)\,l, & n = -2, -3, \ldots . \end{cases} \end{cases}$$

Here r_n is an η-period for all $x(t) \in X$, which period lies in the interval $(n\,l, (n+1)\,l)$.

We designate the set of all $\bar{x}(t)$ by X_η. All functions $\bar{x}(t) \in X_\eta$ satisfy the conditions of the theorem of Arzelà on the interval $[-l, l]$. Therefore, X_η is compact in the sense of uniform convergence on this interval. Since, however, $t - r_n \in [-l, l]$, a sequence of functions $\bar{x}(t)$ which converges uniformly on the interval $[-l, l]$ by definition of these functions converges uniformly also on the entire real number line. Consequently the set X_η is compact in the sense of uniform convergence on the entire real number line, i.e., in the sense of the metric of the space M_∞. For arbitrary $x(t) \in X$ and the corresponding $\bar{x}(t) \in X_n$,

$$x(t) - \bar{x}(t) = 0 \text{ if } -l \leqq t \leqq l$$

and

$$x(t) - \bar{x}(t) = x(t) - x(t - r_n) \quad \text{if} \begin{cases} nl < t \leqq (n+1)\,l, & n = 1, 2, 3, \ldots, \\ nl \leqq t < (n+1)\,l, & n = -2, -3, \ldots . \end{cases}$$

Since r_n is an η-period for $x(t)$, we have for an arbitrary t

$$|x(t) - \bar{x}(t)| < \eta.$$

Hence the compact set X_η forms an η-net for X in the space M_∞. According to Corollary 1 to Theorem 3 §9, X is compact and therefore, we have shown that the conditions 1 and 2 are sufficient [39].

Let a function $x(t)$ be given, $-\infty < t < \infty$. We put

$$x_s(t) = x(t + s).$$

We call a continuous function $x(t)$ *almost periodic* if the set of functions $x_s(t)$, $-\infty < s < \infty$, is compact in the sense of the metric of M_∞.

THEOREM. This definition of an almost periodic function is equivalent to that of Bohr.

For, let $x(t)$ be almost periodic according to Bohr. Obviously the set of functions $x_s(t) = x(t + s)$ satisfies both conditions of the preceding theorem. Therefore it is compact.

Now, let $x(t)$ be almost periodic in the sense of the other definition and let $\varepsilon > 0$ be given arbitrarily. A compact set $\{x_s(t)\}$ contains a finite ε-net $[x_{s_1}(t), x_{s_2}(t), \ldots, x_{s_n}(t)]$. We order the $x_s(t)$ according to increasing subscripts: $s_1 < s_2 < \ldots < s_n$. For every $x_s(t)$ there exists an $x_{s_i}(t)$ such that

$$\rho(x_s, x_{s_i}) = \sup_i |x_s(t) - x_{s_i}(t)| < \varepsilon$$

or

$$|x(t + s) - x(t + s_i)| < \varepsilon$$

for all t. If we put $t + s_i = t'$, then

$$|x(t' + s - s_i) - x(t')| < \varepsilon, -\infty < t' < \infty.$$

Consequently, for arbitrary real s, one of the numbers $s - s_i, i = 1, 2, \ldots, n$, is an ε-period. It follows that every interval $(\alpha, \alpha + s_n - s_1)$ contains an ε-period. For, if we put $\alpha + s_n = s$, then we obtain one of the numbers $\alpha + s_n - s_i$, $i = 1, 2, \ldots, n$ as an ε-period. Because $s_1 \leqq s_i \leqq s_n$,

$$\alpha + s_n - s_i \in [\alpha, \alpha + s_n - s_1].$$

Thus the equivalence of the two definitions is proved.[13]

In the preceding, we considered only the simplest properties of compactness and several applications. There exist many others in abstract spaces. V. Kondrashov [25] and A. G. Sigalov [49] have sought important properties. Since compactness offers wide-spread possibilities of applications to the methods of the calculus of variations and theoretical physics, this concept plays a leading role.

---

[13] The advantage of the second definition lies in the fact that it admits a generalization. We can speak about almost periodic functions $y = x(s)$ where the s are elements of an arbitrary topological group and y elements of a metric space.

CHAPTER II

LINEAR SPACES AND LINEAR OPERATORS

§12. ALGEBRAIC OPERATIONS WITH ARBITRARY SETS

GROUPS. In mathematics, addition and multiplication are fundamental operations. They were first defined for numbers, and introduced later for many mathematical objects. We add and multiply vectors, functions, matrices, mappings, and so forth.

In group theory, we study the operations of addition and multiplication starting with axiomatic definitions. It admits of many concrete applications.

We now define several concepts.

A set G is called a *group*, if with every ordered pair x and y of elements of G, there is associated a unique element of the same set called the *sum* of the elements x and y written $x + y$, and such that

1. Addition is *associative*, i.e.

$$(x + y) + z = x + (y + z).$$

2. There exists at least one element θ, the *right zero element*, such that for all $x \in G$,

$$x + \theta = x.$$

3. For every element $x \in G$ there exists at least one element $-x$, the *right negative* of x, for which

$$x + (-x) = \theta.$$

These three conditions are called the group axioms and addition of elements is the group operation. This group-theoretic operation is often interpreted as *multiplication*, and we write $x \cdot y$. Then instead of a zero element we speak of a *unit element* and designate it by e. Instead of the negative element we have the *inverse element* which is written x^{-1}.

A subset G_1 of a group G, which is also a group with respect to the group operation in G, is called a *subgroup* of G.

Examples. 1. Let G be the set of all real numbers, and let the group operation be the addition of numbers in the usual sense. Evidently axioms 1 through 3 are fulfilled. Hence G is a group. The set of integers is a subgroup.

2. Let G be the set of all positive real numbers and the group operation the usual multiplication. Evidently G is again a group. The set of all positive rational numbers is a subgroup of G.

3. Let G be the set of all continuous functions defined on the interval $[0, 1]$ and let the group operation be the usual addition of functions. Again G is a group. The zero element is the function $x(t) \equiv 0$. The negative element to $x(t)$ is the function $-x(t)$.

4. Let G be the set of all elements of the space l_p. For two arbitrary sequences $x = \{\xi_1, \xi_2, \ldots, \xi_n, \ldots\}$ and $\{y = \eta_1, \eta_2, \ldots, \eta_n, \ldots\}$ of G, we put

$$x + y = \{\xi_1 + \eta_1, \xi_2 + \eta_2, \ldots, \xi_n + \eta_n, \ldots\}.$$

Then $x+y$ is an element of G too.[14]

The group axioms 1–3 are fulfilled. The zero element is the element $\theta=\{0,0,\ldots,0,\ldots\}$, the negative to x is $-x=\{-\xi_1,-\xi_2,\ldots,-\xi_n,\ldots\}$.

If G is a group, then in general

$$x+y\neq y+x,$$

i.e., the addition is not commutative. If, however, for every $x,y\in G$,

$$x+y=y+x,$$

then the group is called *commutative* or *Abelian*. We recognize that the group operation is usually called *addition* and designated by the symbol $+$, only if the group is an Abelian one. With non-commutative groups, we customarily designate the group operation by multiplication.

5. Let G be the set of all non-singular square matrices of nth order. Let the group operation be the usual matrix multiplication. Evidently G is a non-commutative group. The groups in the Examples 1–4 are commutative.

We mention several very simple consequences of the group axioms.

Theorem 1. *A right negative element to $x\in G$ is also a left negative to x.*

We have $x+(-x)=\theta$. If we add to both sides of the equality the element $-x$ on the left and use the associativity of addition, then we obtain $(-x)+x+(-x)=(-x)+\theta=-x$. If we now add to both sides of the equation the element $[-(-x)]$ on the right, then we obtain

$$(-x)+x+\theta=\theta, \text{ therefore } (-x)+x=\theta,$$

q.e.d.

Theorem 2. *A right zero element is also a left zero element. There exists exactly one zero element.*

We have $$\theta+x=x+(-x)+x=x+\theta=x.$$

Let θ_1 be another zero element. Then $\theta_1+\theta=\theta_1$ and $\theta_1+\theta=\theta$. Consequently, $\theta_1=\theta$ and the theorem is proved.

Theorem 3. *For every x, there is exactly one negative element.*

Suppose y is another negative element to x. Then

$$y=y+\theta=y+x+(-x)=-x,$$

q.e.d.

In the following we shall always write $x-y$ instead of $x+(-y)$.

Theorem 4. *The equations $x+a=b$ and $a+x=b$ are, for arbitrary $a\in G$, $b\in G$, uniquely solvable.*

Let us consider the first of these equations (the second one may be handled in the same way). We put $x_0=b-a$. Then x_0 is a solution of this equation. It follows that $x_0+a=b-a+a=b$. Let x_1 be another solution. Then $x_1+a=b$. If we add to both sides $-a$ on the right, we obtain $x_1=b-a=x_0$, q.e.d.

Remark. Sometimes for the definition of a group, the unique solvability of the equations $x+a=b$ and $a+x=b$ for every a and b, is required instead of Axioms 2 and 3. It can be shown that from these requirements 2 and 3 follow.

---

[14] See Appendix I.

THEOREM 5. *From each of the equalities* $x+a=y+a$ *and* $a+x=a+y$, *it follows that* $x=y$.

The proof is left to the reader.

In the following we shall consider only Abelian (commutative) groups.

FACTOR GROUPS. Let G_1 be a subgroup of the group G. We define a new group, the so-called factor group of G with respect to the subgroup G_1, and designated by G/G_1.

We form subsets $\{g\}$ of elements of the group G according to the following prescription: two elements x and x' of G are put in the same subset $\{g\}$ if $x-x' \in G_1$. These subsets are called co-sets of the group G with respect to the subgroup G_1.

By the sum $g+g_1$ of the co-sets g and g_1 we mean the set of all sums $x+x_1$ where $x \in g$ and $x_1 \in g_1$.

The set $g+g_1$ is a co-set.

For, if y and y' are two elements of $g+g_1$, then by definition, $y=x+x_1$, $y'=x'+x_1'$ where $x \in g, x' \in g$ and $x_1 \in g_1, x_1' \in g_1'$. We have

$$y-y'=(x+x_1)-(x'+x_1')=(x-x')+(x_1-x_1'). \qquad (*)$$

Since x and x' are in g, and since g is a co-set of G with respect to G_1, $x-x' \in G_1$. Analogously, $x_1-x_1' \in G_1$. G_1, however, is a subgroup and therefore (*) implies $y-y' \in G_1$. Hence $g+g_1$ is a co-set.

As we easily see, the set of co-sets forms an additive group.

We designate it by G/G_1 and call it the *factor group of G with respect to* G_1. G_1 is the zero element of this group. From the commutativity of G, the commutativity of its factor group follows.

Examples. 1. Let the group G be the set of all measurable, almost everywhere bounded functions. Addition is defined in the usual sense. Let the subgroup G_1 be the set of all functions which are almost everywhere equal to 0. Elements of the factor group G/G_1 are the elements of the space M.

2. Let G be the set of all functions $x(t)$ for which $\int_0^1 |x(t)|^p dt$ exists. Let the subgroup G_1 be the set of all functions which are almost everywhere equal to 0. Elements of the factor group G/G_1 are the elements of the space $L_p(0, 1)$.

Topological groups play a particularly important role in analysis, i.e., those groups which are also topological spaces. The group operation is continuous in the topology.

The theory of topological groups was developed during the last decade by the Soviet mathematicians A. N. Kolmogorov, L. S. Pontriagin, A. A. Markov, A. I. Maltsev, I. M. Gel'fand, M. A. Neumark, the Viennese mathematician Haar and others.

RINGS. A set R of elements $x, y, z, \ldots, a, b, c, \ldots$ is called a ring, if the following conditions are fulfilled:

I. R is an *additive Abelian group*.

II. Besides there is also defined in R a multiplication, which associates with every ordered pair of elements $x \in R, y \in R$, a third element of the set R, the *product*. It is designated by $x \cdot y$, and there must hold:

1) the *associative law* of multiplication, i.e.,

$$x(y \cdot z) = (x \cdot y)z;$$

2) the *distributive laws*, i.e.,

$$(x+y)z = xz + yz, \quad z(x+y) = zx + zy.$$

Examples of rings are the integers, the set of all functions continuous on the interval $[0, 1]$ with the definitions of addition and multiplication usual for these sets.

The distributive laws can be extended to subtraction. For,

$$x(y-z) + xz = x(y-z+z) = xy,$$

therefore

$$x(y-z) = xy - xz.$$

Analogously

$$(y-z)x = yx - zx.$$

In particular, for every $x \in R$

$$x\theta = x(x-x) = xx - xx = \theta,$$

i.e., if one of the factors is equal to zero, then the product is equal to zero too. The converse is not always valid, as we can see from the example of the ring of continuous functions. If $x \neq 0, y \neq 0$, and

$$x \cdot y = 0,$$

then x and y are called *divisors of zero*.

If there exist elements e and e' in the ring R such that $xe = x$ and $e'x = x$ for all $x \in R$, then $e' = e'e = e$. This implies $xe = ex = x$ for all $x \in R$. This element e is called the *unit element* of the ring R, and R is called a *ring with unit element*.

An example of a ring with unit element is the ring of all square matrices of nth order. The unit element is the identity matrix.

If x is an arbitrary element of a ring R with unit element, then an element x^{-1} for which

$$x^{-1}x = e$$

holds, is called *left-inverse element* to x. Analogously, an element $\widetilde{x^{-1}}$, for which

$$x\widetilde{x^{-1}} = e$$

holds, is called *right inverse element* to x. If x^{-1} and $\widetilde{x^{-1}}$ exist, then we easily see that $x^{-1} = \widetilde{x^{-1}}$; x^{-1} is called the *inverse element* to x. Obviously, there exists exactly one inverse element.

If for two arbitrary elements x and y of a ring,

$$xy = yx$$

always holds, i.e., if the multiplication is commutative, then R is called a *commutative ring*.

Analogously to the topological group, we define a topological ring.

LINEAR SYSTEMS. A set E is called a *linear system* if:

I. E is an *additive Abelian group*.

II. There is defined a commutative multiplication of the elements $x, y, z, \ldots$

of E with the elements $\lambda, \mu, \nu, \ldots$ of some ring R with unit element. Here the following conditions have to be fulfilled:

1) If $x \in E, \lambda \in R$, then $\lambda x \in E$.

2) $\lambda(\mu x) = (\lambda\mu)x$, Associativity of multiplication.

3) $\begin{cases} (\lambda + \mu)x = \lambda x + \mu x, \\ \lambda(x + y) = \lambda x + \lambda y, \end{cases}$ Distributivity.

4) $ex = x$.

5) $0x = \theta$, where 0 is the zero element of the ring R and θ the zero element of the group E.

6) If $\lambda x = \theta$ and $x \neq \theta$, then $\lambda = 0$.

The set of n-dimensional vectors with the real numbers as ring multipliers forms the simplest example of a linear system.

Examples. 1. The set E of real solutions of an ordinary linear differential equation of nth order is a linear system with the real numbers as ring multipliers.

2. The complex solutions of an ordinary linear differential equation of nth order form a linear system with the complex numbers as ring multipliers.

3. The real solutions of a linear difference equation of nth order with step h form a linear system with the set of periodic functions with the period h as ring multipliers.

In the examples 1 through 3, all elements $y \in E$ can be expressed linearly in terms of fixed elements $y_1, y_2, \ldots, y_n$:

$$y = \sum_{i=1}^{n} \lambda_i y_i, \qquad (*)$$

where the λ_1 are real numbers in Example 1, complex numbers in Example 2. In Example 3, they are periodic functions with period h.

In (2), n cannot be decreased.

4. The set E of all real (complex) solutions of a homogeneous linear partial differential equation forms a linear system with the real (complex) numbers as ring multipliers.

In this case, there does not exist a finite system of elements of E, by which all other elements can be expressed as in Formula (*).

5. The elements of the real (complex) spaces m, c, l_p and others, form a linear system with the real (complex) numbers as ring multipliers.

By the sum $x + y$ of the elements $x = \{\xi_1, \xi_2, \ldots, \xi_n, \ldots\}$ and $y = \{\eta_1, \eta_2, \ldots, \eta_n, \ldots\}$ we mean the element

$$\{\xi_1 + \eta_1, \xi_2 + \eta_2, \ldots, \xi_n + \eta_n, \ldots\}.$$

The product λx is the element

$$\{\lambda\xi_1, \lambda\xi_2, \ldots, \lambda\xi_n, \ldots\}.$$

6. Let $[e_1, e_2, \ldots, e_k]$ be an n-dimensional simplex of a complex Euclidean space. We consider all algebraic complexes

$$u = \sum_{i=1}^{k} c_i e_i,$$

where the c_i are positive or negative integers. Then the set of all algebraic expressions of this kind forms a linear system with the integers as ring multipliers. This linear system plays an important role in combinatorial topology.

We offer several consequences of the axioms for linear systems:

1. $(-e)x = -x$, since

$$(-e)x + x = (-e + e)x = 0x = \theta$$

holds.

2. $\lambda\theta = \theta$ because
$$\lambda\theta = \lambda(x - x) = \lambda x - \lambda x = \theta.$$

3. If $\lambda x = \mu x$ and $x \neq \theta$, then $\lambda = \mu$. For, if $\lambda x = \mu x$, then $\lambda x - \mu x = \theta$ or $(\lambda - \mu)x = \theta$. Consequently, by II. 6, $\lambda - \mu = 0$, i.e., $\lambda = \mu$.

Note: If E is a linear system then the commutativity of addition is a consequence of the other axioms, for
$$(x + y) - (y + x) = (x + y) + (-e)(y + x) = (x + y) + (-e)y + (-e)x$$
$$= x + (y + (-e)y) + (-e)x = x + (-e)x = \theta.$$

Therefore the axioms given for a linear system are not independent of each other.

In the following we take for the ring R the field of real or complex numbers.

ISOMORPHISMS OF ALGEBRAIC MAPPINGS. Two groups G and G' are *isomorphic* if we can establish between the elements of these groups a one-to-one correspondence such that for two arbitrary elements $x \in G, y \in G$, which correspond to the elements $x' \in G', y' \in G'$, the element $x + y$ corresponds to the element $x' + y'$.

It follows from the definition that in an isomorphism of the groups G and G' there is associated with the zero element of G the zero element of G' and with the negative elements in G the negative elements in G'. Isomorphic groups can differ only by the kind of elements. However, relations between the elements of one group derived from the group operation, hold also in the other isomorphic group. Therefore, for investigations which involve only relations between the elements, isomorphic groups are not to be considered as different.

Analogously, we define *isomorphic rings*. Two rings R and R' are called isomorphic if there can be established between their elements an isomorphic correspondence in the above sense with respect to both operations.

Isomorphic rings are, in particular, isomorphic groups with the ring addition as the group operation.

Example. We associate with every complex number z its conjugate $\bar{z}$. This correspondence is an isomorphism of the ring of complex numbers with itself.

Two linear systems E and E_1 with the ring R as the domain of multipliers are called *linear-isomorphic* if a one-to-one correspondence between their elements x and x', which again is valid in the above sense for addition and multiplication, can be established. Linear-isomorphisms are group isomorphisms with addition as group operation.

Example. The linear systems of Examples 1 and 2 on p. 52 are linear-isomorphic.

Remark. With the definition of linear-isomorphism of two systems E and E_1 we need not require coincidence of the ring multipliers but only isomorphism of the rings.

In the following we shall consider algebraic domains which simultaneously are topological spaces with continuous algebraic operations (topological groups, topological rings, topological linear systems or linear spaces).

Just as with the one-to-one correspondence between the elements of isomorphic algebraic domains, we want to consider this correspondence as being continuous.

In particular, we want to consider a one-to-one correspondence which is a linear isomorphism of two linear systems and which is bi-continuous.

§13. LINEAR SPACES

LINEAR SPACES. Linear spaces are a generalization of n-dimensional vector spaces. The foundations of the general theory of linear spaces were established by F. Riesz and S. Banach.

A linear space E is a linear system with a ring R as domain of operators, where there is introduced a topology in E and R, with respect to which addition and operator multiplication are continuous.

In the following the ring R will be the set of real or complex numbers.

DEFINITIONS. A set E is called a *real* (*complex*) *linear space* if the following conditions are fulfilled:

A. E is a linear system with the real (complex) numbers as multipliers.

B. E is an abstract space, i.e., there is defined a limit concept for sequences of elements in E. It is assumed that Theorems 1 and 2, §2, hold for convergent sequences. In particular, E can be a metric space.

C. 1) If $x_n \to x$ and $x_n, x \in E$, then $\alpha x_n \to \alpha x$ for every number α.

2) If $\alpha_n \to \alpha$ and $x \in E$ is arbitrary, then $\alpha_n x \to \alpha x$.

3) If $x_n \in E$, $x \in E, y_n \in E, y \in E$ and if $x_n \to x, y_n \to y$, then $x_n + y_n \to x + y$.

It follows immediately from Axiom C. 1) that: if $x_n \to x$, then $-x_n \to -x$.

We offer several examples of linear spaces.

Examples. 1. The real function spaces $C[0, 1]$, $L_p(0, 1)$ and the sequence spaces c, l_p are linear spaces with the real numbers as multipliers.

2. The complex spaces $C[0, 1]$, $L_p(0, 1)$, c, l_p are linear spaces with complex multipliers.

3. The Examples 1, 2, 4 of linear systems (p. 52) become linear spaces if we introduce uniform convergence as a limit.

We can introduce in linear spaces the concept of *linear dependence* and *linear independence* of elements. The elements

$$x_1, x_2, \ldots, x_n$$

of a linear space are called *linearly independent* if the equality

$$\lambda_1 x_1 + \lambda_2 x_2 + \ldots + \lambda_n x_n = \theta$$

implies

$$\lambda_1 = \lambda_2 = \ldots = \lambda_n = 0.$$

If, conversely, there exist numbers

$$\lambda_1, \lambda_2, \ldots, \lambda_n$$

not all equal to 0, such that

$$\lambda_1 x_1 + \lambda_2 x_2 + \ldots + \lambda_n x_n = \theta$$

then the elements $x_1, x_2, \ldots, x_n$ are called *linearly dependent*. If, for example, $\lambda_n \neq 0$, then

$$x_n = -\frac{\lambda_1}{\lambda_n} x_1 - \frac{\lambda_2}{\lambda_n} x_2 - \ldots - \frac{\lambda_{n-1}}{\lambda_n} x_{n-1}$$

or if we put $-\dfrac{\lambda_i}{\lambda_n} = \alpha_i$,

$$x_n = \alpha_1 x_1 + \alpha_2 x_2 + \ldots + \alpha_{n-1} x_{n-1}.$$[15]

In this case we say that the element x_n is a *linear combination* of the elements $x_1, x_2, \ldots, x_{n-1}$.

LINEAR MANIFOLDS AND SUBSPACES. A non-empty set L of elements of a linear space E is called a *linear manifold* if L contains with the elements $x_1, x_2, \ldots, x_n$ every linear combination $\alpha_1 x_1 + \alpha_2 x_2 + \ldots + \alpha_n x_n$ of these elements. If L is closed, then L is called *closed linear manifold* or *subspace* of the linear space E.

We note that every linear manifold contains the zero element θ. Since L is not empty, L contains an element x. Because L is a linear manifold, it contains also the element $-x = (-1)x$ and consequently also the element $x + (-x) = \theta$.

We consider the elements $x_1, x_2, \ldots, x_k$ of a linear space E. The set of all sums $\sum_{i=1}^{k} \alpha_i x_i$ obviously forms a linear manifold L_0 in E. For, if the elements y_j have the form $y_j = \sum_{i=1}^{k} \alpha_i^{(j)} x_i$, then every linear combination

$$\lambda_1 y_1 + \lambda_2 y_2 + \ldots + \lambda_n y_n = \sum_{i=1}^{k} \beta_i x_i$$

has the same form. The linear manifold L_0 thus constructed is clearly the *smallest linear manifold* which contains the elements $x_1, x_2, \ldots, x_k$. That is, every other linear manifold containing $x_1, x_2, \ldots, x_k$ has L_0 as a subset. L_0 is also the *smallest linear subspace* containing $x_1, x_2, \ldots, x_k$.

The definition of the smallest linear manifold and the smallest linear subspace which contain given elements can easily be generalized to the case of an infinite set, for instance, a denumerable set of elements. Let $\{x_1, x_2, \ldots x_k, \ldots\}$ be a denumerable set of elements of E. The smallest linear manifold L_0 containing this set evidently is the set of all sums of the form $\sum_{i=1}^{k} \lambda_i x_i$ where the λ_i are arbitrary and the k are natural numbers. The smallest linear subspace which contains the given elements is $\bar{L}$, the closure of the union of all the L_0. The smallest linear manifold (the smallest linear subspace) containing the given elements is also called the *linear manifold* (the linear subspace) *defined by the given elements*. We can show: If L_0 is a linear manifold defined by a finite number of elements $x_1, x_2, \ldots, x_k$ of a linear space E, then $\bar{L}_0 = L_0$ (see remark on p. 70).

If L_0 is determined by infinitely many elements, then in general, $\bar{L}_0 \neq L_0$.

Let *e.g.*, $E = C[0, 1]$ and

$$x_1 = 1, x_2 = t, x_3 = t^2, \ldots, x_n = t^{n-1}, \ldots$$

Then L_0 is the set of all polynomials, $\bar{L}_0 = C[0, 1]$ and $\bar{L}_0 \neq L_0$.

If a subspace G of a space E is defined by a finite number of elements, then it is called of *finite dimension*. If G is defined by linearly independent elements

---

[15] The last formulation can be applied only if the ring of multipliers is a field, i.e., if there exists for every element x the inverse x^{-1}. L. S. Pontriagin has proved the following theorem [45]: *Every commutative connected locally compact topological field with second denumerability axiom is isomorphic to the ring of real or complex numbers.* If we ignore the commutativity, it is the field of quaternions.

$x_1, x_2, \ldots, x_n$, then n is called the *dimension* of the subspace G. In this case, we call the set of the elements $x_1, \ldots, x_n$ a *basis* of the subspace G. If $x_1, \ldots, x_n$ are linearly dependent, then the dimension of the subspace G is the maximal number of linearly independent elements of the set $\{x_1, \ldots, x_n\}$.

The set of all elements of the form

$$y = tx, \; -\infty < t < \infty$$

is a *straight line*, and the set of all elements of the form

$$y = tx_1 + (1-t)x_2,$$

where t traverses all real numbers of the interval $[0, 1]$, is called the *segment* joining the points x_1 and x_2. A set K of a space E is called *convex*, if all segments joining any two points of the set K are contained in this set. For instance, every subspace of a space E is a convex set.

DIRECT SUM. We now introduce a representation of a linear space as a direct sum of two or more subspaces. Let E be a linear space and $L_1, L_2, \ldots, L_n$ subspaces of E. If we can write every element $x \in E$ uniquely in the form

$$x = x_1 + x_2 + \ldots + x_n, \; x_i \in L_i, \; i = 1, 2, \ldots, n \tag{1}$$

then we say that the space E is a direct sum of the subspaces $L_1, \ldots, L_n$ and (1) is called a *representation of x by the elements of the subspaces $L_1, \ldots, L_n$*.

We shall write

$$E = L_1 \dot{+} L_2 \dot{+} \ldots \dot{+} L_n \text{ or } E = \sum_{i=1}^{n} \dot{+} L_i.$$

If

$$E = \sum_{i=1}^{n} \dot{+} L_i, \text{ and } L_i = \sum_{k=1}^{m_i} \dot{+} L_k^{(i)},$$

then

$$E = \sum_{i=1}^{n} \sum_{k=1}^{m_i} \dot{+} L_k^{(i)},$$

as we easily see. Then we can represent every element $x \in E$ in the form

$$x = \sum_{i=1}^{n} x_i = \sum_{i=1}^{n} (x_1^{(i)} + x_2^{(i)} + \ldots + x_{m_i}^{(i)}), x_i \in L_i, x_k^{(i)} \in L_k^{(i)}$$

and this representation is unique. For, if

$$x = \sum_{i=1}^{n} \tilde{x}_i = \sum_{i=1}^{n} (\tilde{x}_1^{(i)} + \tilde{x}_2^{(i)} + \ldots + \tilde{x}_{m_i}^{(i)})$$

is another one, then we have, because of the uniqueness of the representation of $x \in E$ by the elements of $L_1, \ldots, L_n$,

$$x_i = x_1^{(i)} + x_2^{(i)} + \ldots + x_{m_i}^{(i)} = \tilde{x}_1^{(i)} + \tilde{x}_2^{(i)} + \ldots + \tilde{x}_{m_i}^{(i)} = \tilde{x}_i,$$

and because of the unique representation of the $x_i \in L_i$ by the elements of the $L_1^{(i)}, \ldots, L_{m_i}^{(i)}$,

$$x_k^{(i)} = \tilde{x}_k^{(i)}, i = 1, 2, \ldots, n, \; k = 1, 2, \ldots, m_i.$$

If $E = L_1 \dot{+} L_2$, then the subspaces L_1 and L_2 have only the zero element θ of the space in common.

For, if L_1 and L_2 would contain an element $u \neq \theta$, then for $x \in E$ with

$$x = y + z, \; y \in L_1, z \in L_2$$

there would also hold the representation

$$x = (y - u) + (z + u), \; y - u \in L_1, \; z + u \in L_2,$$

which is different from the first one. This, however, is impossible.

If, conversely, every element $x \in E$ *can be represented in the form*

$$x = y + z, \; y \in L_1, \; z \in L_2 \tag{2}$$

and if $L_1 \cdot L_2 = \theta$, *then* $E = L_1 \dot{+} L_2$.

To prove this theorem, it is sufficient to prove the uniqueness of the representation (2). If

$$x = y + z = y_1 + z_1, \; y, y_1 \in L_1, \; z, z_1 \in L_2,$$

then

$$y - y_1 = z_1 - z, \; y - y_1 \in L_1, \; z_1 - z \in L_2.$$

By virtue of our hypothesis, it follows that

$$y - y_1 = z - z_1 = \theta, \text{ i.e., } y = y_1, z = z_1,$$

q.e.d.

In many cases, the concept of the direct sum of two or of several spaces happens to be useful. Let $E_1, \ldots, E_n$ be linear spaces. We consider the set of all ordered n-tuples $x = (x_1, x_2, \ldots, x_n)$ of elements of the given space where $x_i \in E_i, i = 1, 2 \ldots, n$. If $x = (x_1, \ldots, x_n), y = (y_1, \ldots, y_n)$ and a scalar λ are given, then we put,

$$x + y = (x_1 + y_1, x_2 + y_2, \ldots, x_n + y_n)$$

and

$$\lambda x = (\lambda x_1, \lambda x_2, \ldots, \lambda x_n).$$

Finally, if $x_k = (x_1^{(k)}, x_2^{(k)}, \ldots, x_n^{(k)})$, then we define the convergence $x_k \to x$ if $x_i^{(k)} \to x_i$ for all i.

We easily see that the operations of addition, scalar multiplication, and taking limits thus defined fulfill all axioms of a linear space. Therefore the set x of ordered n-tuples is a linear space.

If all spaces E_n are metric spaces, then we can metrize X by putting, for example

$$\rho(x, y) = \sum_{i=1}^{n} \rho(x_i, y_i)$$

or

$$\rho(x, y) = \sqrt{\sum_{i=1}^{n} \rho^2(x_i, y_i)}.$$

Here $\rho(x_i, y_i)$ means the distance between the points x_i and y_i of the space E_i.

Example. Let E_i be a real number line for every i. Then $\sum_{i=1}^{n} \dot{+} E_i$ is the n-dimensional Euclidean space.

FACTOR SPACES. We consider a linear space E and one of its subspaces L_0. The space E, being an additive group, decomposes into co-sets L of L_0 in such a way that two elements x_1 and x_2 belong to the same L if and only if $x_1 - x_2$ belongs to L_0. If x' is an arbitrary element of L, then every x of L can be represented in the form $x = x' + x_0$ where $x_0 \in L_0$. Therefore L is formed from the subspace $\bar{L}_0$ by means of x'. We consider the factor group E/L_0, the group of co-sets. The operation of addition in E/L_0 is defined in §12. Let L_1 and L_2 be elements of E/L_0. Then by the sum $L_1 + L_2$, we mean the co-set formed by all sums $x_1 + x_2$ where $x_1 \in L_1, x_2 \in L_2$. E/L_0 is an additive commutative group.

Analogously, we designate by λL the set of all elements λx where $x \in L$. We can see that $\lambda L \in E/L_0$. For, if x_1 and $x_2 \in \lambda L$, then

$$x_1 = \lambda y_1, x_2 = \lambda y_2 \text{ with } y_1, y_2 \in L.$$

Consequently
$$x_1 - x_2 = \lambda(y_1 - y_2),$$

where $y_1 - y_2 \in L_0$, because L is a co-set with respect to L_0. L_0, however, is a subspace. Therefore $\lambda(y_1 - y_2)$ is contained in L_0 too. Hence λL is a co-set, i.e., $\lambda L \in E/L_0$.

We say that a sequence L_n converges to L if there exists a sequence $\{x_1, x_2, \ldots, x_n, \ldots\}$, $x_n \in L_n$, such that $x_n \to x$ and $x \in L$.

We easily see that the operations of addition and scalar multiplication defined in this way and taking limits in E/L_0, make E/L_0 a linear space, the so-called *factor space* of the space E with respect to the subspace L_0. Here L_0 plays the role of the zero elements of E/L_0. If $L \in E/L_0$ contains the zero element θ of the space E, then L coincides with L_0 because in this case every element $x \in L$ has the form $x = \theta + x_0 = x_0, x_0 \in L_0$.

Example. In $C[0, 1]$ we consider the subspace C_0 of all continuous functions which assume the value zero for $x = \frac{1}{2}$. The corresponding factor space is isomorphic with the set of all real numbers.

If $E = E_1 \dot{+} E_2$, then E/E_1 is linearly isomorphic with E_2.

CONNECTION BETWEEN THE REAL AND THE COMPLEX SPACES. For complex numbers, besides the algebraic operations, taking the conjugate complex number

$$\overline{a + ib} = a - ib$$

is meaningful.

Such an operation we call an *involution*.

More generally, we define:

A mapping which associates with every a $\in E$ an $\bar{a} \in E$, where E is a linear complex space, is called an *involution* if it has the properties:

1. $\overline{a + b} = \bar{a} + \bar{b}$.
2. $\overline{\lambda a} = \bar{\lambda} \cdot \bar{a}$ (λ complex).
3. $\bar{\bar{a}} = a$.
4. If $a_n \to a$, then $\bar{a}_n \to \bar{a}$.

All elements a $\in E$ for which $a = \bar{a}$ are called *real*. The elements $a \in E$ for which $\bar{a} = -a$ are called *pure imaginary*. Obviously: If a is real, then ia is pure imaginary, and if c is pure imaginary, then $\frac{1}{i} c$ is real. Therefore, the set of all pure imaginary elements coincides with the set of all elements of the form ia where a is real.

Every element $x \in E$ can be represented uniquely in the form $x = a + ib$ with real a and b.

For, if we put $a = \frac{x+\bar{x}}{2}$, $b = \frac{x-\bar{x}}{2i}$, then $x = a + ib$,

$$\bar{a} = \frac{1}{2}\overline{(x+\bar{x})} = \frac{1}{2}(\bar{x}+\bar{\bar{x}}) = \frac{1}{2}(x+\bar{x}) = a,$$

and

$$\bar{b} = -\frac{1}{2i}\overline{(x-\bar{x})} = -\frac{1}{2i}(\bar{x}-\bar{\bar{x}}) = \frac{1}{2i}(x-\bar{x}) = b,$$

i.e., a and b are real.

The representation of every element $x \in E$ in the form $x = a + ib$ is unique: if

$$a + ib = c + id, \tag{*}$$

then $a = c$ and $b = d$.

For, because of (*) $a - c = i(d - b)$ with a, b, c, d real.

Furthermore,

$$\overline{a-c} = \bar{a} - \bar{c} = a - c,$$

$$\overline{i(d-b)} = -i(\bar{d} - \bar{b}) = -i(d-b)$$

which implies

$$a - c = -i(d-b),$$

i.e.,

$$i(d-b) = -i(d-b),$$

and therefore

$$d - b = \theta, d = b.$$

Hence $a - c = \theta$ and $a = c$.

Thus we have proved that the space E is the direct sum of the subspaces of the real and the pure imaginary elements of E. Therefore many investigations on complex spaces can be reduced to considerations of real spaces. In the same way, an n-dimensional complex space can be interpreted as a direct sum of two n-dimensional real spaces.

In the sequel, when we speak of linear spaces, we shall mean real linear spaces.

§14. LINEAR OPERATORS

Let two linear spaces E and E_1 be given, and let $y = A(x)$ be an operator defined on E with a range lying in E_1. We shall also write $y = Ax$. The element y is called the *image* of x and x the *inverse image* of y.

DEFINITION. The operator A is called *linear*, if:

1) A is *additive*, i.e.,

$$A(x_1 + x_2) = Ax_1 + Ax_2; \tag{1}$$

2) A is *continuous*, i.e., if $x_n \to x$ in the sense of convergence in E, then

$$Ax_n \to Ax$$

in the sense of convergence in E_1.

Obviously, in the case of a metric space, continuity of A means that there exists for every $\varepsilon > 0$ a $\delta > 0$ such that the images of elements from the sphere $S(x, \delta)$ lie in $S(Ax, \varepsilon)$.

By $(E \to E_1)$ we shall designate the set of all linear operators which map E into E_1.

Examples. 1. We consider a square matrix $(a_{ik})_{i,\, k=1,\dots,n}$ of nth order. The equations

$$\eta_i = \sum_{k=1}^{n} a_{ik}\xi_k, \quad i = 1, 2, \dots, n$$

obviously define an operator A, which transforms an element $x = \{\xi_1, \dots, \xi_n\}$ of the n-dimensional Euclidean space E_n into an element $y = \{\eta_1, \dots, \eta_n\}$ of the same space. A is a linear operator.

The additivity of A follows from the equality

$$\sum_{k=1}^{n} a_{ik}(\xi_k + \xi'_k) = \sum_{k=1}^{n} a_{ik}\,\xi_k + \sum_{k=1}^{n} a_{ik}\,\xi'_k \quad i = 1, 2, \dots, n,$$

which is equivalent to $A(x + x') = Ax + Ax'$. The continuity follows from

$$\sqrt{\sum_{i=1}^{n} (\eta_i^{(k)} - \eta_i)^2} \leq \sqrt{\sum_{i=1}^{n} \sum_{k=1}^{n} a_{ik}^2}\, \sqrt{\sum_{k=1}^{n} (\xi_k^{(n)} - \xi_k)^2},$$

which we obtain by means of the Schwarz inequality, where

$$x_k = \{\xi_1^{(k)}, \dots, \xi_n^{(k)}\}, y_k = \{\eta_1^{(k)}, \dots, \eta_n^{(k)}\}, x_k \to x$$

is assumed.

2. We put

$$y(s) = \int_0^1 K(s, t)\, x(t)\, dt.$$

Here $K(s, t)$ is a function continuous on the square $0 \leq s, t \leq 1$. If $x(t) \in C[0,1]$, then obviously $y(s) \in C[0, 1]$ also. Consequently, the operator $y = Ax$ maps the space $C[0, 1]$ into itself. As we easily see, A is a linear operator because

1) $$A(x_1 + x_2) = \int_0^1 K(s, t)\,[x_1(t) + x_2(t)]\,dt$$

$$= \int_0^1 K(s, t)\, x_1(t)\, dt + \int_0^1 K(s, t)\, x_2(t)\, dt = Ax_1 + Ax_2,$$

and the additivity is fulfilled.

2) Let $\{x_n(t)\}$ converge to $x(t)$ in the sense of convergence in $C[0, 1]$, i.e., uniformly on $[0, 1]$. Because in the case of uniform convergence we can take the limit under the integration sign,

$$\lim_{n \to \infty} \int_0^1 K(s, t)\, x_n(t)\, dt = \int_0^1 K(s, t)\, x(t)\, dt,$$

i.e., $\lim_n Ax_n = Ax$, and continuity of the operator A is proved.

3. Let E be the set of functions defined and continuous on a closed plane curve Γ with continuous curvature. Let the metric be given by the equality $\rho(x, y) = \max_\Gamma |x(t) - y(t)|$. Let further E_1 be a space of functions of two variables defined and continuous on the region G bounded by the curve Γ. Here, let the distance be defined by the equality $\rho(u, v) = \max_G |u(\xi, \eta) - v(\xi, \eta)|$. With every function $x(t) \in E$ we associate that function $u(\xi, \eta) \in E_1$ which is solution of the Dirichlet problem for the region G with the boundary condition $x(t)$. As is well-known, this problem is solvable uniquely under the given hypotheses. This correspondence defines a certain operator $u = Ax$. By virtue of the well-known properties of harmonic functions, A is a linear operator defined on E with a range lying in E_1.

4. Let $E = C[0, 1]$. In this space we consider the operator A defined by

$$y(t) = \int_0^t x(\tau)\, d\tau.$$

Obviously, A is a linear operator defined on the whole of E. Let, furthermore, another operator B on E be given by

$$y(t) = \frac{d}{dt} x(t).$$

B is not defined for all $x(t) \in C[0, 1]$, and if Bx exists, then $y(t)$ is not always $\in C[0, 1]$. If, however, we take for the domain of definition of B a linear manifold of functions which have continuous derivatives (these functions are everywhere dense in $C[0, 1]$), then the range of B is also contained in $C[0, 1]$.

The operator B is clearly additive. It is, however, not continuous, because the derivative of a limit element of a uniformly convergent sequence of functions need not be equal to the limit of the derivative of these functions.

SIMPLEST PROPERTIES OF LINEAR OPERATORS. Let A be a linear operator. We put $y = x + z$, then $z = y - x$. We obtain $Ay = Ax + Az = Ax + A(y - x)$, which implies

$$A(y - x) = Ay - Ax. \tag{2}$$

If in (2), $y = x$, then $A\theta = Ax - Ax = \theta_1$, where θ_1 is the zero element of the space in which the range of the operator A lies.

If we put $y = \theta$ in (2), then we get $A(-x) = -Ax$.

THEOREM 1. *Let A be an additive operator defined on a linear space E with range lying in a linear space E_1. If A is continuous at $x_0 \in E$, then A is continuous on the whole space and therefore linear.*

Let x be an arbitrary point of E, and let $x_n \to x$. Then $x_n - x + x_0 \to x_0$, and since A is continuous at the point x_0, we have

$$\lim_n A(x_n - x + x_0) = Ax_0.$$

However,

$$A(x_n - x + x_0) = Ax_n - Ax + Ax_0$$

because of the additivity of A. Therefore

$$\lim_n Ax_n - Ax + Ax_0 = Ax_0,$$

and consequently

$$\lim_n Ax_n = Ax.$$

THEOREM 2. *A linear operator is homogeneous, i.e., for every real t, $A(tx) = tAx$. (A scalar factor can be taken from under the operator sign.)*

First let $t = n$ be a natural number. Then

$$A(nx) = A(x + x \ldots + x) = Ax + Ax + \ldots + Ax = nAx.$$

If $t = m$ is a negative integer, then

$$A(mx) = -A(-mx) = -(-m)Ax = mAx.$$

If $t = \dfrac{m}{n}$ is a rational number, then we obtain

$$A\left(\frac{m}{n}x\right) = mA\left(\frac{1}{n}x\right).$$

We put $\frac{1}{n}x = y$. Then $x = ny$ and

$$Ax = A(ny) = nAy = nA\left(\frac{1}{n}x\right),$$

which implies $A\left(\frac{1}{n}x\right) = \frac{1}{n}Ax$, consequently

$$A\left(\frac{m}{n}x\right) = mA\left(\frac{1}{n}x\right) = \frac{m}{n}Ax.$$

Finally, let t be irrational. There exists a sequence $\{r_n\}$ of rational numbers for which $\lim_{n\to\infty} r_n = t$. Hence $\lim_{n\to\infty} r_n x = tx$, and because A is continuous,

$$A(tx) = A(\lim_{n\to\infty} r_n x) = \lim_{n\to\infty} A(r_n x) = \lim_{n\to\infty} r_n Ax = tAx.$$

We can introduce algebraic operations in the set of linear operators defined on a linear space E whose range lies in a linear space E_1. Let A and B be such operators. We define addition by

$$(A+B)x = Ax + Bx$$

and multiplication by a scalar by

$$(\lambda A)x = \lambda Ax.$$

The considered set of linear operators is a linear system, for with these definitions all the required axioms are fulfilled. In particular, the zero element is an operator 0, such that $0x = \theta$ holds for every $x \in E$. The operators A_n converge to the operator $A (A_n \to A)$ if for every $x \in E, A_n x \to Ax$. Then the set of linear operators is a linear space. We shall treat this space in more detail with additional assumptions about E and E_1 later on.

THE RING OF LINEAR OPERATORS. We now consider an arbitrary linear space E and the set $(E \to E)$ of all linear operators defined on E whose range lies in E. As we have shown above, these operators form a linear system. We define the product of two linear operators A and B by

$$(AB)x = A(Bx).$$

As we easily see, this is again a linear operator. By induction, we define the product of a finite number of linear operators. In particular, we write

$$AA = A^2, A^2A = A^3, \text{ and so on.}$$

We note that $(AB)C = A(BC)$, $(A+B)C = AC + BC$ and also $C(A+B) = CA + CB$. Furthermore, there exists a unit operator I defined by $Ix = x$ for every x. For all operators $A, AI = IA = A$.

Thus the set $(E \to E)$ forms a non-commutative ring (with unit element) because in general $AB \neq BA$.

Example. Let $E = C[0, 1]$. We consider the operators

$$y(s) = \int_0^1 stx(t)\,dt = Ax \text{ and } y(s) = sx(s) = Bx.$$

Then

$$ABx = \int_0^1 st \cdot tx(t)\,dt = s\int_0^1 t^2x(t)\,dt,$$

$$BAx = s\int_0^1 st\,x(t)\,dt = s^2\int_0^1 tx(t)\,dt.$$

Therefore $AB \neq BA$.

The concept of the *inverse operator* is fundamental. Analogous to the definition of the inverse element in a ring, the operator B is called *left inverse* of A, if $BA = I$ and C is *right inverse* of A, if $AC = I$. If $B = C$, then we say that the operator A has an inverse designated by A^{-1}. If, therefore, A^{-1} exists, then by definition $A^{-1}A = AA^{-1} = I$. We shall return later to the concept of an inverse operator.

FUNCTIONS OF OPERATORS. The operator $A^n = A \cdot A \ldots A$ represents the simplest example of an operator-function. It is a special case of a more general function, namely of a polynomial operator

$$P_n(A) = a_0I + a_1A + a_2A^2 + \ldots + a_nA^n.$$

We can consider even more general operator functions $f(A)$.

For instance, let E be an n-dimensional vector space and A an operator transforming E into itself and given by a symmetric matrix $\mathfrak{A}$. By means of a unitary transformation U we transform $\mathfrak{A}$ to diagonal form

$$\Lambda = U\mathfrak{A}U^{-1},$$

where

$$\Lambda = \begin{Vmatrix} \lambda_1 0 0 \ldots 0 \\ 0 \lambda_2 0 \ldots 0 \\ \ldots\ldots \\ 0 0 0 \ldots \lambda_n \end{Vmatrix}.$$

Let $f(t)$ be an arbitrary function of a real variable t, defined on the interval $[m, M]$ with $m = \inf_i \lambda_i$ and $M = \sup_i \lambda_i$. We put

$$f(\Lambda) = \begin{Vmatrix} f(\lambda_1) & 0 & 0 \ldots 0 \\ 0 & f(\lambda_2) & 0 \ldots 0 \\ \ldots & \ldots & \ldots \\ 0 & 0 & 0 \ldots f(\lambda_n) \end{Vmatrix}$$

and

$$f(\mathfrak{A}) = U^{-1}f(\Lambda)U.$$

Therefore, there is associated with every function $f(t)$ of a real variable t defined on $[m, M]$ a function $f(\mathfrak{A})$ of the matrix $\mathfrak{A}$. Obviously, there corresponds to the function $f(t) \equiv 0$ the zero matrix, to the function $f(t) \equiv 1$ the unit matrix, and to $f(t) = t^n$ the matrix $\mathfrak{A}^n$. If, furthermore, $f(t) = f_1(t) + f_2(t)$, then $f(\mathfrak{A}) = f_1(\mathfrak{A}) + f_2(\mathfrak{A})$, and if $\varphi(t) = f_1(t) \cdot f_2(t)$, then $\varphi(\mathfrak{A}) = f_1(\mathfrak{A}) \cdot f_2(\mathfrak{A})$. These equalities follow from the relations

$$U(B + C)U^{-1} = UBU^{-1} + UCU^{-1}$$

$$U(BC)U^{-1} = (UBU^{-1})(UCU^{-1})$$

and

$$f(\Lambda) = f_1(\Lambda) + f_2(\Lambda),\ \varphi(\Lambda) = f_1(\Lambda) \cdot f_2(\Lambda),$$

which hold for arbitrary matrices B and C. We can build up the theory of functions of matrices also in another way, going over from polynomials of matrices to power series of matrices. So doing, we can define, however, only analytic functions of a matrix. Far-reaching research in this direction was done by A. I. Lappo-Danilevskij. This theory of analytic functions of matrices was used for investigations in systems of differential equations [37]. In spectral theory, the construction of matrix functions by reduction of the matrix to diagonal form was carried out for infinite matrices and then for arbitrary self-adjoint operators in Hilbert space (see §37).

As a second example, we consider the differential operators. Let $E = C[0, 1]$ and $D = \frac{d}{dt}$ the operator which assigns to every continuously differentiable function $x(t)$ the derivative of this function:

$$Dx(t) = \frac{dx}{dt}.$$

For n-times differentiable functions

$$P_n(D)\,x(t) = a_0 x(t) + a_1 \frac{dx}{dt} + a_2 \frac{d^2x}{dt^2} + \ldots + a_n \frac{d^n x}{dt^n}$$

is meaningful. Here $P_n(s)$ is an arbitrary polynomial of nth degree in the argument s. For infinitely differentiable functions

$$\sum_{n=0}^{\infty} a_n D^n x(t) = \sum_{n=0}^{\infty} a_n \frac{d^n x}{dt^n}$$

has meaning, provided the series on the right converges and its sum belongs to $C[0, 1]$. In particular, this is the case if $x(t)$ is a polynomial of degree m, since then $D^n x(t) = 0$ for $n > m$ and the series becomes a finite sum.

Polynomials of differential operators find application in the theory of linear differential equations. The simplest example is the so-called symbolic method for solving a differential equation with constant coefficients. Further applications to linear differential equations—ordinary and partial—will be found in the calculus of operators [7].

Formal operations with series and functions of a differential operator defined by means of series were used in the first half of the nineteenth century in order to obtain certain formulas. We explain this by way of examples.

First we remark that the operator

$$I + hD + \frac{h^2 D^2}{2!} + \frac{h^3 D^3}{3!} + \ldots + \frac{h^n D^n}{n!} + \ldots = e^{hD}$$

applied to $x(t)$ furnishes the function $x(t+h)$. For,

$$e^{hD} x(t) = x(t) + hx'(t) + \frac{h^2 x''(t)}{2!} + \ldots + \frac{h^n x^{(n)}(t)}{n!} + \ldots = x(t+h).$$

If we therefore designate by Δ_h the difference operator

$$\Delta_h x(t) = x(t+h) - x(t)$$

then

$$e^{hD}x(t) = x(t+h) = x(t) + \Delta_h x(t) = (I + \Delta_h)\, x(t)$$

i.e.,

$$I + \Delta_h = e^{hD}.$$

This implies

$$hD = \ln(I + \Delta_h) = \sum_{n=1}^{\infty} (-1)^n \frac{\Delta_h^n}{n},$$

if we develop formally in a power series. This is the formula of Gregory, expressing the differential operator by difference operators.

We consider further the operators

$$Jx(t) = \int_0^1 x(t+\tau)\, d\tau \quad \text{and} \quad Sx(t) = \sum_{i=1}^{n} c_i x(t+\tau),$$

where $\sum_{i=1}^{n} c_i = 1$. We can easily convince ourselves that the operators J and S can be expressed as functions of the operator D, for we have

$$J = \int_0^1 e^{\tau D} d\tau = \frac{e^D - I}{D}, \quad S = \sum_{i=1}^{n} c_i e^{\tau_i D}.$$

This implies

$$J - S = J[I - J^{-1}S] = J\left[I - \sum_{i=1}^{n} c_i \frac{De^{\tau_i D}}{e^D - I}\right].$$

As known, the Bernoulli polynomials are defined by

$$\frac{ze^{az}}{e^z - 1} = \sum_{k=0}^{\infty} \frac{z^k}{k!} B_k(a).$$

Therefore, the preceding formula becomes

$$J - S = J\left[I - \sum_{i=1}^{n} c_i \left(\sum_{k=0}^{\infty} \frac{D^k}{k!} B_k(\tau_i)\right)\right] = -J\left[\sum_{i=1}^{n} c_i \left(\sum_{k=1}^{\infty} \frac{D^k}{k!} B_k(\tau_i)\right)\right],$$

because $\sum_{i=1}^{n} c_i B_0(\tau_i) = \sum_{i=1}^{n} c_i = 1$.

Since, in addition

$$J[D^k x(t)] = \int_0^1 x^{(k)}(t+\tau)\, d\tau = x^{(k-1)}(t+1) - x^{(k-1)}(t),$$

we arrive at the generalized Euler-MacLaurin summation formula

$$\int_0^1 x(t+\tau)\, d\tau = \sum_{i=1}^{n} c_i x(t+\tau_i) - \sum_{k=1}^{\infty} \left[\sum_{i=1}^{n} c_i \frac{B_k(\tau_i)}{k!}\right] [x^{(k-1)}(t+1) - x^{(k-1)}(t)].$$

In particular, for $t = 0$, we obtain

$$\int_0^1 x(\tau)\, d\tau = \sum_{i=1}^{n} c_i x(\tau_i) - \sum_{k=1}^{\infty} \left[\sum_{i=1}^{n} c_i \frac{B_k(\tau_i)}{k!}\right] [x^{(k-1)}(1) - x^{(k-1)}(0)].$$

This derivation of the formulas of Gregory and Euler-MacLaurin is characteristic. If $x(t)$ is a polynomial, then the infinite series arising become finite sums. For sufficiently often differentiable functions, the formulas of Gregory and Euler-MacLaurin can be proved by estimating the remainder terms.

§15. NORMED LINEAR SPACES

An important class of linear spaces are the spaces of type B (Banach spaces). A set E is called a *normed linear space*, if

1. E is a linear system with real (complex) numbers as ring scalars, and if
2. With every element x of the linear system E there is associated a unique real number, called the norm of this element, and designated by $\|x\|$. The norm of an element has to satisfy the following conditions (*norm axioms*):

1) $\|x\| \geqq 0$ and $\|x\| = 0$, only if $x = \theta$;
2) $\|x+y\| \leqq \|x\| + \|y\|$;
3) $\|\lambda x\| = |\lambda| \, \|x\|$.

In a normed linear space we can introduce a metric by

$$\rho(x, y) = \|x - y\|.$$

We easily show that the introduced distance satisfies all metric axioms. Having introduced the metric, we define

$$x = \lim_n x_n \text{ or } x_n \to x \text{ by } \|x_n - x\| \to 0$$

as $n \to \infty$. This convergence in a normed linear space is called *convergence in norm.*

If a normed linear space is complete in the sense of the convergence in norm, then it is called a *Banach space* or a *space of type B.*

Examples. 1. The n-dimensional vector space E_n is a space of type B.

If we define the sum of elements and the product of elements with a scalar as in Example 5 on p. 52 and the norm by

$$\|x\| = \left(\sum_{i=1}^{n} \xi_i^2 \right)^{\frac{1}{2}},$$

then we see that E_n is of type B. Here the metric is the one introduced earlier in E_n.

2. $C[0, 1]$ is a Banach space. We define addition of functions and multiplication of functions by a real number in the usual manner. We put further

$$\|x\| = \max_{0 \leq t \leq 1} |x(t)|.$$

The metric coincides with the one introduced earlier in $C[0, 1]$.

3. l_p is a space of type B, if we define addition and multiplication by a real number by

$$x + y = \{\xi_i + \eta_i\},$$

$$\lambda x = \{\lambda \xi_i\}$$

and put

$$\|x\| = \left(\sum_{i=1}^{\infty} |\xi_i|^p \right)^{\frac{1}{p}}.$$

The corresponding metric is again the earlier one.

4. $L_p(0, 1)$ is a Banach space. We put, for $x(t) \in L_p(0, 1)$,

$$\|x\| = \left(\int_0^1 |x(t)|^p dt \right)^{\frac{1}{p}}.$$

Again we obtain the old metric.

5. m is a space of type B. For, putting for $x = \{\xi_i\}$, $\|x\| = \sup_i |\xi_i|$, we obtain again a Banach space with the metric already known.

6. The spaces s and $S(0, 1)$ are not Banach spaces.

From the relations

$$\|(x+y)-(x_n+y_n)\| \leqq \|x-x_n\|+\|y-y_n\|,$$

$$\|\lambda x-\lambda x_n\| = |\lambda|\,\|x-x_n\|$$

and

$$\|\lambda_n x-\lambda x\| = |\lambda_n-\lambda|\,\|x\|$$

there follows: If

$$\lambda_n \to \lambda,\ x_n \to x \text{ and } y_n \to y$$

then

$$x_n+y_n \to x+y,\ \lambda x_n \to \lambda x \text{ and } \lambda_n x \to \lambda x.$$

Furthermore,

$$\|x\| = \|y+(x-y)\| \leqq \|y\|+\|x-y\|$$

or

$$\|x\|-\|y\| \leqq \|x-y\|.$$

Interchanging the roles of x and y, we obtain

$$\|y\|-\|x\| \leqq \|x-y\|$$

and therefore

$$|\,\|x\|-\|y\|\,| \leqq \|x-y\|.$$

In particular, it follows that if $x_n \to x$, then

$$\|x_n\| \to \|x\|.$$

Since Banach spaces are complete metric linear spaces all concepts such as sphere, bounded set, separability, compactness, linear dependence of elements, linear manifold, subspaces, etc. are meaningful because they were introduced in complete metric and linear spaces.

As we easily see, a sphere in a normed linear space is a convex set. Let $x_1, x_2 \in \bar{S}(a, r)$, i.e.,

$$\|x_1-a\| \leqq r,\ \|x_2-a\| \leqq r.$$

We choose an arbitrary element of the form

$$y = tx_2+(1-t)x_1,\ t \in [0,1].$$

Then we have

$$\begin{aligned}\|y-a\| &= \|tx_2+(1-t)x_1-a\|\\ &= \|tx_2+(1-t)x_1-ta-(1-t)a\|\\ &\leqq \|tx_2-ta\|+\|(1-t)x_1-(1-t)a\|\\ &= t\|x_2-a\|+(1-t)\|x_1-a\| \leqq tr+(1-t)r = r.\end{aligned}$$

Hence

$$\|y-a\| \leqq r$$

and therefore

$$y \in \bar{S}(a, r).$$

NORMING. The introduction of a norm in a linear space is called *norming*.

Conditions for normability of linear spaces were established by A. N. Kolmogorov [35].

Example. Let E be a linear normal space, L_0 a subspace of E, E/L_0 the factor group. E/L_0 can be normed by

$$\|L\| = \inf_{x \in L} \|x\|$$

for every $L \in E/L_0$.

We shall show that $\|L\|$ satisfies all conditions of a norm.

1) We have $\|L\| \geqq 0$. Here $\|L\| = 0$ only if $L = L_0$. The first inequality is immediate. Furthermore L is a closed set. For, if $\{x_n\}$ is a sequence of elements in L converging to $x \in E$, then we have for every $n, m, x_n - x_m \in L_0$ and as $m \to \infty$,

$$x_n - x_m \to x_n - x.$$

Since L_0 is closed, it follows that $x_n - x \in L_0$. Then, however, x together with x_n is contained in L.

Now let

$$\|L\| = \inf_{x \in L} \|x\| = 0.$$

Then there exists in L a sequence $\{x_n'\}$ such that $\|x_n'\| \to 0$, i.e., $x_n' \to \theta$. Because of closedness, L must contain θ; then, however, $L = L_0$. We see immediately that, conversely, $\|L_0\| = 0$. Consequently, the first axiom is fulfilled.

2) $$\|L_1 + L_2\| \leqq \|L_1\| + \|L_2\|, \quad \text{for}$$
$$\|L_1 + L_2\| = \inf_{x_1 \in L_1, x_2 \in L_2} \|x_1 + x_2\| \leqq \inf_{x_1 \in L_1, x_2 \in L_2} (\|x_1\| + \|x_2\|)$$
$$= \inf_{x_1 \in L_1} \|x_1\| + \inf_{x_2 \in L_2} \|x_2\| = \|L_1\| + \|L_2\|.$$

3) $$\|\lambda L\| = |\lambda| \, \|L\|, \quad \text{since}$$
$$\|\lambda L\| = \inf_{x \in L} \|\lambda x\| = |\lambda| \inf_{x \in L} \|x\| = |\lambda| \, \|L\|.$$

Finally we remark: If E is a complete space, then E/L_0 is also complete.

In conclusion, we show that the concept of a limit introduced by this norm is equivalent to the earlier one.

According to the definition given in §13 (p. 58),
$$L_n \to L,$$
if there exists a sequence $x_n \in L_n$ such that $x_n \to x$ where $x \in L$.

Then, however,
$$\|x_n - x\| \to 0,$$
with
$$x_n - x \in L_n - L \text{ and } \|L_n - L\| \leqq \|x_n - x\|.$$
A fortiori,
$$\|L_n - L\| \to 0.$$
If conversely
$$\|L_n - L\| \to 0,$$
i.e., if
$$\|L_n - L\| = \varepsilon_n \text{ and } \varepsilon_n \to 0$$
then $L_n - L$ contains an element $x_n - x_n'$ such that $x_n \in L_n, x_n' \in L$, and $\|x_n - x_n'\| < 2\varepsilon_n$. Here we can take for x_n', any fixed (not dependent on n) element $x_0 \in L$. For, if
$$\|x_n' - x_n\| < 2\varepsilon_n$$
where $x_n' \in L$, $x_n \in L_n$, then
$$\|x_0 - (x_0 - x_n' + x_n)\| < 2\varepsilon_n,$$
and since $x_0 \in L$, $x_n' \in L$, it follows that $x_0 - x_n' \in L_0$ and also
$$y_n = x_0 - x_n' + x_n \in L_n.$$
Thus we have constructed for an arbitrary element $x_0 \in L$, a sequence $y_n \in L_n$ such that $y_n \to x_0$. This is, however, the convergence of L_n to L in §13.

§16. SPACES AND SUBSPACES OF FINITE DIMENSION

FINITE DIMENSION AND COMPACTNESS. As we know, in n-dimensional Euclidean space, every bounded set is compact. We shall prove that compactness of bounded sets is a characteristic property of normed linear spaces with finite dimension.

LEMMA. *Let L be a proper subspace of a normed linear space E. We can then find for every given $\varepsilon > 0$ a $y \in E$ with $\|y\| = 1$ such that for all $x \in E$, the inequality $\|x - y\| > 1 - \varepsilon$ is fulfilled.*

Proof. Let $y_0 \in E$, but not $\in L$ and
$$d = \inf_{x \in L} \|y_0 - x\|.$$
Necessarily $d > 0$, for otherwise y_0 would be a limit point of L and therefore, contradicting the hypothesis, would belong to L. For every $\eta > 0$, we can find an $x_0 \in L$ with

$$d \leqq \|y_0 - x_0\| < d + \eta.$$

Then the element

$$y = \frac{y_0 - x_0}{\|y_0 - x_0\|}$$

does not belong to L, for contrariwise y_0 would belong to L. In addition $\|y\| = 1$ and

$$\|y - x\| = \left\| \frac{y_0 - x_0}{\|y_0 - x_0\|} - x \right\| = \frac{1}{\|y_0 - x_0\|} \|y_0 - x'\|$$

$$> \frac{1}{d+\eta} \|y_0 - x'\| \geqq \frac{d}{d+\eta} = 1 - \frac{\eta}{d+\eta},$$

where

$$x' = x_0 + \|y_0 - x_0\| x \in L$$

for every $x \in L$. If we now choose η such that $\frac{\eta}{d+\eta} < \varepsilon$, then

$$\|y - x\| > 1 - \varepsilon$$

implies the assertion.

THEOREM. *A subspace L of a normed linear space E is of finite dimension if and only if every bounded subset of L is compact.*

Necessity. Let L be n-dimensional. Then every $x \in L$ can be represented in the form

$$x = \lambda_1 x_1 + \lambda_2 x_2 + \ldots + \lambda_n x_n$$

where $x_1, x_2, \ldots, x_n$ are arbitrary but fixed linearly independent elements of L. The elements y_k of an arbitrary bounded sequence $\{y_k\}$ from L can, in particular, be written

$$y_k = \lambda_1^{(k)} x_1 + \lambda_2^{(k)} x_2 + \ldots + \lambda_n^{(k)} x_n, \; k = 1, 2, \ldots \; .$$

Because of the boundedness of $\{y_k\}$, there exists an M such that $\|y_k\| \leqq M$, $k = 1, 2, \ldots$. In order to prove that the sequence of sums

$$|\lambda_1^{(k)}| + |\lambda_2^{(k)}| + \ldots + |\lambda_n^{(k)}|, k = 1, 2, \ldots$$

is bounded, let us assume the contrary. We would have a sequence of indices $k_1, k_2, \ldots, k_m, \ldots$ with

$$|\lambda_1^{(k_m)}| + |\lambda_1^{(k_m)}| + \ldots + |\lambda_n^{(k_m)}| = c_m \geqq m.$$

For the elements

$$y_{k_m}^* = \frac{1}{c_m} y_{k_m}$$

it then follows that

$$\|y_{k_m}^*\| = \frac{1}{c_m} \|y_{k_m}\| \leqq \frac{1}{c_m} M \leqq \frac{M}{m}.$$

Therefore $y_{k_m}^* \to \theta$ as $m \to \infty$. On the other hand,

$$y_{k_m}^* = \mu_1^{(k_m)} \, x_1 + \mu_1^{(k_m)} \, x_2 + \ldots + \mu_n^{(k_m)} \, x_n,$$

where

$$\mu_i^{(k_m)} = \frac{1}{c_m} \lambda_i^{(k_m)} .$$

Because of $$|\mu_1^{(km)}| + |\mu_2^{(km)}| + \ldots + |\mu_n^{(km)}| = 1$$
the $\{\mu_i^{(k_1)}, \mu_i^{(k_2)}, \ldots\}$ are bounded sequences of real numbers for all i. Hence, we can select subsequences $\{\mu_i^{(k_{m_1})}, \mu_i^{(k_{m_2})}, \ldots\}$ converging to $\mu_i^{(0)}$.

We choose the sequences $\{k_{m_j}\}$ of indices in such a way that for $i = 1, \ldots, n$, convergence results.

Obviously
$$|\mu_1^{(0)}| + |\mu_2^{(0)}| + \ldots + |\mu_n^{(0)}| = 1.$$
If we put
$$y_0 = \mu_1^{(0)} x_1 + \mu_2^{(0)} x_2 + \ldots + \mu_n^{(0)} x_n,$$
then
$$\|y_o - y^*_{k_{mj}}\| \leqq |\mu_1^{(0)} - \mu_1^{(k_{mj})}| \, \|x_1\| + |\mu_2^{(0)} - \mu_2^{(k_{mj})}| \, \|x_2\| + \ldots + |\mu_n^{(0)} - \mu_n^{(k_{mj})}| \, \|x_n\| \to 0$$
as $m_j \to \infty$, i.e., $y^*_{k_{mj}} \to y_0$. Because, on the other hand, $y^*_{k_{mj}} \to \theta$, it follows that $y_0 = \theta$. This cannot be, however, because of the assumed linear independence of the $x_1, x_2, \ldots, x_n$ for surely all of the numbers $\mu_1^{(0)}, \mu_2^{(0)}, \ldots, \mu_n^{(0)}$ do not vanish.

Hence all sums
$$|\lambda_1^{(k)}| + |\lambda_2^{(k)}| + \ldots + |\lambda_n^{(k)}|$$
and therefore all sequences $\{\lambda_i^{(1)}\ \lambda_i^{(2)}, \ldots\}$ are bounded. Consequently, we can select a sequence of indices $k_1, k_2, \ldots, k_j, \ldots$ such that
$$\lambda_i^{(k_j)} \to \lambda_i^{(0)}$$
as $k_j \to \infty$ and i takes the values 1 to n. Then, however,
$$y_{k_j} \to y_0 = \lambda_1^{(0)} x_1 + \ldots + \lambda_n^{(0)} x_n.$$
We had selected an arbitrary bounded sequence $\{y_k\}$ from L, hence every bounded set of a space with finite dimension is compact.

Remark. By this proof we have shown simultaneously: If $\{y_k\}$ lies in a linear manifold L of finite dimension and $y_k \to y$, then y is also contained in L. Therefore $L = \bar{L}$.

For, if $y_k \to y$, then $\{y_k\}$ is surely a bounded sequence. As we have proved, we can select a convergent subsequence $\{y_{k_j}\}$ from $\{y_k\}$ such that
$$y_{k_j} \to y_0 = \lambda_1^{(0)} x_1 + \lambda_2^{(0)} x_2 + \ldots + \lambda_n^{(0)} x_n \in L.$$
Since $\{y_{k_j}\}$ on the other hand, is a subsequence of the sequence $\{y_k\}$ which converges to y,
$$y_{k_j} \to y,$$
$y = y_0 \in L$.

Sufficiency. Let every bounded set of elements of L be compact. Let x_1 be an arbitrary element of L with $\|x_1\| = 1$ and let L_1 be the subspace generated by x_1. If $L = L_1$, then the theorem is proved. Contrariwise, applying the lemma, there exists according to the remark above, an $x_2 \in L$ with $\|x_2\| = 1$ and $\|x_1 - x_2\| \geqq \frac{1}{2}$. The subspace L_2 of L generated by x_1, x_2 is then either equal to L or there exists, by virtue of our lemma, an x_3 with
$$\|x_3\| = 1, \ \|x_1 - x_3\| \geqq \tfrac{1}{2}, \ \|x_2 - x_3\| \geqq \tfrac{1}{2}.$$

If we continue this process, two possibilities arise: Either there exists an n with $L_n = L$ and our theorem is proved, or we obtain an infinite sequence $\{x_n\}$, for whose elements $\|x_n\| = 1$ and $\|x_n - x_m\| \geqq \frac{1}{2}$ holds for arbitrary $m, n, m \neq n$. The second possibility, however, contradicts the hypothesis, for otherwise there would exist a bounded ($\|x_n\| = 1$) and non-compact ($\|x_n - x_m\| \geqq \frac{1}{2}$ for $m \neq n$) set.

THE PROBLEM OF BEST APPROXIMATION. P. L. Chebyshev sought thc approximation of arbitrary functions by linear combinations of given ones. We can consider approximations in the spaces C, $L_2^{(\rho)}$,[16] L, etc.

We shall deal with the problem of the best approximation in a normed space E by linear combinations of a given finite system of linearly independent elements $x_1, x_2, \ldots, x_n$ of E[2].

THEOREM. *There exist real numbers* $\lambda_1^{(0)}, \lambda_2^{(0)}, \ldots \lambda_n^{(0)}$ *such that*

$$\varphi(\lambda_1, \lambda_2, \ldots, \lambda_n) = \|y - \lambda_1 x_1 - \lambda_2 x_2 - \ldots - \lambda_n x_n\|$$

attains its minimum for $\lambda_1 = \lambda_1^{(0)}, \lambda_2 = \lambda_2^{(0)}, \ldots, \lambda_n = \lambda_n^{(0)}$.[17]

For the proof, we shall assume that y does not lie in the subspace generated by $x_1, x_2, \ldots, x_n$. According to the following inequality, $\varphi(\lambda_1, \ldots, \lambda_n)$ is a continuous function of its arguments $\lambda_1, \ldots, \lambda_n$:

$$|\varphi(\lambda_1, \lambda_2, \ldots, \lambda_n) - \varphi(\mu_1, \mu_2, \ldots, \mu_n)| = \left| \left\| y - \sum_{i=1}^{n} \lambda_i x_i \right\| - \left\| y - \sum_{i=1}^{n} \mu_i x_i \right\| \right|$$

$$\leqq \left\| \sum_{i=1}^{n} (\lambda_i - \mu_i) x_i \right\| \leqq \sum_{i=1}^{n} |\lambda_i - \mu_i| \, \|x_i\| \leqq \max_i |\lambda_i - \mu_i| \sum_{i=1}^{n} \|x_i\|.$$

LEMMA. *If* $\sum_{i=1}^{n} \lambda_i^{(2)}$ *increases without bound, then* $\varphi(\lambda_1, \ldots, \lambda_n) \to \infty$.

Proof. We have

$$\varphi(\lambda_1, \lambda_2, \ldots, \lambda_n) \geqq \|\lambda_1 x_1 + \lambda_2 x_2 + \ldots + \lambda_n x_n\| - \|y\| .$$

The continuous function

$$\psi(\lambda_1, \lambda_2, \ldots, \lambda_n) = \|\lambda_1 x_1 + \lambda_2 x_2 + \ldots + \lambda_n x_n\|$$

of the parameters $\lambda_1, \lambda_2, \ldots, \lambda_n$ assumes its minimum (which because of the linear independence of the $x_1, x_2, \ldots, x_n$ is greater than 0) on the sphere

$$\sum_{i=1}^{n} \lambda_i^2 = 1$$

of the n-dimensional Euclidean space (this sphere is a set compact in itself). If now

$$\sqrt{\sum_{i=1}^{n} \lambda_i^2} > \frac{1}{\mu} (K + \|y\|),$$

[16] $L_2^{(\rho)}(a, b)$ is the space of measurable functions whose squares are integrable with weight $\rho(t) > 0$, i.e.,

$$\|x\| = \int_a^b |x(t)|^2 \rho(t) \, dt < \infty.$$

[17] In other words, in the subspace of finite dimension generated by the elements $x_1, x_2, \ldots, x_n$ there exists an element which is at the smallest distance from y.

where μ is the minimum, then it follows immediately that

$$\varphi(\lambda_1, \lambda_2, \ldots, \lambda_n) \geqq \left\| \sum_{k=1}^{n} \lambda_k x_k \right\| - \|y\|$$

$$= \sqrt{\sum_{i=1}^{n} \lambda_i^2} \left\| \sum_{k=1}^{n} \frac{\lambda_k}{\sqrt{\sum_{i=1}^{n} \lambda_i^2}} x_k \right\| - \|y\| \geqq \sqrt{\sum_{i=1}^{n} \lambda_i^2} \cdot \mu - \|y\| > K,$$

and this proves our lemma.

Hence $\varphi(\lambda_1, \lambda_2, \ldots, \lambda_n) > \|y\|$ outside of a certain sphere $\sum_{i=1}^{n} \lambda_i^2 \leqq r^2$. Being a continuous function, $\varphi(\lambda_1, \lambda_2, \ldots, \lambda_n)$ assumes its minimum ν on the sphere compact in itself at the point $(\lambda_1^{(0)}, \lambda_2^{(0)}, \ldots, \lambda_n^{(0)})$. Now $\varphi(0, 0, \ldots, 0) = \|y\|$. Hence $\nu \leqq \|y\|$, and ν is simultaneously the minimum on the whole space. This proves our theorem.

In general, the approximation problem is not uniquely solvable. In order to have uniqueness, we have to impose further conditions on the space. For this reason, in $C[0, 1]$ we consider systems of functions satisfying the Tchebychev condition. There also exist spaces, however, where the approximation problem has a unique solution.

A space E is called (*strongly*) *normed* if the equality $\|x + y\| = \|x\| + \|y\|$ for $x \neq \theta, y \neq \theta$ holds only if $y = \alpha x, \alpha > 0$.

In (strongly) normed spaces the best approximation is determined uniquely. If we have two linear combinations $\sum_{i=1}^{n} \lambda_i' x_i$ and $\sum_{i=1}^{n} \lambda_i'' x_i$ such that

$$\left\| y - \sum_{i=1}^{n} \lambda_i' x_i \right\| = \left\| y - \sum_{i=1}^{n} \lambda_i'' x_i \right\| = \omega$$

where $\omega = \min_{\lambda_i} \left\| y - \sum_{i=1}^{n} \lambda_i x_i \right\| > 0$, then

$$\left\| y - \sum_{i=1}^{n} \frac{\lambda_i' + \lambda_i''}{2} x_i \right\| \leqq \frac{1}{2} \left\| y - \sum_{i=1}^{n} \lambda_i' x_i \right\| + \frac{1}{2} \left\| y - \sum_{i=1}^{n} \lambda_i'' x_i \right\| = \frac{1}{2} \omega + \frac{1}{2} \omega = \omega.$$

Because of

$$\left\| y - \sum_{i=1}^{n} \frac{\lambda_i' + \lambda_i''}{2} x_i \right\| \geqq \omega,$$

$$\left\| y - \sum_{i=1}^{n} \frac{\lambda_i' + \lambda_i''}{2} x_i \right\| = \omega$$

and consequently

$$\left\| y - \sum_{i=1}^{n} \frac{\lambda_i' + \lambda_i''}{2} x_i \right\| = \left\| \frac{1}{2} y - \sum_{i=1}^{n} \frac{\lambda_i'}{2} x_i \right\| + \left\| \frac{1}{2} y - \sum_{i=1}^{n} \frac{\lambda_i''}{2} x_i \right\|$$

must hold. Because our space is (strongly) normed,

$$y - \sum_{i=1}^{n} \lambda_i' x_i = \alpha \left\{ y - \sum_{i=1}^{n} \lambda_i'' x_i \right\}$$

follows. Now if $\alpha \neq 1$, then y would have to be a linear combination of $x_1, x_2, \ldots, x_n$; this, however, is excluded by the hypothesis. Hence $\alpha = 1$ and therefore

$$\sum_{i=1}^{n} (\lambda_i' - \lambda_i'') x_i = 0,$$

which, by virtue of the linear independence of $x_1, x_2, \ldots, x_n$, finally implies

$$\lambda_i' = \lambda_i'', \ i = 1, 2, \ldots, n, \text{ q.e.d.}$$

For example, $L_p(0, 1)$ and $l_p(p > 1)$ are (strongly) normed spaces. On the other hand $C[0, 1]$ is not (strongly) normed. In order to see this, it is sufficient to consider two non-negative, linearly independent functions $x(t), y(t) \in C[0, 1]$ which assume their minima at the same point of the interval $[0, 1]$. In spite of the fact that $x \neq \alpha y$,

$$\| x + y \| = \| x \| + \| y \|$$

obviously holds.

§17. ABSTRACT HILBERT SPACE

INTRODUCTORY REMARKS. Besides addition of vectors and multiplication of a vector by a real (complex) number, we introduce in the n-dimensional vector space E_n the *scalar* or *inner product* of two vectors

$$x = \{ \xi_1, \xi_2, \ldots, \xi_n \} \text{ and } y = \{ \eta_1, \eta_2, \ldots, \eta_n \}$$

of this space by

$$(x, y) = \sum_{i=1}^{n} \xi_i \bar{\eta}_i.$$

By means of this scalar product the length of a vector $x = \{ \xi_1, \xi_2, \ldots, \xi_n \}$ can be given as

$$\| x \| = \sqrt{\sum_{i=1}^{n} | \xi_i |^2} = \sqrt{(x, x)} .$$

In analysis, we frequently consider the scalar product of functions.

It is therefore natural to consider the class of those linear spaces where an inner product is defined. Such spaces are called Hilbert spaces and are defined by the following axioms.

AXIOMS OF ABSTRACT HILBERT SPACE. Let H consist of elements $x, y, z, \ldots$ and suppose there hold:

1. H is a *linear system* over the set of complex numbers.
2. With every pair of elements x, y there is associated a complex number (x, y), the *scalar product* of x and y, with the properties:

a) $(x, y) = \overline{(y, x)}$[18] (in particular (x, x) is real);
b) $(x_1 + x_2, y) = (x_1, y) + (x_2, y)$;
c) $(\lambda x, y) = \lambda (x, y)$ for every complex λ;
d) $(x, x) \geqq 0$, where $(x, x) = 0$ if and only if $x = \theta$.

---

[18] $\bar{a}$ denotes the complex conjugate of a.

The number $\|x\| = \sqrt{(x, x)}$ is called the *norm* of x.

3. With respect to the metric $\rho(x, y) = \|x - y\|$, H is complete.

If these axioms are fulfilled, the set H is called an *abstract complex space with quadratic metric* or, shorter, *a unitary space*. An n-dimensional unitary space is a complex Euclidean space.

4. For every natural number n, there exist n linearly independent elements.

5. H is separable.

If H satisfies axioms 1 through 5, then H is called an *abstract complex separable Hilbert space*. In the following we shall say, for short, a (separable) Hilbert space.

Examples. 1. The complex space l_2 becomes a Hilbert space if we put for every two of its elements $x = \{\xi_1, \xi_2, \ldots, \xi_n, \ldots\}$ and $y = \{\eta_1, \eta_2, \ldots, \eta_n, \ldots\}$,

$$(x, y) = \sum \xi_i \bar{\eta}_i$$

2. The complex space $L_2^{(\rho)}$ (0, 1) becomes a Hilbert space, if the scalar product of the elements $x = x(t)$ and $y = y(t)$ is defined by

$$(x, y) = \int_0^1 x(t)\overline{y(t)}\, \rho(t)\, dt.$$

In particular, $L_2(0, 1)$ is a Hilbert space, if we put

$$(x, y) = \int_0^1 x(t)\, \bar{y}(t)\, dt.$$

Real Hilbert space is analogously defined. There we require that the inner product of two elements be real. The real spaces $l_2, L_2, L_2^{(\rho)}$ are real Hilbert spaces.

Let us briefly consider the simplest properties of Hilbert space. First of all, it is easy to deduce from Axioms 1 through 3 the relations

$$(x, y_1 + y_2) = (x, y_1) + (x, y_2), \quad (x, \lambda y) = \bar{\lambda}(x, y).$$

From the second one, in particular, it follows that

$$\|\lambda x\| = |\lambda|\, \|x\|. \tag{1}$$

We shall now prove the Schwarz inequality for the scalar product. For an arbitrary complex λ, we have

$$(x + \lambda y, x + \lambda y) \geqq 0$$

or

$$(x, x) + \bar{\lambda}(x, y) + \lambda(y, x) + |\lambda|^2 (y, y) \geqq 0.$$

Putting

$$\lambda = -\frac{(x, y)}{(y, y)},$$

we obtain

$$(x, x) - \frac{|(x, y)|^2}{(y, y)} \geqq 0$$

or

$$|(x, y)| \leqq \|x\|\, \|y\|. \tag{2}$$

This is the desired inequality.

Furthermore, we get

$$\begin{aligned} \|x + y\|^2 &= (x + y, x + y) = (x, x) + (y, x) + (x, y) + (y, y) \\ &\leqq \|x\|^2 + 2\|x\|\, \|y\| + \|y\|^2 = (\|x\| + \|y\|)^2 \end{aligned}$$

or

$$\|x + y\| \leqq \|x\| + \|y\|. \tag{3}$$

Axiom 2d) and relations (1) and (3) show that the norm introduced by means of the scalar product satisfies all axioms of a norm of a normed linear space and consequently, the distance introduced by this norm satisfies all axioms of a metric space.

We easily deduce

LEMMA 1. *The scalar product is a continuous function with respect to norm convergence.*

If $x_n \to x$ and $y_n \to y$, then the numbers $\|x_n\|, \|y_n\|$ are bounded. Let M be their upper bound. Then

$$\begin{aligned} |(x_n, y_n) - (x, y)| &\leqq |(x_n, y_n) - (x_n, y)| + |(x_n, y) - (x, y)| \\ &= |(xn, y_n - y)| + |(x_n - x, y)| \leqq \|x_n\| \; \|y_n - y\| + \|x_n - x\| \; \|y\| \\ &\leqq M\|y_n - y\| + \|y\| \; \|x_n - x\|. \end{aligned}$$

Now $\|y_n - y\| \to 0$ and $\|x_n - x\| \to 0$ as $n \to \infty$, and therefore $|(x_n, y_n) - (x, y)| \to 0$ as $n \to \infty$, q.e.d.

ORTHOGONALITY. Two elements x and y in H are called *orthogonal* $(x \perp y)$, if $(x, y) = 0$. The element x is said to be orthogonal to the subspace $L \subset H (x \perp L)$, if x is orthogonal to every $y \in L$. There holds

LEMMA 2. *If $x \in H$ and L is a subspace of H, then there exists a unique representation.*

$$x = y + z \tag{4}$$

where $y \in L$ and $z \perp L$.

Proof. For $x \in L$, we obviously have $y = x$ and $z = \theta$. Therefore we can assume $x \bar{\in} L$. Let $d = \inf_{y' \in L} \|x - y'\|^2$ and let $\{y_n\}$ be a sequence in L such that $d_n = \|x_n - y_n\|^2 \to d$ as $n \to \infty$. Furthermore, let h be an arbitrary element in L. Then, for arbitrary complex ε, we have $y_n + \varepsilon h \in L$ and therefore $\|x - (y_n + \varepsilon h)\|^2 \geqq d$, i.e.,

$$\|x - y_n\|^2 - \bar{\varepsilon}(x - y_n, h) - \varepsilon(h, x - y_n) + |\varepsilon|^2 \|h\|^2 \geqq d.$$

For

$$\varepsilon = \frac{(x - y_n, h)}{\|h\|^2}$$

we obtain

$$\|x - y_n\|^2 - \frac{|(x - y_n, h)|^2}{\|h\|^2} \geqq d$$

from which

$$|(x - y_n, h)|^2 \leqq \|h\|^2 (d_n - d)$$

or

$$|(x - y_n, h)| \leqq \|h\| \sqrt{d_n - d}. \tag{5}$$

This implies

$$|(y_n - y_m, h)| \leqq |(y_n - x, h)| + |(x - y_m \; h)| \leqq \left(\sqrt{d_n - d} + \sqrt{d_m - d}\right) \|h\|.$$

For $h = y_n - y_m$, we get

$$\|y_n - y_m\| \leqq \sqrt{d_n - d} + \sqrt{d_m - d}.$$

Consequently $\{y_n\}$ is a fundamental sequence. Because of the completeness of H, there exists a $y \in H$ towards which $\{y_n\}$ converges. y also belongs to L because L is closed.

Taking the limit in inequality (5) we get $(x - y, h) = 0$ and since h is an arbitrary element of the subspace L, it follows that $x-y \perp L$. For $x-y=z$ we obtain the desired representation

$$x=y+z.$$

There remains to prove the uniqueness of the representation. Let

$$x=y+z=y'+z',$$

where $y, y' \in L$ and $z, z' \perp L$. Then $y-y'=z'-z$ and

$$\|y-y'\|^2=(z'-z, y-y')=0, \tag{6}$$

because of $\qquad y-y' \in L$ and $z'-z \perp L$.

Equality (6) means $y=y'$ and this implies $z=z'$, and the lemma is proved.

The element y in the representation (4) is called the *projection* of the element x on the subspace L. Obviously the totality M of all elements orthogonal to a given subspace constitute again a subspace. It is clear that this collection is a linear manifold. Because of the continuity of the scalar product, it is closed. Therefore the element z of the representation above can be called the projection of x on M. M is called the *orthogonal complement* of the subspace L and is designated by $H \dot{-} L$. Furthermore H is called the *orthogonal sum* of L and M and we write $H=L+M$. Obviously, the orthogonal sum is a special case of the direct sum. Lemma 2 therefore achieves the decomposition of an element into its projections onto two complementary orthogonal subspaces.

LEMMA 3. *The linear manifold M is everywhere dense in H if and only if there exists no element different from zero which is orthogonal to all elements of M.*

Necessity. First it is clear that $x \perp M$ implies $x \perp \bar{M}$. If now $\bar{M}=H$, then $x \perp H$, in particular, therefore $x \perp x$, i.e., $x=\theta$.

Sufficiency. Let us assume that M is not everywhere dense in H. Then $\bar{M} \neq H$ and there exists an $x \bar{\in} M$ and $x \in H$. According to Lemma 2, $x=y+z$, where $y \in \bar{M}$ and $z \perp M$. Because $x \bar{\in} M, z \neq 0$, which contradicts our assumption.

ORTHONORMAL SYSTEMS. A system $e_1, e_2, \ldots, e_n, \ldots$ of elements of the space H is called *orthonormal* if

$$(e_i, e_j)=\delta_{ij}$$

where δ_{ij} is the Kronecker symbol ($\delta_{ij}=1$ for $i=j$ and $\delta_{ij}=0$ for $i \neq j$).

For example, in the complex space $L_2(0,1)$ the functions $e^{2\pi int}$, $n=0, \pm 1, \pm 2, \ldots$, are an orthonormal system on the interval $[0,1]$.

Every system $h_1, h_2, \ldots, h_n$, of linearly independent elements can be changed into an orthonormal one by means of the following *orthogonalization process.*[19]

We put $e_1=\dfrac{h_1}{\|h_1\|}$. Let $g_2=h_2-c_{21}e_1$, then we choose the number c_{21} such that g_2 is orthogonal to e_1. Obviously $c_{21}=(h_2, e_1)$. Furthermore, we put $e_2=\dfrac{g_2}{\|g_2\|}$. Here $\|g_2\| \neq 0$, for otherwise $g_2=\theta$ and then h_1 and h_2 would be linearly dependent and that contradicts the hypothesis. If $e_1, e_2, \ldots, e_{k-1}$ are already constructed, then we take $g_k=h_k-\sum_{i=1}^{k-1} c_{ki}e_i$ and choose the c_{ki}'s, putting $c_{ki}=(h_k, e_i)$,

[19] The process is called the Gram-Schmidt process.

such that g_k is orthogonal to $e_1, e_2, \ldots, e_{k-1}$. Then $e_k = \frac{g_k}{\|g_k\|}$ is another element of the orthonormal system under construction.

Example. By orthonormalization of the powers

$$1, t, t^2, \ldots, t^n, \ldots$$

in the real space $L_2^{(\rho)}(a, b)$ we arrive at the system of Chebyshev polynomials

$$P_0 = \text{const}, P_1(t), P_2(t), \ldots, P_n(t), \ldots,$$

which are orthonormal with weight $\rho(t)$:

$$\int_a^b \rho(t) P_i(t) P_j(t)\, dt = \delta_{ij}.$$

For $\rho(t) \equiv 1$, $a = -1$, $b = 1$ we obtain the Legendre polynomials to within constant factors; for $\rho(t) = e^{-t^2}$, $a = -\infty$, $b = +\infty$ the Chebyshev-Hermite polynomials and for $\rho(t) = e^{-t}$, $a = 0$, $b = \infty$ the system of Chebyshev-Laguerre polynomials.

Let L be the subspace generated by the orthonormal system $e_1, e_2, \ldots, e_n, \ldots$ and let $x \in L$. Then there exists for every $\varepsilon > 0$ a linear combination $\sum_{i=1}^{n} \alpha_i e_i$ such that

$$\left\| x - \sum_{i=1}^{n} \alpha_i e_i \right\| < \varepsilon.$$

Moreover,

$$\left\| x - \sum_{i=1}^{n} \alpha_i e_i \right\|^2 = \|x\|^2 - \sum_{i=1}^{n} \bar{\alpha}_i (x, e_i) - \sum_{i=1}^{n} \alpha_i (e_i, x) + \sum_{i=1}^{n} \sum_{j=1}^{n} \alpha_i \bar{\alpha}_j (e_i, e_j)$$

$$= \|x\|^2 - \sum_{i=1}^{n} \bar{\alpha}_i c_i - \sum_{i=1}^{n} \alpha_i \bar{c}_i + \sum_{i=1}^{n} |\alpha_i|^2,$$

where

$$c_i = (x, e_i).$$

The numbers c_i are called *Fourier coefficients* of x with respect to the orthonormal system $\{e_i\}$. Furthermore we obtain from the last equality

$$\left\| x - \sum_{i=1}^{n} \alpha_i e_i \right\|^2 = \|x\|^2 - \sum_{i=1}^{n} |c_i|^2 + \sum_{i=1}^{n} |\alpha_i - c_i|^2.$$

We note that the norm of the difference $x - \sum_{i=1}^{n} \alpha_i e_i$ assumes its least value if the α_i are precisely the Fourier coefficients of x with respect to the system $\{e_i\}$. In this case

$$0 \leqq \left\| x - \sum_{i=1}^{n} c_i e_i \right\|^2 = \|x\|^2 - \sum_{i=1}^{n} |c_i|^2 < \varepsilon, \tag{7}$$

and since ε can be chosen arbitrarily small,

$$x = \lim_{n} \sum_{i=1}^{n} c_i e_i = \sum_{i=1}^{\infty} c_i e_i.$$

In addition, (7) implies the convergence of the series $\sum_{i=1}^{\infty} |c_i|^2$, and in particular $\sum_{i=1}^{\infty} |c_i|^2 = \|x\|^2$.

Now let x be an arbitrary element in H. Let the projection of x on L be designated by z. Then $z = \sum_{i=1}^{\infty} c_i e_i$, where $c_i = (z, e_i) = (x, e_i)$ and $\sum_{i=1}^{\infty} |c_i|^2 = \|z^2\|$. As we easily see, $\|z\|^2 \leqq \|x\|^2$ and so

$$\sum_{i=1}^{\infty} |c_i|^2 \leqq \|x\|^2. \tag{8}$$

This is the *Bessel inequality*.

COMPLETENESS IN THE SENSE OF STEKLOV. V. A. Steklov introduces in his investigations of the representation of functions by elements of an orthonormal system the concept of closure of such a system.

Let there be given in the space H an orthonormal system $\{e_i\}$. Then, if for every $x \in H$,

$$\sum_{i=1}^{\infty} |c_i|^2 = \|x\|^2, \tag{9}$$

where $c_i = (x, e_i)$, then the system $\{e_i\}$ is called *complete in the sense of Steklov* and the equality (9) is called the *Parseval-Steklov equality* or *the completeness relation.* If this condition is fulfilled, then there exists no element in H different from 0 and orthogonal to all e_i. For, if $x \perp e_i$ for $i = 1, 2, \ldots$, then $c_i = 0$ for all i. Then, however, according to (9), $\|x\| = 0$, and consequently $x = \theta$.

An orthonormal system is called *complete* if there exists no element different from 0 which is orthogonal to all elements of the system.

We have seen that every complete system is closed. Conversely, every closed system is also complete. For, if there exists no element different from 0 which is orthogonal to all elements of the orthonormal system, then, according to Lemma 3, L lies everywhere dense in H. This is obviously equivalent to the completeness of the orthonormal system.

For example, the trigonometric functions form a complete orthogonal system in the real Hilbert space.

As we can easily show, there exists in every Hilbert space a complete orthonormal system:

Let $G = \{g_1, g_2, \ldots, g_n, \ldots\}$ be an arbitrary denumerable everywhere dense set in H. Let L_1 be the one-dimensional subspace generated by

$$e_1 = \frac{g_1}{\|g_1\|}.$$

Let g_{n_2} be the first element of G which does not belong to L_1 and let h_2 be the projection of g_{n_2} on $H \dot{-} L_1$. Furthermore let

$$e_2 = \frac{h_2}{\|h_2\|}$$

and e_1 generate the subspace L_2 and let g_{n_3} be the first element of the set G which does not belong to L_2. Taking h_3 for the projection of g_{n_3} on $H \dot{-} L_2$, we put

$$e_3 = \frac{h_3}{\|h_3\|}.$$

Continuing this procedure, we obtain an orthonormal system $e_1, e_2, \ldots, e_n, \ldots$. Since, according to the construction, each g_m belongs to an L_n, the subspace generated by the system $\{e_i\}$ coincides with the subspace generated by $\{g_i\}$, i.e., with H. The system $\{e_i\}$ cannot be finite. If it had only p elements, then there would exist in H, as is known from linear algebra, no $p+1$ linearly independent elements. This contradicts Axiom 4.

If $\{e_i\}$ is a complete orthonormal system and if $x, y \in H$ with the Fourier coefficients c_i, d_i, respectively, $i = 1, 2, \ldots$, then we easily show the validity of

$$(x, y) = \sum_{i=1}^{\infty} c_i d_i.$$

ISOMORPHISMS OF SEPARABLE HILBERT SPACE. Let H be an arbitrary Hilbert space with the complete orthonormal system $\{e_i\}$. Then every $x \in H$ determines a sequence $\{c_1, c_2, \ldots, c_n, \ldots\}$ of numbers consisting of its Fourier coefficients and the series

$$\sum_{i=1}^{\infty} |c_i|^2$$

converges, as we have shown above. We can therefore look upon $\{c_1, c_2, \ldots, c_n, \ldots\}$ as an element of the complex space l_2. Each $x \in H$ is therefore assigned an $\tilde{x} \in l_2$. Because the completeness of $\{e_i\}$ has been required, we have furthermore

$$\|x\|_H = \|\tilde{x}\|_{l_2}. \tag{10}$$

Here the indices H and l_2 shall suggest that one time we mean the norm introduced in H, the other time that one defined in l_2. If there correspond to the elements $\tilde{x} \in H, y \in H$ the elements $\tilde{x} \in l_2, \tilde{y} \in l_2$, respectively, then obviously $x \pm y$ is assigned $\tilde{x} \pm \tilde{y}$. This and (10) imply

$$\|x - y\|_H = \|\tilde{x} - \tilde{y}\|_{l_2}. \tag{11}$$

Now let $\tilde{z} = \zeta_i$, an arbitrary element in l_2. We have then for the elements $z_n = \sum_{i=1}^{n} \zeta_i e_i$, $n = 1, 2, \ldots$ of H,

$$\|z_m - z_n\|^2 = \left\| \sum_{i=n+1}^{m} \zeta_i e_i \right\|^2 = \sum_{i=n+1}^{m} |\zeta_i|^2$$

and therefore

$$\|z_m - z_n\| \to 0,$$

as $m, n \to \infty$. Consequently, the sequence $\{z_n\}$ converges in the sense of the metric of H. Because of the completeness of H, $\{z_n\}$ must converge to a $z \in H$. Since

$$(z, e_i) = \lim_n (z_n, e_i) = \zeta_i$$

the ζ_i are just the Fourier coefficients of z with respect to the orthonormal system $\{e_i\}$. Therefore, there corresponds to every $\tilde{z}$ in l_2 a certain $z \in H$. Hence there exists a one-to-one correspondence between the elements of H and those of l_2.

Formula (11) shows that the correspondence between the spaces H and l_2 is isometric. The two spaces are also isomorphic: If x and $\tilde{x}$ are associated with each other, then so are λx and $\lambda \tilde{x}$. This, together with what has been shown above implies the validity of

THEOREM 1. *Every complex (real) separable Hilbert space is isomorphic and isometric with the complex (real) space l_2. In particular, all complex (real) separable Hilbert spaces are isomorphic and isometric to each other.*

As a corollary to this theorem, we have

THEOREM 2 (Riesz-Fischer). *The real spaces $L_2(0, 1)$ and l_2 are isomorphic and isometric.*

§18. LINEAR OPERATORS IN NORMED LINEAR SPACES

Let E, E_1 be two normed linear spaces. Because every normed linear space is a fortiori a linear space, the above definition of a linear operator on E with range in E_1 and the definition of the totality $(E \to E_1)$ of operators on normed linear spaces remains valid. Also the theorems proved in §14 remain valid.

We only remark that the continuity of an operator A means the following:

$$\| Ax_n - Ax \| \to 0 \text{ as } \| x_n - x \| \to 0.$$

The operators considered in Examples 1 and 2 of §14 are linear and transform the normed linear spaces E_n, $C[0, 1]$, respectively into themselves. The operator of Example 3 is also linear. By means of it, the normed linear space of the functions defined and continuous on Γ is transformed into the normed linear space of the harmonic functions in G, if the norms in these spaces are defined by

$$\|x\| = \max_{\Gamma} |x(t)|, \quad \|u\| = \max_{G} |u(\xi, \eta)|, \text{ respectively.}$$

As a further example, we consider the infinite matrix (a_{ik}), $i, k = 1, 2, \ldots$ with

$$\sum_{i=1}^{\infty} \sum_{k=1}^{\infty} |a_{ik}|^q < \infty, \ q > 1.$$

Then the system of equalities

$$\eta_i = \sum_{i=1}^{\infty} a_{ik}\xi_k, \quad i = 1, 2, 3, \ldots,$$

defines an operator A linear in l_p. The range of A lies in l_q, where q is given by

$$\frac{1}{p} + \frac{1}{q} = 1.$$

For, by means of the above system, each $x = \{\xi_i\} \in l_p$ is associated with a $y = \{\eta_i\} \in l_q$, i.e.,

$$A \in (l_p \to l_q).$$

First we show that indeed $y \in l_q$ for $x \in l_p$. According to Hölder's inequality for sums,[20] we have

$$\sum_{i=1}^{N} |\eta_i|^q = \sum_{i=1}^{N} \left| \sum_{k=1}^{\infty} a_{ik}\xi_k \right|^q \leqq \sum_{i=1}^{N} \left\{ \left(\sum_{k=1}^{\infty} |\xi_k|^p \right)^{\frac{1}{p}} \left(\sum_{k=1}^{\infty} |a_{ik}|^q \right)^{\frac{1}{q}} \right\}^q$$

$$= \|x\|^q \sum_{i=1}^{N} \sum_{k=1}^{\infty} |a_{ik}|^q \leqq \|x\|^q \sum_{i=1}^{\infty} \sum_{k=1}^{\infty} |a_{ik}|^q.$$

Since this inequality holds for arbitrary N, we can pass to the limit and obtain

$$\|y\|^q = \sum_{i=1}^{\infty} |\eta_i|^q \leqq \|x\|^q \sum_{i=1}^{\infty} \sum_{k=1}^{\infty} |a_{ik}|^q < \infty,$$

i.e., $y \in l_q$.

---

[20] See Appendix I.

Putting

$$x_1 = \{\xi_i^{(1)}\} \in l_p \text{ and } x_2 = \{\xi_i^{(2)}\} \in l_p,$$

$$\sum_{k=1}^{\infty} a_{ik}(\xi_k^{(1)} + \xi_k^{(2)}) = \sum_{k=1}^{\infty} a_{ik}\xi_k^{(1)} + \sum_{k=1}^{\infty} a_{ik}\xi_k^{(2)}$$

implies the validity of $\qquad A(x_1 + x_2) = Ax_1 + Ax_2,$

that is, the additivity of the operator A. Now let

$$x_n = \{\xi_i^{(n)}\} \in l_p, \ x = \{\xi_i\} \in l_p,$$

$$Ax_n = y_n, \ Ax = y, \ y_n = \{\eta_i^{(n)}\}, \ y = \{\eta_i\}.$$

Then

$$\|y - y_n\| = \left(\sum_{i=1}^{\infty} |\eta_i^{(n)} - \eta_i|^q\right)^{\frac{1}{q}} = \left(\sum_{i=1}^{\infty} \left|\sum_{k=1}^{\infty} a_{ik}(\xi_k^{(n)} - \xi_k)\right|^q\right)^{\frac{1}{q}}$$

$$\leqq \left(\sum_{k=1}^{\infty} |\xi_k^{(n)} - \xi_k|^p\right)^{\frac{1}{p}} \left(\sum_{i=1}^{\infty}\sum_{k=1}^{\infty} |a_{ik}|^q\right)^{\frac{1}{q}} = \|x_n - x\| \left(\sum_{i=1}^{\infty}\sum_{k=1}^{\infty} |a_{ik}|^q\right)^{\frac{1}{q}}$$

and therefore

$$\|y_n - y\| \to 0$$

as $\|x_n - x\| \to 0$. A is therefore continuous, which, together with the already proved additivity of A, implies the linearity of our operator.

An operator A is called *bounded*, if there exists a constant M such that $\|Ax\| \leqq M\|x\|$ for all $x \in E$. ($\|Ax\|$ means the norm in the space E_1, whereas $\|x\|$ is the norm in E.)

If A is bounded, then the boundedness of the pre-image set $\{x\} \subset E$ implies the boundedness of the image set $\{Ax\} \subset E_1$.

Examples. 1. Let $\qquad Ax = \int_0^1 K(s,t)x(t)\,dt,$

where $K(s,t)$ is continuous on the square $0 \leqq s,\ t \leqq 1$ and $x(s) \in C[0,1]$. Then

$$Ax = y(s) \in C[0,1].$$

For $M = \max_{0 \leqq s,\, t \leqq 1} |K(s,t)|$,

$$\|y\| = \max_{0 \leqq s \leqq 1} |y(s)| \leqq M \max_{0 \leqq t \leqq 1} |x(t)| = M\|x\|,$$

i.e., our integral operator $y = Ax$ is bounded.

2. In $C[0,1]$ we consider the operator

$$Ax = \frac{d}{ds}(x(s)),$$

defined on the set of continuously differentiable functions $x(s) \in C[0,1]$. We shall show the unboundedness of A. In order to do so, we put

$$x_n(s) = \sqrt{2}\sin n\pi s.$$

Obviously, $\|x_n\| = 1$ for all n and

$$Ax_n = (\sqrt{2}\sin n\pi s)' = \sqrt{2}\, n\pi \cos n\pi s$$

where $\|Ax_n\| = n\pi$. Because $\|x_n\| = 1$ and Ax_n increases without bound as $n \to \infty$, there exists no constant M such that $\|Ax\| \leqq M\|x\|$ for all x of the domain of definition. Finally, we remark that our operator is not defined everywhere in $C[0,1]$.

THEOREM 3. *An additive operator is continuous if and only if it is bounded.*[21]

[21] Therefore the operators linear in our sense are often called bounded linear operators in contrast with "linear" additive operators, which are perhaps unbounded and which need not be defined on the whole of E.

Necessity. Let A be linear and let us assume the unboundedness of A. There then exists a sequence $\{x_n\}$ with

$$\| Ax_n \| > n \| x_n \|.$$

If $x_n' = \dfrac{x_n}{n \| x_n \|}$ then $x_n' \to \theta$, because

$$\| x_n' \| = \frac{1}{n} \to 0 \text{ as } n \to \infty.$$

On the other hand
$$\| Ax_n' \| = \frac{1}{n \| x_n \|} \| Ax_n \| > 1$$

and therefore

$$\| Ax_n' \| \text{ does not tend to } \| A\theta \| = \| \theta \| = 0 \text{ as } n \to \infty.$$

Therefore A cannot be continuous at the origin, and this contradicts the hypothesis.

Sufficiency. Let A be additive and bounded, therefore $\| A_x \| \leqq M \| x \|$. If $x_n \to x$, i.e., $\| x_n - x \| \to 0$, then $\| Ax_n - Ax \| = \| A(x_n - x) \| \leqq M \| x_n - x \|$ implies $Ax_n \to Ax$, q.e.d.

Remark. In proving the necessity we have used only the additivity and the homogeneity of A. Consequently, an additive homogeneous operator defined on a linear manifold cannot be continuous if it isn't bounded there. In particular, there follows the discontinuity of the differential operator considered above.

THE NORM OF AN OPERATOR. Let A be a linear operator. The smallest number M, for which $\| Ax \| \leqq M \| x \|$ for all x, is called the *norm* of A and designated by $\| A \|$.

According to the definition, $\| A \|$ has the following properties.

For arbitrary x, we have

$$\| Ax \| \leqq \| A \| \| x \| \tag{1}$$

and for every $\varepsilon > 0$, there exists an x' such that

$$\| Ax' \| > (\| A \| - \varepsilon) \| x' \|.$$

One can show without difficulty that

$$\| A \| = \sup_{\| x \| \leqq 1} \| Ax \|. \tag{2}$$

For, if $\| x \| \leqq 1$, then

$$\| Ax \| \leqq \| A \| \| x \| \leqq \| A \|$$

and consequently
$$\sup_{\| x \| \leqq 1} \| Ax \| \leqq \| A \|. \tag{3}$$

On the other hand, there exists for every $\varepsilon > 0$ an x' such that

$$\| Ax' \| > (\| A \| - \varepsilon) \| x' \|.$$

Putting
$$x_1 = \frac{x'}{\| x' \|},$$

we get
$$\| A x_1 \| = \frac{1}{\| x' \|} \| A x' \| > \frac{1}{\| x' \|} (\| A \| - \varepsilon) \| x' \| = \| A \| - \varepsilon.$$

Because $\| x_1 \| = 1$,
$$\sup_{\| x \| \leqq 1} \| A x \| \geqq \| A x_1 \| > \| A \| - \varepsilon$$

and therefore
$$\sup_{\|x\| \leqq 1} \|Ax\| \geqq \|A\|. \tag{4}$$

The assertion (2) follows from (3) and (4).

As an example, we shall determine the norm of the linear operator

$$y(s) = \int_0^1 K(s, t)\, x(t)\, dt$$

with continuous kernel. Obviously this operator maps the space $C[0, 1]$ into itself. Putting

$$Ax = \int_0^1 K(s, t)\, x(t)\, dt,$$

there results

$$\|Ax\| = \max_s \left| \int_0^1 K(s, t)\, x(t)\, dt \right|$$
$$\leqq \max_t |x(t)| \max_s \int_0^1 |K(s, t)|\, dt = \|x\| \max_s \int_0^1 |K(s, t)|\, dt$$

and
$$\|A\| \leqq \max_s \int_0^1 |K(s, t)|\, dt. \tag{α}$$

As a continuous function of s, the integral $\int_0^1 |K(s, t)|\, dt$ must assume its maximum at a point s_0 of the interval $[0, 1]$. We put

$$z_0(t) = \operatorname{sgn} K(s_0, t).$$

Let $x_n(t)$ be continuous functions approximating the functions $z_0(t)$ such that always $|x_n(t)| \leqq 1$ and in addition everywhere except on a set M_{ε_n} of measure ε_n, $x_n(t) = z_0(t)$ holds. Here $\varepsilon_n < \dfrac{1}{2Mn}$ where $M = \max\limits_{s,t} |K(s, t)|$. Then on M_{ε_n}, $|x_n(t) - z_0(t)| \leqq 2$. Furthermore

$$\left| \int_0^1 K(s, t)\, z_0(t)\, dt - \int_0^1 K(s, t)\, x_n(t)\, dt \right| \leqq \int_0^1 |K(s, t)|\, |x_n(t) - z_0(t)|\, dt$$
$$= \int_{M_{\varepsilon_n}} |K(s, t)|\, |x_n(t) - z_0(t)|\, dt \leqq 2 \max_{s,t} |K(s, t)| \frac{1}{2Mn} = \frac{1}{n}$$

for every $s \in [0, 1]$.

Consequently, for all $s \in [0, 1]$,

$$\int_0^1 K(s, t)\, z_0(t)\, dt \leqq \int_0^1 K(s, t)\, x_n(t)\, dt + \frac{1}{n} \leqq \|A\|\, \|x_n\| + \frac{1}{n}.$$

Now, putting $s = s_0$, we obtain

$$\int_0^1 |K(s_0, t)|\, dt \leqq \|A\|\, \|x_n\| + \frac{1}{n}.$$

Since $\|x_n\| < 1$, there results from the above inequality as $n \to \infty$

$$\int_0^1 |K(s_0, t)|\, dt \leqq \|A\|,$$

i.e.,
$$\max_s \int_0^1 |K(s, t)|\, dt \leqq \|A\|. \tag{β}$$

It follows from (α) and (β), that

$$\|A\| = \max_s \int_0^1 |K(s,t)|\,dt.$$

§19. THE SPACE OF LINEAR OPERATORS

STRONG CONVERGENCE. We have seen in §14 that all operators defined on a linear space E and with ranges in a linear space E_1 themselves form a linear space $(E \to E_1)$. If E and E_1 are normed, then we can define a norm in $(E \to E_1)$.

There was introduced in §18 a norm for each linear operator A which maps E into E_1. We have only to show that it satisfies the axioms for a norm.

1) Obviously $\|A\| = \sup_{\|x\| \leqq 1} \|Ax\| \geqq 0$. If $\|A\| = 0$, i.e., $\sup_{\|x\| \leqq 1} \|Ax\| = 0$, then $\|Ax\| = 0$ and therefore $Ax = \theta$ for all x with $\|x\| \leqq 1$. According to the homogeneity of A, $Ax = \theta$ for all x and consequently $A = \theta$.

2) $\|\lambda A\| = \sup_{\|x\| \leqq 1} \|\lambda A x\| = |\lambda| \sup_{\|x\| \leqq 1} \|Ax\| = |\lambda|\,\|A\|$.

3) $\|A + B\| = \sup_{\|x\| \leqq 1} \|Ax + Bx\| \leqq \sup_{\|x\| \leqq 1} \|Ax\| + \sup_{\|x\| \leqq 1} \|Bx\| = \|A\| + \|B\|$.

Therefore *the linear operators form a normed linear space.*

If $\|A_n - A\| \to 0$ (where the norm is that of $(E \to E_1)$), then we say that the sequence $\{A_n\}$ of operators converges to A, and we speak of *convergence in norm* or *strong convergence.*

We now prove that

Whenever E_1 is complete, so is the space of linear operators and therefore it is a Banach space.

In order to prove this, let $\{A_n\}$ be a sequence of linear operators which converges in the sense of the norm of $(E \to E_1)$, i.e., let $\|A_n - A_m\| \to 0$ as $n, m \to \infty$. For an arbitrary x,

$$\|A_m x - A_n x\| = \|(A_m - A_n)x\| \leqq \|A_m - A_n\|\,\|x\| \to 0$$

as $n, m \to \infty$.

Therefore, the sequence $\{A_n x\}$ of elements of the space E_1 is convergent for every fixed x. Because E_1 is complete, $\{A_n x\}$ has a limit y in E_1.

Thus with every $x \in E$ there is associated a $y \in E_1$ and we obtain an additive and bounded operator defined by the equation $Ax = y$:

Additivity of A follows immediately from

$$A(x_1 + x_2) = \lim_n A_n(x_1 + x_2) = \lim_n A_n x_1 + \lim_n A_n x_2 = Ax_1 + Ax_2.$$

By hypothesis,

$$\|A_n - A_m\| \to 0$$

as $n, m \to 0$. Therefore

$$|\,\|A_n\| - \|A_m\|\,| \to 0$$

as $n, m \to \infty$, i.e., the numerical sequence $\{\|A_n\|\}$ converges and therefore is bounded. Consequently, there exists a K such that $\|A_n\| \leqq K$ for all n. Hence $\|A_n x\| \leqq K\|x\|$ for all n and so

$$\| Ax \| = \lim_n \| A_n x \| \leqq K \| x \|,$$

which proves the boundedness of A.

Since A is also additive, A is a linear operator.

Now we shall prove that A is the limit of the sequence $\{A_n\}$ in the sense of norm convergence in $(E \to E_1)$. We can find for every $\varepsilon > 0$ an n_0 such that the inequality

$$\| A_{n+p} x - A_n x \| < \varepsilon \tag{1}$$

is fulfilled if $n \geqq n_0$, $p > 0$ and for all x with $\| x \| \leqq 1$. From (1), passing to the limit as $p \to \infty$, there results

$$\| Ax - A_n x \| \leqq \varepsilon$$

if $n \geqq n_0$ and for all x with $\| x \| \leqq 1$. This implies for $n \geqq n_0$

$$\| A - A_n \| = \sup_{\|x\| \leqq 1} \| (A - A_n) x \| = \sup_{\|x\| \leqq 1} \| A x - A_n x \| \leqq \varepsilon,$$

i.e.,

$$A = \lim_n A_n$$

in the sense of convergence in the norm in the space of linear operators. Therefore, the above assertion holds: $(E \to E_1)$ is complete whenever E_1 is.

We shall call convergence of a sequence of linear operators in the sense of convergence in the norm in the space of linear operators *uniform convergence of a sequence of linear operators.*

STRONG CONVERGENCE. In the space $(E \to E_1)$ we can define another convergence of sequences $\{A_n\}$ of linear operators.

$\{A_n\}$ is called *strongly convergent* to the linear operator A (in itself) if for every x, the sequence $\{A_n x\}$ converges to Ax (in itself).

Obviously, uniform convergence of a sequence $\{A_n\}$ implies strong convergence of this sequence.

We have that if E_1 is complete, then the space of linear operators is also complete in the sense of strong convergence.

Because, for every x the sequence $\{A_n x\}$ is a fundamental sequence, the limit $y = \lim_n A_n x$ exists for every x. Thus we obtain an operator $Ax = y$ defined on E with range in E_1. Additivity can be proved as above. The boundedness of A follows from the following

THEOREM (BANACH). *If a sequence of linear operators converges (strongly) at each point x of space E, then the sequence of norms $\{\| A_n \|\}$ of this operator is bounded.*

Proof. We assume the contrary. Then the set $\{\| A_n x \|\}$ is not bounded on any closed sphere $\| x - x_0 \| \leqq \varepsilon$. For, if

$$\| A_n x \| \leqq K$$

were valid for all n and all $x \in \bar{S}(x_0, \varepsilon)$, then for every y $(y \neq \theta)$ the element

$$y' = \frac{\varepsilon}{\|y\|} y + x_0$$

would have to belong to this sphere and we would have

$$\| A_n y' \| \leqq K$$

or
$$\frac{\varepsilon}{\|y\|}\|A_n y\| - \|A_n x_n\| \leqq \left\|\frac{\varepsilon}{\|y\|}A_n y + A_n x\right\| \leqq K.$$

This implies
$$\|A_n y\| \leqq \frac{K + \|A_n x_0\|}{\varepsilon}\|y\| = K_1\|y\|.$$

By virtue of the convergence of $\{A_n x_0\}$, the sequence $\{\|A_n x_0\|\}$ is bounded. Therefore
$$\|A_n y\| \leqq K_1\|y\|$$
and $\|A_n\| \leqq K_1$. This contradicts the hypothesis.

Now let $\bar{S}_0$ be an arbitrary closed sphere in E. Because, according to the preceding, $\{\|A_n x\|\}$ is not bounded on it, there exists in particular an operator A_{n_1} and an $x_1 \in \bar{S}_0$ such that
$$\|A_{n_1} x_1\| > 1.$$
By the continuity of A_{n_1} this inequality holds on a sphere $\bar{S}_1(x_1, \varepsilon_1) \subset S_0$. On $\bar{S}_1$, $\{\|A_n x\|\}$ again is not bounded and there exists therefore an operator A_{n_2}, $n_2 > n_1$, and an $x_2 \in \bar{S}_1$ such that
$$\|A_{n_2} x_2\| > 2.$$
Because A_{n_2} is continuous, this inequality must be valid on a sphere $\bar{S}_2(x_2, \varepsilon_2) \subset \bar{S}_1$. Continuing this reasoning and letting ε_n tend to 0 as $n \to \infty$, then because of the completeness, there exists a point $\bar{x}$, belonging to all spheres $\bar{S}_n(x_n, \varepsilon_n)$ and satisfying the inequality
$$\|A_{n_k}\bar{x}\| \geqq k.$$
This, however, cannot be, because $\{A_n x\}$ converges for every x. Consequently, the theorem of Banach is valid.

We return now to the operator
$$A = \lim_n A_n.$$
According to the inequality
$$\|A_n x\| \leqq M\|x\|$$
obtained from the theorem of Banach, we have $\|Ax\| \leqq M\|x\|$ if we let $n \to \infty$. Therefore A is bounded, hence linear. Consequently, every weakly Cauchy convergent sequence of linear operators has a well determined limit. Because it is itself a linear operator, the space of linear operators is complete in the sense of weak convergence.

APPLICATIONS TO INTERPOLATION THEORY. The theorem of Banach may be applied in many ways. As an example, we shall prove the following theorem [43].

Let there be given by the triangular array

$$\begin{matrix} t_1^{(1)} & & & \\ t_1^{(2)} & t_2^{(2)} & & \\ t_1^{(3)} & t_2^{(3)} & t_3^{(3)} & \\ \cdot\ \cdot\ \cdot & \cdot\ \cdot & \cdot\ \cdot & \\ t_1^{(n)} & t_2^{(n)} & \ldots & t_n^{(n)} \\ \cdot\ \cdot\ \cdot & \cdot\ \cdot\ \cdot & \cdot\ \cdot\ \cdot & \cdot\ \cdot\ \cdot \end{matrix} \tag{T}$$

certain points in the interval [0, 1]. There can then be found, for every function $x(t)$ defined on [0, 1] the Lagrange interpolation polynomials $L_n x$, whose partition points are the points of the nth row of (T):

$$L_n x = \sum_{k=1}^{n} x(t_k^{(n)})\, l_k^{(n)}(t)$$

with

$$l_k^{(n)}(t) = \frac{\omega_n(t)}{\omega_n'(t_k^{(n)})\,(t - t_k^{(n)})}, \quad \omega_n(t) = \prod_{k=1}^{n} (t - t_k^{(n)}).$$

We have the

THEOREM. *For an arbitrary array (T) there always exists a continuous function $x(t)$ such that the $L_n x$ do not converge to $x(t)$ as $n \to \infty$.*

We consider the $L_n x$ as operators mapping the functions $x(t) \in C[0, 1]$ into elements of the same space and we put

$$\lambda_n = \max_t \lambda_n(t), \text{ where } \lambda_n(t) = \sum_{k=1}^{n} | l_k^{(n)}(t) |.$$

Then we can easily show [43] that

$$\|L_n\| = \lambda_n.$$

On the other hand, there holds the inequality of S. N. Bernstein [43]:

$$\lambda_n > \frac{\ln n}{8\sqrt{\pi}}.$$

Consequently,

$$\|L_n\| \to \infty$$

as $n \to \infty$. This implies immediately the theorem formulated above. For, if $L_n x \to x$ for all $x \in C[0, 1]$, then the norms $\|L_n\|$ would have to be bounded.

§20. INVERSE OPERATORS

With the concept of the inverse operator introduced earlier, there is connected the question as to the existence and uniqueness of the solution of the functional equation

$$Ax = y, \tag{1}$$

where y is a given element of the linear space E and x the element of E sought. Since by (1) there can be expressed systems of linear algebraic equations, linear differential and integral equations, etc., we recognize the great importance of the inverse operator.

If we assume that A in (1) has an inverse operator A^{-1}, then we obtain by substituting $x = A^{-1}y$ in (1) the identity

$$AA^{-1}y = y,$$

i.e., $y = y$. Consequently

$$x = A^{-1}y$$

is a solution of (1).

Let x_1 be another solution of (1), i.e., assume $Ax_1 = y$. Multiplying both sides of this equality by A^{-1} we get $x_1 = A^{-1}y$ and therefore the uniqueness of the solution $x = A^{-1}y$ of (1). If A has a right inverse operator C, then we recognize immediately that $x = Cy$ satisfies (1); the question as to the uniqueness, however, remains open. If A has a left inverse operator B, the opposite prevails. If in this case there exists at all a solution $\bar{x}$, then this is the only one, because $A\bar{x} = y$ implies by left-sided multiplication by B that $\bar{x} = By$. Whether there exists a solution at all, however, cannot be decided in general.

THEOREM OF THE INVERSE OPERATOR.

THEOREM 1 (BANACH). *For every linear operator A which maps a Banach space E biuniquely onto a Banach space E_1, there exists a linear inverse operator A^{-1} which maps E_1 onto E.*

Proof. By hypothesis, $Ax = y$ determines a one-to-one correspondence $x \leftrightarrow y$ between the $x \in E$ and $y \in E_1$. Because of the linearity of A, θ must be associated with θ_1 (θ_1 is the zero element of E_1). This correspondence arises not only from A but one can think of it also as created in the other direction by the inverse operator $A^{-1}y = x$. It remains to show that the thus interpreted operator A^{-1} is linear and bounded.

Let $y_1, y_2 \in E_1$ and let us put

$$z = A^{-1}(y_1 + y_2) - A^{-1}y_1 - A^{-1}y_2.$$

Then $Az = A\,[A^{-1}(y_1 + y_2) - A^{-1}y_1 - A^{-1}y_2]$

$$= AA^{-1}(y_1 + y_2) - AA^{-1}y_1 - AA^{-1}y_2 = (y_1 + y_2) - y_1 - y_2 = \theta_1$$

and therefore $$z = \theta,$$

i.e., $$A^{-1}(y_1 + y_2) = A^{-1}y_1 + A^{-1}y_2.$$

Exactly in the same way one can show the validity of

$$A^{-1}(\lambda y) = \lambda A^{-1}y.$$

Hence the operator A^{-1} is additive.

Let Y_n be the set of elements $y \in E_1$ with $\|A^{-1}y\| \leqq n\|y\|$. Certainly Y_n is not empty, because $\theta_1 \in Y_n$ for all n. Also every $y \in E_1$ belongs to one of these sets, just choose n as the smallest integer $n \geqq \dfrac{\|A^{-1}y\|}{\|y\|}$. One can therefore write

$$E_1 = \sum_{n=1}^{\infty} Y_n.$$

Since E_1 is complete, E_1 cannot be the union of a denumerable number of nowhere dense sets (Theorem 6 §2). Therefore, there certainly exists a set, say Y_{n_0}, such that the set $S(y, r) \cap Y_{n_0}$ is everywhere dense in a certain sphere $S(y, r)$. Then let $\bar{S}(y_1, r_1)$ be a sphere entirely in $S(y, r)$ with $y_1 \in Y_{n_0}$. If we choose an arbitrary y with $\|y\| = r_1$, then $y + y_1 \in \bar{S}(y_1, r_1)$ because $\|(y + y_1) - y_1\| = r_1$. By virtue of $\bar{S}(y_1, r_1) \subset \bar{Y}_{n_0}$, there can be found a sequence $\{z^{(k)}\}$ of elements of $Y_{n_0} \cdot \bar{S}(y_1, r_1)$ such that $z^{(k)} \to y + y_1$ as $k \to \infty$. (This sequence can be stationary if $y + y_1 \in Y_{n_0}$). From $z^{(k)} \to y + y_1$, there follows the convergence of $y^{(k)} = z^{(k)} - y_1 \to y$. Since $y^{(k)} \to y$ and $\|y\| = r_1$, $\dfrac{r_1}{2} \leqq \|y^{(k)}\|$ and $\|y^{(k)}\| = r_1$.

Because $z^{(k)}, y_1 \in Y_{n_0}$,

$$\|A^{-1}y^{(k)}\| = \|A^{-1}z^{(k)} - A^{-1}y_1\| \leqq \|A^{-1}z^{(k)}\| + \|A^{-1}y_1\| \leqq n_0(\|z^{(k)}\| + \|y_1\|)$$

and

$$\|z^{(k)}\| = \|y^{(k)} + y_1\| \leqq \|y^{(k)}\| + \|y_1\| \leqq r_1 + \|y_1\|.$$

These inequalities imply

$$\|A^{-1}y\| \leqq n_0(r_1 + 2\|y_1\|) \leqq \frac{2n_0(r_1 + 2\|y_1\|)}{r_1}\|y^{(k)}\|.$$

For the smallest integer N which is not smaller than

$$\frac{2n_0(r_1 + 2\|y_1\|)}{r_1}$$

we have

$$\|A^{-1}y^{(k)}\| \leqq N\|y^{(k)}\|,$$

i.e., $y^{(k)} \in Y_N$ for all k.

Therefore an arbitrary element with norm equal to r_1 can be approximated by elements Y_N. Now let y be arbitrary in E_1 and

$$y' = \frac{r_1}{\|y\|}y.$$

Then

$$\|y'\| = r_1.$$

According to what we have just proved, there exists a sequence $\{y'^{(k)}\} \subset Y_N$ converging to y'.

With $y'^{(k)} \to y'$,

$$y^{(k)} = y'^{(k)}\frac{\|y\|}{r_1} \to y$$

and

$$\|A^{-1}y^{(k)}\| = \frac{\|y\|}{r_1}\|A^{-1}y'^{(k)}\| \leqq \frac{\|y\|}{r_1}N\|y'^{(k)}\| = N\|y^{(k)}\|,$$

therefore

$$y^{(k)} \in Y_N.$$

Consequently Y_N is everywhere dense in E_1. Again we choose an arbitrary $y \in E_1$ with $\|y\| = l$, however. For this y there exists a $y_1 \in Y_N$ such that

$$\|y - y_1\| \leqq \frac{l}{2}, \quad \|y_1\| \leqq l.$$

(Such an element does exist, because $\bar{S}(\theta_1, l) \cdot Y_N$ is everywhere dense in $\bar{S}(\theta_1, l)$ and $y \in \bar{S}(\theta_1, l)$). Furthermore, there exists a $y_2 \in Y_N$ such that

$$\|(y - y_1) - y_2\| \leqq \frac{l}{2^2}, \quad \|y_2\| \leqq \frac{l}{2}.$$

Continuing this procedure, we construct in general $y_n \in Y_N$ in such a way that

$$\|y - (y_1 + y_2 + \ldots + y_n)\| \leqq \frac{l}{2_n} \text{ and } \|y_n\| \leqq \frac{l}{2^{n-1}}.$$

We then obtain

$$y = \lim_n \sum_{k=1}^{n} y_k.$$

For $x_k = A^{-1}y_k$ we have

$$\|x_k\| \leqq N\|y_k\| \leqq \frac{Nl}{2^{k-1}}.$$

Because
$$\|S_{n+p} - S_n\| = \left\| \sum_{k=n+1}^{n+p} x_k \right\| < \frac{Nl}{2^{n-1}},$$

and because E is complete, the sequence $\{S_n\}$, $S_n = \sum_{k=1}^{n} x_k$ converges to a limit $x \in E$, i.e.,

$$x = \lim_n \sum_{k=1}^{n} x_k = \sum_{k=1}^{\infty} x_k.$$

From
$$Ax = A\left(\lim_n \sum_{k=1}^{n} x_k\right) = \lim_n \sum_{k=1}^{n} Ax_k = \lim_n \sum_{k=1}^{n} y_k = y$$

there follows finally

$$\|A^{-1}y\| = \|x\| = \lim_n \left\| \sum_{k=1}^{n} x_k \right\| \leqq \lim_n \sum_{k=1}^{n} \|x_k\|$$

$$\leqq \lim_n \sum_{k=1}^{n} \frac{Nl}{2^{k-1}} = 2\,Nl = 2N\|y\|,$$

and consequently, the inverse operator A^{-1} is bounded because y was an arbitrary element of E_1, q.e.d.

Up until now, we were only concerned with linear bounded operators A and with linear bounded inverse operators A^{-1}. Frequently, the inverse of a bounded operator is additive and homogeneous but it is, however, not defined on the whole of E_1, rather only on a certain manifold everywhere dense in E_1 and not bounded there. If, conversely, A is linear and unbounded on a manifold everywhere dense in E, then A^{-1} may be bounded and defined on all of E. A more detailed discussion of these different cases would be outside the scope of this book. Therefore, we limit ourselves to two examples.

1. Let $E = E_1 = C[0, 1]$ and

$$Ax = \int_0^t x(\tau)\, d\tau.$$

Obviously, A is bounded whereas the inverse operator

$$A^{-1}y = \frac{d}{dt} y(t)$$

is not bounded and defined only on the everywhere dense manifold of continuously differentiable functions in $C[0, 1]$.

2. Let $E = E_1 = C[0, 1]$ and $Lx = \dfrac{d}{dt}\left\{p(t)\dfrac{dx}{dt}\right\} + q(t)x$, the unbounded Sturm-Liouville operator, which is defined on the linear manifold of twice continuously differentiable functions. The inverse operator of Lx,

$$A^{-1}y = \int_0^1 G(t, \tau)\, y(\tau)\, d\tau$$

($G(t, \tau)$ is the Green function) is bounded, linear and defined on all of $C[0, 1]$.

NORMED RINGS, RINGS OF OPERATORS. A normed linear complex (real) space E is called a *normed ring*, if there is defined a multiplication for its elements $a, b, c, \ldots$ which satisfies the following conditions.

1. With every pair $a \in E$, $b \in E$ there is associated an element $ab \in E$ such that
$$(ab)c = a(bc).$$
2. For every complex (real) λ
$$(\lambda ab) = \lambda(ab),$$
$$a(\lambda b) = \lambda(ab).$$
3. $a(b + c) = ab + ac,$
$(a + b)c = ac + bc.$
4. $\lambda(a + b) = \lambda a + \lambda b.$
5. $\| ab \| \leqq \| a \| \| b \|.$

Continuity of multiplication follows from 5.

Besides this, if $ab = ba$ for arbitrary $a, b \in E$, then the ring is called *commutative.* The theory of such rings was founded by the Soviet mathematicians I. M. Gelfand, V. A. Ditkin, G. E. Shilov and D. A. Raikov.

For the following, the study of the non-commutative normed ring $(E \to E)$ of all operators $(A, B, \dots)$, mapping a normed linear space E into itself is necessary.

Let A and $B \in (E \to E)$. We have
$$\| A B x \| = \| A (Bx) \| \leqq \| A \| \| B x \| \leqq \| A \| \| B \| \| x \|,$$
therefore, according to the definition of the norm of an operator,
$$\| AB \| \leqq \| A \| \| B \|$$
and $(E \to E)$ fulfills condition 5. The remaining conditions can be proved immediately.

For normed rings with unit e, there holds

THEOREM 2. *If* $\| a \| < 1$, *then* $e + a$ *has an inverse element* $(e + a)^{-1}$ *with* $\| (e + a)^{-1} \| \leqq \dfrac{1}{1 - \| a \|}$.

Proof. By 5, $\| a^n \| \leqq \| a \|^n$ and therefore the series
$$e - a + a^2 - \dots + (-1)^n a^n + \dots$$
converges absolutely. From
$$(e + a)(e - a + a^2 - \dots + (-1)^n a^n) = e + (-1)^n a^{n+1},$$
it follows that
$$(e + a)(e - a + a^2 - \dots) = e,$$
because $a^{n+1} \to \theta$ as $n \to \infty$, i.e.,
$$e - a + a^2 - \dots = (e + a)^{-1}$$
and therefore
$$\| (e + a)^{-1} \| \leqq \sum_{n=0}^{\infty} \| a^n \| \leqq \sum_{n=0}^{\infty} \| a \|^n = \frac{1}{1 - \| a \|}.$$

THEOREM 3. *If x and Δx are elements of a normed ring, and if x has an inverse element x^{-1} and $\| \Delta x \| < \| x^{-1} \|^{-1}$, then $x + \Delta x$ also has an inverse.*

First $x + \Delta x = x(e + x^{-1} \Delta x)$, where $\| x^{-1} \Delta x \| \leqq \| x^{-1} \| \| \Delta x \| < 1$. Therefore
$$(e + x^{-1} \Delta x)^{-1} = \sum_{n=0}^{\infty} (-x^{-1} \Delta x)^n$$

is the inverse element of $e+x^{-1}\Delta x$ and consequently $(e+x^{-1}\Delta x)^{-1}x^{-1}$ is the inverse of $x(e+x^{-1}\Delta x)=x+\Delta x$.

Furthermore

$$\|(x+\Delta x)^{-1}-x^{-1}\| \leqq \|(e+x^{-1}\Delta x)^{-1}-e\|\,\|x^{-1}\|$$
$$=\left\|\sum_{n=1}^{\infty}(-x^{-1}\Delta x)^n\right\|\|x^{-1}\| \leqq \frac{\|x^{-1}\Delta x\|}{1-\|x^{-1}\Delta x\|}\|x^{-1}\| < \frac{\|x^{-1}\|}{1-\|x^{-1}\Delta x\|}.$$

Theorem 3 could also be obtained from the principle of the contracting mapping (§7). Theorem 3 implies

THEOREM 3′. *If $A\in(E\to E)$ has an inverse operator A^{-1} and if the operator ΔA satisfies the inequality $\|\Delta A\|<\|A^{-1}\|^{-1}$, then $B=A+\Delta A$ also has an inverse operator B^{-1} and*

$$\|B^{-1}-A^{-1}\| < \frac{\|A^{-1}\|}{1-\|A^{-1}\|\,\|\Delta A\|}.$$

OPERATORS DEPENDING ON A PARAMETER. Equations of the form

$$Ax-\lambda x=y \text{ or } (A-\lambda I)x=y \tag{2}$$

frequently occur in different branches of mathematics. Here A shall be a linear operator and λ a parameter. A special case of (2) is

$$Ax-\lambda x=\theta \text{ or } (A-\lambda I)x=\theta. \tag{3}$$

This is the eigenvalue equation belonging to (2). The always-existing solution $x=\theta$ of (3) is called the *trivial solution.*

Let us assume that the operator $A-\lambda I$ has an inverse operator $(A-\lambda I)^{-1}=R_\lambda$ for certain λ. R_λ is called the *resolvent* of (2). If, therefore, R_λ exists, then there exists for this λ and arbitrary y the unique solution

$$x=R_\lambda y.$$

The eigenvalue equation (3) then has only the trivial solution $x=\theta$.

Those λ for which (2) has a unique solution for all y are called *regular* values of (2) (or of A). If (3) has a non-trivial solution for a λ, then λ is called an *eigenvalue* or *characteristic number* of (2) (or of A). The non-trivial solution is called an *eigenelement* of A for the eigenvalue λ. If λ is an eigenvalue of (2) and if (2) has a solution for certain y, then this solution is not unique, as we can easily see.

The totality of non-regular values of λ is called the *spectrum* of the operator A. In particular, all the eignenvalues belong to the spectrum.

From Theorem 3′ there follows

THEOREM. *If $\frac{1}{|\lambda|}\|A\|=q<1$, then $A-\lambda I$ has an inverse operator.*

If λ is a regular value, then so is $\lambda+\Delta\lambda$ for $|\Delta\lambda|<\|(A-\lambda I)^{-1}\|^{-1}$. Therefore, the totality of all regular values forms an open set and consequently the spectrum is closed.

Examples. In $C[0,1]$, we consider the integral equation

$$x(s)=y(s)+\lambda\int_0^1 K(s,t)\,x(t)\,dt, \tag{*}$$

where $K(s,t)$ is continuous on the square $0\leqq s,t\leqq 1$. We put $\frac{1}{\lambda}=\mu$. Then, (*) becomes

$$\int_0^1 K(s,t)\,x(t)\,dt - \mu x(s) = -\mu y(s).$$

For $Ax = \int_0^1 K(s,t)\,x(t)\,dt$, we get

$$Ax - \mu x = -\mu y,$$

and according to Theorem 2

$$R_\mu = -\frac{1}{\mu}\left(I + \frac{A}{\mu} + \frac{A^2}{\mu^2} + \ldots\right) = -\lambda(I + \lambda A + \lambda^2 A^2 + \ldots).$$

Here $A^p z = \int_0^1 K_p(s,t)\,z(t)\,dt$, where $K_p(s,t)$ is the pth iteration of the kernel $K(s,t)$. Thus we have

$$R_{\frac{1}{\lambda}} z = R^\mu z = -\lambda z(s) - \lambda^2 \int_0^1 K(s,t)\,z(t)\,dt - \lambda^3 \int_0^1 K_2(s,t)\,z(t)\,dt - \ldots .$$

Therefore, the solution of (*) becomes

$$x(s) = R_{\frac{1}{\lambda}}(-\mu y) = R_{\frac{1}{\lambda}}\left(-\frac{1}{\lambda} y\right)$$

$$= y(s) + \lambda \int_0^1 K(s,t)\,y(t)\,dt + \lambda^2 \int_0^1 K_2(s,t)\,y(t)\,dt + \ldots$$

and we have the same solution

$$x(s) = y(s) + \lambda \int_0^1 R(s,t,\lambda)\,y(t)\,dt$$

which we obtain in the theory of integral equations, where we have for the resolvent $R(s,t,\lambda)$ of the kernel $K(s,t)$

$$R(s,t,\lambda) = K(s,t) + \lambda K_2(s,t) + \lambda^2 K_3(s,t) + \ldots .$$

The equations for the resolvent as given in the theory of integral equations are the conditions for $R_{\frac{1}{\lambda}}$ being a right- and left-sided inverse operator of $\lambda A - I$.

APPROXIMATE SOLUTIONS OF FUNCTIONAL EQUATIONS. Very often a functional equation

$$Ax = y,$$

where $x, y \in E$, cannot be solved immediately because the inverse operator A^{-1} is not known. L. V. Kantorovich [9] has given a procedure by means of which the operator A is approximated by another one A_0 whose inverse operator can be more easily found. If $x_0 = A_0^{-1}y$ and $\Delta = A - A_0$, then

$$\|x - x_0\| = \|A^{-1}y - A_0^{-1}y\| \leqq \|A^{-1} - A_0^{-1}\|\,\|y\| < \frac{\|A_0^{-1}\|\,\|y\|}{1 - \|A_0^{-1}\|\,\|\Delta\|}$$

(under the hypothesis $\|\Delta\|\,\|A_0^{-1}\| < 1$).

Thus we obtain an estimate for the error $\|x - x_0\|$, which arises by substituting the approximate solution for the exact one.

As an example of this, the approximate solution of a Fredholm integral equation may serve, if the kernel is replaced by the degenerated kernel.

A more general procedure [9] allows us to approximate not only the operator but also in the known sense the space itself. Using this, the most important approximation methods of linear analysis can be considered from a unified point of view.

CHAPTER III

LINEAR FUNCTIONALS

§21. LINEAR FUNCTIONAL ON NORMED LINEAR SPACES

BASIC CONCEPTS. If the range of an operator consists of real numbers, then the operator is called a *functional*. Since the set R of real numbers is a Banach space, all earlier definitions and theorems on linear operators which map one linear or normed linear space E onto another one remain valid for functionals $f(x)$. According to the general definition (for operators), a function $f(x)$ is *linear* if

1) $f(x_1+x_2)=f(x_1)+f(x_2)$ and
2) $f(x_n)\to f(x)$ as $x_n\to x$ in the sense of convergence in E.

Examples of linear functionals.

Let $E=L_2(0,1)$. Then $f(x)=\int_0^1 x(t)\,dt$ is a linear functional. From the inequality,

$$\left|\int_0^1 x(t)\,dt\right| \leqq \int_0^1 |x(t)|\,dt \leqq \left(\int_0^1 x^2(t)\,dt\right)^{\frac{1}{2}}$$

there follows the existence of

$$\int_0^1 x(t)\,dt$$

for arbitrary $x(t)\in L_2(0,1)$, and

$$|f(x)| \leqq \|x\|.$$

Obviously, $f(x)$ is additive. If $x_n\to x$, i.e., $\|x_n-x\|\to 0$, then

$$|f(x_n)-f(x)| = |f(x_n-x)| \leqq \|x_n-x\|\to 0,$$

i.e., $f(x)$ is continuous.

As for linear operators, the following theorems hold for linear functionals.

THEOREM 1. *If an additive functional is continuous at a point of the linear space E, then it is continuous on all of E and consequently it is linear.*

THEOREM 2. *Linear functionals are homogeneous.*

THEOREM 3. *An additive functional defined on a normed linear space E is linear if and only if there exists a constant M such that* $|f(x)|\leqq M\|x\|$ *for all* $x\in E$.

The smallest of these constants is called the *norm* of the functional $f(x)$ and is designated by $\|f\|$. For an arbitrary linear functional $f(x)$ and every x, there consequently holds the inequality

$$|f(x)| \leqq \|f\|\,\|x\|.$$

THEOREM 4. $\|f\| = \sup\limits_{\|x\|=1} |f(x)|$.

The norm of a linear functional can be interpreted geometrically. The totality of all points x satisfying the equation $f(x)=c$ where f is a linear functional is called a *plane* in the space E. The planes defined by $f(x)=c_0$ and $f(x)=c_1$ are called parallel. $f(x)=c$ divides the space into two parts, the totality of all points $f(x)$ with $f(x)\geqq c$ and that of all points with $f(x)<c$. The first of these sets is said to be on the right of $f(x)=c$, the second one on the left of $f(x)=c$. The plane $f(x)=\|f\|$ has the property that the sphere $\|x\|\leqq 1$ is on the left of it (because for the points of the sphere $\|x\|\leqq 1$, $f(x)\leqq\|f\|$). On the other hand, none of the parallel planes $f(x)=\|f\|-\varepsilon$ for $\varepsilon>0$ has this property. In analogy with the theory of convex bodies, the plane $f(x)=\|f\|$ is called the *supporting plane* for the sphere $\|x\|\leqq 1$.

We define the sum of two linear functionals f_1 and f_2 and the product of a linear functional f by a real number λ by the equalities

$$(f_1+f_2)x=f_1(x)+f_2(x),$$

$$(\lambda f)(x)=\lambda f(x),$$

respectively.

If we observe that 1) $\|f\|\geqq 0$ and $\|f\|=0$ if and only if $f(x)\equiv 0$, 2) $\|f_1+f_2\|\leqq\|f_1\|+\|f_2\|$, 3) $\|\lambda f\|=|\lambda|\,\|f\|$, then, as was done with the operators, we can consider the totality of all linear functionals defined on E as a normed linear space. This space is even a Banach space if E is also of type B, as was shown earlier for operators in general.

The space of linear functionals defined on E is called the *conjugate space of* E and is designated by $\bar{E}$. Hence $\bar{E}=(E\rightarrow R)$.

THEOREM ON EXTENSION OF A LINEAR FUNCTIONAL. We shall now prove theorems which hold especially for functionals. In order to do so several definitions are necessary.

Let E be a normed linear space and G a linear manifold of E. We study first linear functionals which are defined only for elements x in G. We put

$$\|f\|_G=\sup_{x\in G,\ \|x\|\leqq 1}|f(x)|.$$

Then the linear functional $F(x)$ defined on all of E is called an *extension* of the linear functional $f(x)$ onto the whole space E if $f(x)=F(x)$ for all $x\in G$.

THEOREM 5 (HAHN-BANACH). *Each linear functional $f(x)$ defined on a linear manifold G of a normed linear space E can be extended onto the whole space with preservation of the norm. That is, there can be found a linear functional $F(x)$ defined on all of E, such that*

1) $F(x)=f(x)$ *for* $x\in G$ *and*
2) $\|f\|_G=\|F\|_E$.

Proof. Let $x_0\in G$ and consider the set $\{G+x_0\}$ of all elements $x+tx_0$ where $x\in G$ and t is an arbitrary real number.

Obviously, $\{G+x_0\}$ is a linear manifold. Also, each element of $\{G+x_0\}$ is uniquely representable in the form $x+tx_0$. Let us assume there are two representations

$$y=x_1+t_1x_0 \text{ and } y=x_2+t_2x_0.$$

Here we can assume $t_1\neq t_2$ for otherwise $x_1+t_1x_0=x_2+t_1x_0$ would imply $x_1=x_2$ and the representation still would be unique. Assume that

$$x_1 - x_2 = (t_2 - t_1)x_0 \text{ and } x_0 = \frac{x_1 - x_2}{t_2 - t_1}.$$

This is impossible, however, because $x_0 \bar{\in} G$ and $x_1, x_2 \in G$. Hence $t_1 = t_2$ and therefore $x_1 = x_2$, which proves the uniqueness.

For two elements x' and x'' we have

$$f(x') - f(x'') = f(x' - x'') \leqq \|f\| \|x' - x''\| \leqq \|f\| \{\|x' + x_0\| + \|x'' + x_0\|\}$$

and therefore also

$$f(x') - \|f\| \|x' + x_0\| \leqq f(x'') + \|f\| \|x'' + x_0\|.$$

Since x' and x'' were arbitrary from G,

$$\sup_{x \in G} \{f(x) - \|f\| \|x + x_0\|\} \leqq \inf_{x \in G} \{f(x) + \|f\| \|x + x_0\|\}.$$

Therefore there exists a real number c, satisfying the inequality

$$\sup_{x \in G} \{f(x) - \|f\| \|x + x_0\| \{ \leqq c \leqq \inf_{x \in G} \{f(x) + \|f\| \|x + x_0\|\}. \tag{1}$$

Now let y be an arbitrary element of $\{G + x_0\}$. According to what has been shown above, y is of the form

$$y = x + tx_0$$

with $x \in G$ and a uniquely determined real t. We introduce a new functional defined for all elements $y = x + tx_0$ by the equality

$$\varphi(y) = f(x) - tc,$$

where c is a fixed real number satisfying (1).

Obviously, f and φ coincide on G. φ is additive, because f has this property. We will prove that $\varphi(y)$ is bounded and that it has the same norm as $f(x)$.

We distinguish two cases:

1) $t > 0$. $\frac{x}{t} \in G$ and (1) then imply

$$\varphi(y) = t\left\{f\left(\frac{x}{t}\right) - c\right\} \leqq t\left\{\|f\| \left\|\frac{x}{t} + x_0\right\|\right\} = \|f\| \|x + t x_0\| = \|f\| \|y\|,$$

therefore

$$\varphi(y) \leqq \|f\| \|y\|. \tag{2}$$

2) $t < 0$. (1) implies

$$f\left(\frac{x}{t}\right) - c \geqq - \|f\| \left\|\frac{x}{t} + x_0\right\| = -\frac{1}{|t|} \|f\| \|y\| = \frac{1}{t} \|f\| \|y\|$$

and therefore $\quad \varphi(y) = t\left\{f\left(\frac{x}{t}\right) - c\right\} \leqq t\frac{1}{t} \|f\| \|y\| = \|f\| \|y\|,$

consequently, we have (2) again.

Hence inequality (2) holds for all $y \in \{G + x_0\}$. Substituting $-y$ for y in (2), we have

$$-\varphi(y) \leqq \|f\| \|y\|.$$

Together with (2) we obtain

$$|\varphi(y)| \leqq \|f\| \|y\|$$

and therefore
$$\|\varphi\| \leqq \|f\|.$$
Since the functional φ is, however, an extension of f from G to $\{G+x_0\}$,
$$\|\varphi\| \geqq \|f\|$$
and consequently
$$\|\varphi\| = \|f\|.$$
(Here we have determined the norm of the functional φ with respect to that linear manifold on which φ is defined.) Therefore, the functional $f(x)$ is indeed extended with preservation of norm.

We consider the elements of the set $E-G$ to be well-ordered according to the Zermelo well-ordering theorem, i.e., arranged into a transfinite sequence $x_0, x_1, \ldots, x_\omega, x_{\omega+1}, \ldots, x, \ldots$. We extend the functional $f(x)$ onto the sets
$$\{G+x_0\} = G_1, \{G_1+x_1\} = G_2, \ldots .$$
Because each time the norm is unchanged we shall obtain after an infinite number of steps, a functional $F(x)$ satisfying all requirements of the theorem.

Remark. If E is separable, then the proof of the Hahn-Banach theorem can be constructed without using the Zermelo theorem and transfinite induction. For then there exists in E a countable everywhere dense set
$$x^{(1)}, x^{(2)}, \ldots, x^{(n)}, \ldots .$$
We choose those elements of this set which belong to $E-G$ and arrange them into a sequence
$$x_0, x_1, x_2, \ldots .$$
If we now extend $f(x)$ onto
$$\{G+x_0\} = G_1,\ \{G_1+x_1\} = G_2, \text{ and so on,}$$
then we arrive at a functional $\varphi_\omega(x)$ which is defined on the linear manifold G_ω, everywhere dense in E and containing all G_n and the relation $\|\varphi_\omega\| = \|f\|$ is fulfilled. Now let $y \bar{\in} G_\omega$, then there can be found a sequence $\{y_n\}, y_n \in G_\omega$, with $y = \lim_n y_n$. We define
$$F(y) = \lim_n \varphi_\omega(y_n). \tag{3}$$
Because, in particular, we can put for the elements $y \in G_\omega$,
$$y_1 = y_2 = \ldots = y,$$
the functional $F(y)$ is the extension of φ_ω onto E. Obviously $F(y)$ is additive. From
$$|F(y)| = \lim_n |\varphi_\omega(y_n)| \leqq \|\varphi_\omega\| \lim_n \|y_n\| = \|f\|\,\|y\|,$$
we conclude
$$\|F\| \leqq \|f\|.$$
Since, however, $\|F\|$ cannot be smaller than $\|f\|$, it follows that
$$\|F\| = \|f\|.$$
Naturally $F(x) = f(x)$ holds on G.

This last step, the extension of a functional, from an everywhere dense set onto the whole space using a limiting process is called *extension by means of the continuity of the functional.*

G. A. Sukhomlinov has generalized the extension theorem for spaces over the complex numbers and quarternions [50].[22]

[22] See also H. Bohnenblust - A. Sobczyk — Extensions of Functionals on complex linear spaces, Bulletin American Mathematical Society, 44 (1938), 91–93. [Trans.]

COROLLARY 1. *Let E be a normed linear space and x_0 an arbitrary but fixed element in E. Then there exists a linear functional $f(x)$ defined on E and having the following properties*

1) $\|f\|=1$ and
2) $f(x_0)=\|x_0\|$.

Proof. We consider the set $\{tx_0\}=G$, where t is an arbitrary real number. G is a subspace of E generated by x_0. Now

$$\varphi(x)=t\|x_0\|$$

for $x=tx_0$. Obviously

1) $\varphi(x_0)=\|x_0\|$ and
2) $|\varphi(x)|=|t|\|x_0\|=\|tx_0\|=\|x\|$, i.e., $\|\varphi\|=1$.

If we now extend the functional $\varphi(x)$ onto the whole space without changing the norm, then we arrive at a functional $f(x)$ with the following properties.

COROLLARY 2. *In a normed linear space E let there be given a linear manifold G and an element $y_0 \overline{\in} G$. Let $d>0$ be the distance of y_0 from G ($d=\inf_{x\in G}\|y_0-x\|$). Then there exists a functional $f(x)$ defined everywhere on E and it has the properties*

1) $f(x)=0$ *for* $x\in G$,
2) $f(y_0)=1$ *and*
3) $\|f\|=\dfrac{1}{d}$.

Proof. As we have already seen, every element y of the set $\{G+y_0\}$ is uniquely representable in the form $y=x+ty_0$ with $x\in G$ and t real. We define the functional $\varphi(y)$ in the following manner. Let $\varphi(y)=t$ for $y=x+ty_0$. Then, obviously, $\varphi(x)=0$ for $x\in G$ and $\varphi(y_0)=1$. Now we determine $\|\varphi\|$. We have

$$|\varphi(y)|=|t|=\frac{|t|\,\|y\|}{\|y\|}=\frac{|t|\,\|y\|}{\|x+ty_0\|}=\frac{\|y\|}{\left\|\dfrac{x}{t}+y_0\right\|}=\frac{\|y\|}{\left\|y_0-\left(-\dfrac{x}{t}\right)\right\|}\leqq\frac{\|y\|}{d}$$

and consequently

$$\|\varphi\|\leqq\frac{1}{d}. \tag{4}$$

Furthermore, there exists a sequence $\{x_n\}$, $x_n\in G$ with

$$\lim_n\|x_n-y_0\|=d.$$

We have

$$|\varphi(x_n-y_0)|\leqq\|\varphi\|\,\|x_n-y_0\|,$$

by virtue of

$$|\varphi(x_n-y_0)|=|\varphi(x_n)-\varphi(y_0)|=1,$$

$$1\leqq\|\varphi\|\,\|x_n-y_0\|.$$

By taking the limit, this implies

$$1\leqq\|\varphi\|d$$

or

$$\|\varphi\|\geqq\frac{1}{d}. \tag{5}$$

From (4) and (5), we obtain

$$\|\varphi\|=\frac{1}{d}.$$

By extension of $\varphi(x)$ onto the whole space with preservation of the norm, we then obtain a functional $f(x)$ with the required properties. The first corollary assures the existence of a functional on an arbitrary normed linear space which does not vanish everywhere. On the other hand, this corollary tells us that $x=\theta$ if for a fixed $x \in E$ the equality $f(x)=0$ is fulfilled for every linear functional of $\bar{E}$.

This corollary can be given the following geometric wording:

Through each point x_0 on the surface of the sphere $\|x\| \leqq 1$ ($\|x_0\|=1$) there can be passed a supporting plane.

This theorem is a generalization of a proposition proved for n-dimensional space by G. Minkowski.

The equation of the supporting plane for the unit sphere is $f(x)=\|f\|$. For x_0 there exists, however, a functional f_0 with $\|f_0\|=1$ and

$$f_0(x_0)=\|x_0\|=1. \tag{6}$$

Because of (6), the supporting plane

$$f_0(x)=\|f_0\|=1$$

passes through x_0.

The second corollary is interesting for the problem: to approximate an element y_0 by linear combinations of other elements $x_1, x_2, \ldots$ of E. The second corollary implies the

THEOREM. *y_0 is the limit of a sequence of linear combinations $\sum_{i=1}^{n} c_i x_i$ if and only if $f(y_0)=0$ for all functionals vanishing for $x_1, x_2, \ldots$.*

Proof. Let $f(x_i)=0, i=1,2,\ldots,$ imply $f(y_0)=0$. Then y cannot have a distance $d>0$ from the linear manifold generated by the elements x_i, for otherwise, according to the second corollary, there would exist a functional f_0 with $f_0(x_i)=0, i=1,2,\ldots,$ but $f_0(y_0)=1$. $d=0$ means, however, that y_0 is either a point of accumulation of G or that it itself belongs to G, i.e., y must be approximable by elements of the form $\sum_{i=1}^{n} c_i x_i$.

Conversely, let y_0 be the limit of a sequence of elements of G and let $f(x_i)=0$ for all functionals f. Putting

$$z_n=\sum_{i=1}^{n} c_i x_i,$$

we find that

$$f(z_n)=\sum_{i=1}^{n} c_i f(x_i)=0$$

and consequently

$$f(y_0)=f(\lim_n z_n)=\lim_n f(z_n)=0.$$

§22. THE GENERAL FORM OF LINEAR FUNCTIONALS IN SOME FUNCTIONAL SPACES

We can find, for several spaces, the general form of the functionals defined on them. For various investigations this appears to be useful.

LINEAR FUNCTIONALS ON THE n-DIMENSIONAL SPACE E_n. As will be shown later (see p. 108), the functionals on E_n are of the form

$$f(x)=\sum_{i=1}^{n} f_i x_i,$$

where the x_i are the components of the vector $x \in E_n$ and the f_i arbitrary numbers. The f_i themselves can be interpreted as components of an n-dimensional vector $f \in \bar{E}_n$.

The conjugate space $\bar{E}_n$ is therefore an n-dimensional space (with a metric which in general is different from the one in E_n). For sake of conformity with the terminology used in tensor algebra, the elements of E_n are called *contravariants*, those of $\bar{E}_n$ *co-variants*. Each linear functional can also be written as a scalar product

$$f(x) = (f, x),$$

where $x \in E_n, f \in \bar{E}_n$.

GENERAL FORM OF LINEAR FUNCTIONALS ON s. Let $f(x)$ be a linear functional given on s (see p. 13). We put $x_n = \{\xi_i^{(n)}\}$ with $\xi_n^{(n)} = 1$ and $\xi_i^{(n)} = 0$ for $i \neq n$. Furthermore, let $f(x_n) = a_n$. Since the convergence in s is coordinate-wise, the equality

$$x = \lim_n \sum_{k=1}^{n} \xi_k x_k = \sum_{k=1}^{\infty} \xi_k x_k,$$

holds for every $x = \{\xi_1, \xi_2, \ldots, \xi_n, \ldots\}$ and it follows from this, because of the continuity of $f(x)$, that

$$f(x) = \sum_{k=1}^{\infty} \xi_k f(x_k) = \sum_{k=1}^{\infty} a_k \xi_k.$$

Now this series must converge for every number sequence $\{\xi_k\}$. This, however, is possible only if the a_k are equal to 0 from a certain index on. Consequently

$$f(x) = \sum_{k=1}^{N} a_k \xi_k.$$

Conversely, each such expression with arbitrary real numbers a_k and natural number N is a linear functional on s. Hence we have the result:

Every linear functional defined on s is of the form

$$f(x) = \sum_{k=1}^{N} a_k \xi_k.$$

It is easy to see that N and $a_k, k = 1, 2, \ldots, N$, are uniquely determined by the functional f.

GENERAL FORM OF LINEAR FUNCTIONALS ON $C[0, 1]$ (Riesz Theorem). Let $f(x)$ be a linear functional on $C[0, 1]$. Since every continuous function is bounded and measurable and since

$$\inf_{\substack{E_0 \subseteq [0,1] \\ m(E_0) = 0}} \left\{ \sup_{t \in [0,1] - E_0} |x(t)| \right\} = \max_{t \in [0,1]} |x(t)|$$

for every continuous functions $x(t)$, $C[0, 1]$ can be considered as a subspace of $M(0, 1)$, because in $M(0, 1)$,

$$\|x\| = \rho(x, \theta) = \inf_{\substack{E_0 \subseteq [0,1] \\ m(E_0) = 0}} \left\{ \sup_{t \in [0,1] - E_0} |x(t)| \right\}.$$

We extend the functional $f(x)$ given on $C[0, 1]$ with preservation of the norm onto all of $M(0, 1)$. Let this extension be called $F(x)$.

For the function

$$u_t(\xi) = \begin{cases} 1 \text{ for } 0 \leqq \xi < t, \\ 0 \text{ for } t \leqq \xi \leqq 1 \end{cases}$$

there obviously holds $u_t(\xi) \in M(0,1)$.
The function $g(t)$ defined by $F[u_t(\xi)] = g(t)$
is of bounded variation. With

$$t_0 = 0 < t_1 < t_2 < \ldots < t_{n-1} < t_n = 1$$

as partition of the interval $[0, 1]$, we consider the sum

$$\sum_{i=1}^{n} |g(t_i) - g(t_{i-1})|.$$

Putting $\varepsilon_i = \operatorname{sgn} [g(t_i) - g(t_{i-1})]$,
we have

$$\sum_{i=1}^{n} |g(t_i) - g(t_{i-1})| = \sum_{i=1}^{n} \varepsilon_i [g(t_i) - g(t_{i-1})]$$

$$= \sum_{i=1}^{n} \varepsilon_i [F(u_{t_i}) - F(u_{t_{i-1}})] = F\left[\sum_{i=1}^{n} \varepsilon_i (u_t - u_{t_{i-1}})\right].$$

This implies $$\sum_{i=1}^{n} |g(t_i) - g(t_{i-1})| \leqq \|F\| \left\| \sum_{i=1}^{n} \varepsilon_i (u_{t_i} - u_{t_{i-1}}) \right\| = \|f\|$$

because of $\|F\| = \|f\|$ and $\left\| \sum_{i=1}^{n} \varepsilon_i (u_{t_i} - u_{t_{i-1}}) \right\| = 1.$

Hence we have finally

$$\left| \sum_{i=1}^{n} [g(t_i) - g(t_{i-1})] \right| \leqq \|f\|,$$

i.e., $g(t)$ is of bounded variation.

Now let $x(t)$ be an arbitrary function continuous on $[0, 1]$. Then the functions

$$z_n(t) = \sum_{k=1}^{n} x\left(\frac{k}{n}\right) \left[u_{\frac{k}{n}}(t) - u_{\frac{k-1}{n}}(t)\right]$$

approximate $x(t)$. From the equality

$$F(z_n) = \sum_{k=1}^{n} x\left(\frac{k}{n}\right) \left[g\left(\frac{k}{n}\right) - g\left(\frac{k-1}{n}\right)\right]$$

there follows $$\lim_n F(z_n) = \lim_n \sum x\left(\frac{k}{n}\right) \left[g\left(\frac{k}{n}\right) - g\left(\frac{k-1}{n}\right)\right] = \int_0^1 x(t)\, dg(t).$$

On the other hand, the sequence $\{z_n(t)\}$ converges strongly to $x(t)$, i.e., $\|z_n - x\| \to 0$. Because $F(x)$ is continuous,

$$F(z_n) \to F(x)$$

and therefore $$F(x) = \int_0^1 x(t)\, dg(t). \tag{1}$$

According to the hypothesis, $x(t)$ is continuous; therefore

$$F(x) = f(x),$$

and

$$f(x) = \int_0^1 x(t)\,dg(t).$$

This leads us to the

THEOREM (F. RIESZ). *Every linear functional on* $C[0,1]$ *can be represented by a Stieltjes integral* (1). *Here* $g(t)$ *is determined by the functional* $f(x)$ *and is of bounded variation.*

Conversely, as we easily see,

$$\varphi(x) = \int_0^1 x(t)\,dh(t),$$

where $h(t)$ is of bounded variation, is a linear functional on $C[0,1]$. Obviously $\varphi(x)$ is additive. Because we can pass to the limit under the Stieltjes integral for uniformly convergent sequences of functions, $\varphi(x)$ is also continuous. Hence (1) is the general form of linear functionals on $C[0,1]$, i.e., if we let $g(t)$ run through all functions of bounded variation, we obtain all linear functionals on $C[0,1]$.

We want to determine the norm of $f(x)$. First

$$\sum_{i=1}^{n} |g(t_i) - g(t_{i-1})| \leqq \|f\|$$

and therefore

$$\operatorname*{Var}_0^1(g) \leqq \|f\|. \tag{2}$$

From (1), on the other hand, there follows

$$|f(x)| = \left| \int_0^1 x(t)\,dg(t) \right| \leqq \max_{t\in[0,1]} |x(t)| \operatorname*{Var}_0^1(g) = \|x\| \operatorname*{Var}_0^1(g).$$

Hence

$$\|f\| \leqq \operatorname*{Var}_0^1(g). \tag{3}$$

(2) and (3) imply

$$\|f\| = \operatorname*{Var}_0^1(g). \tag{4}$$

If two functions of bounded variation are considered to be equal if they differ at all points of continuity by at most an additive constant, then obviously, the above association between the linear functionals on $C[0,1]$ and the functions of bounded variation is one-to-one.

A. A. Markov extended the theorem of Riesz by finding the general form of linear functionals on the space $C(Q)$ of all functions continuous on a compact set Q [42].

GENERAL FORM OF LINEAR FUNCTIONALS ON $L_p(0,1)$. We consider any linear functional $f(x)$ on $L_p(0,1)$ for $p > 1$; we put

$$u_t(\xi) = \begin{cases} 1 \text{ for } 0 \leqq \xi < t, \\ 0 \text{ for } t \leqq \xi \leqq 1 \end{cases}$$

and we shall prove that $g(t) = f[u_t(\xi)]$ is absolutely continuous. For that purpose, let $\delta_i = (\tau_i, t_i)$, $i = 1, 2, \ldots, n$, be an arbitrary system of non-overlapping intervals situated in $[0, 1]$. Besides this, we shall use the ε_i defined on p. 101.

$$\text{Then } \sum_{i=1}^{n} |g(t_i) - g(\tau_i)| = \sum_{i=1}^{n} \varepsilon_i [g(t_i) - g(\tau_i)]$$

$$= f\left\{\sum_{i=1}^{n} \varepsilon_i [u_{t_i}(\xi) - u_{\tau_i}(\xi)]\right\} \leqq \|f\| \left\| \sum_{i=1}^{n} \varepsilon_i [u_{t_i}(\xi) - u_{\tau_i}(\xi)] \right\|$$

$$= \|f\| \left(\int_0^1 \left| \sum_{i=1}^{n} \varepsilon_i [u_{t_i}(\xi) - u_{\tau_i}(\xi)] \right|^p d\xi \right)^{\frac{1}{p}}$$

$$\leqq \|f\| \left(\sum_{i=1}^{n} \int_{\delta_i} d\xi \right)^{\frac{1}{p}} = \|f\| \left(\sum_{i=1}^{n} m\, \delta_i \right)^{\frac{1}{p}},$$

hence $g(t)$ is absolutely continuous. As an absolutely continuous function, $g(t)$ has an almost everywhere L-integrable derivative and $g(t)$ is equal to the Lebesgue integral of this derivative. Putting $\alpha(t) = g'(t)$, we have

$$g(t) = g(0) + \int_0^t \alpha(\tau) d\tau.$$

Now
$$g(0) = f[u_0(\xi)] = 0$$
because
$$u_0(\xi) \equiv 0$$
is the zero element of $L_p(0, 1)$. Consequently

$$g(t) = \int_0^t \alpha(\tau)\, d\tau.$$

With the above defined functions $u_t(\tau)$, we obtain

$$f(u_t(\tau)) = g(t) = \int_0^t \alpha(\tau)\, d\tau = \int_0^1 u_t(\tau)\, \alpha(\tau)\, d\tau,$$

and since f was supposed to be a linear functional,

$$f(z_n) = \int_0^1 z_n(\tau)\, \alpha(\tau)\, d\tau,$$

if we put
$$z_n(\tau) = \sum_{k=1}^{n} c_k \left[u_{\frac{k}{n}}(\tau) - u_{\frac{k-1}{n}}(\tau) \right].$$

If $x(t)$ is an arbitrary bounded and measurable function, then we can find a sequence $\{z_n(t)\}$ of functions such that

$$z_n(t) \to x(t).$$

Here $\{z_n(t)\}$ can be assumed to be uniformly bounded.

According to the Lebesgue theorem on the integration of a bounded sequence, we obtain

$$\lim_n f(z_n) = \lim_n \int_0^1 z_n(\tau)\,\alpha(\tau)\,d\tau = \int_0^1 \lim_n z_n(\tau)\,\alpha(\tau)\,d\tau = \int_0^1 x(\tau)\,\alpha(\tau)\,d\tau.$$

Because, on the other hand,

$$z_n(t) \to x(t),$$

$$\| z_n - x \| = \left(\int_0^1 | z_n(t) - x(t) |^p \, dt \right)^{\frac{1}{p}} \to 0$$

as $n \to \infty$. Therefore $f(z_n) \to f(x)$, and we finally have

$$f(x) = \int_0^1 x(\tau)\,\alpha(\tau)\,d\tau.$$

The functions defined by

$$x_N(t) = \begin{cases} |\alpha(t)|^{q-1} \operatorname{sgn} \alpha(t), & \text{if } |\alpha(t)| \leqq N, \\ N^{q-1} \operatorname{sgn} \alpha(t), & \text{if } |\alpha(t)| > N \end{cases}$$

are bounded and measurable. Here q is the number conjugate to p, i.e., $\frac{1}{p} + \frac{1}{q} = 1.$

Consequently
$$f(x_N) = \int_0^1 x_N(t)\,\alpha(t)\,dt$$

and
$$|f(x_N)| \leqq \|f\| \, \|x_N\| = \|f\| \left(\int_0^1 |x_N(t)|^p dt \right)^{\frac{1}{p}}.$$

On the other hand,

$$|f(x_N)| \geqq f(x_N) = \int_0^1 x_N(t)\,\alpha(t)\,dt = \int_0^1 |x_N(t)|\,|\alpha(t)|\,dt$$

$$\geqq \int_0^1 |x_N(t)|\,|x_N(t)|^{\frac{1}{q-1}} dt = \int_0^1 |x_N(t)|^{\frac{q}{q-1}} dt = \int_0^1 |x_N(t)|^p dt.$$

Therefore
$$\int_0^1 |x_N(t)|^p dt \leqq \|f\| \, \|x_N\| = \|f\| \left(\int_0^1 |x_N(t)^p| dt \right)^{\frac{1}{p}}$$

and
$$\left(\int_0^1 |x_N(t)|^p dt \right)^{\frac{1}{q}} \leqq \|f\|.$$

Obviously,
$$|x_N(t)| \to |\alpha(t)|^{q-1}$$

almost everywhere on $[0, 1]$ as $N \to \infty$, because $\alpha(t)$ is L-integrable and therefore becomes infinite only for a set of points of measure zero. As $N \to \infty$, we obtain

$$\left(\int_0^1 |\alpha(t)|^{(q-1)p}\, dt\right)^{\frac{1}{q}} \leqq \|f\|$$

or

$$\left(\int_0^1 |\alpha(t)^q|\, dt\right)^{\frac{1}{q}} \leqq \|f\|. \tag{5}$$

Hence

$$\alpha(t) \in L_q(0, 1).$$

If now $x(t)$ is an arbitrary function of $L_p(0, 1)$, then there exists

$$\int_0^1 x(t)\,\alpha(t)\,dt$$

because of the Hölder inequality. Furthermore there can be found a sequence $\{x_n(t)\}$ of bounded functions where

$$\int_0^1 |x(t) - x_n(t)|^p\, dt \to 0$$

as $n \to \infty$. According to Hölder's inequality

$$\int_0^1 x_n(t)\,\alpha(t)\,dt \to \int_0^1 x(t)\,\alpha(t)\,dt.$$

Because, however, the $x_n(t)$ were supposed to be bounded measurable functions,

$$\int_0^1 x_n(t)\,\alpha(t)\,dt = f(x_n)$$

and consequently

$$f(x_n) \to \int_0^1 x(t)\,\alpha(t)\,dt$$

as $n \to \infty$. On the other hand,

$$f(x_n) \to f(x).$$

Hence

$$f(x) = \int_0^1 x(t)\,\alpha(t)\,dt. \tag{6}$$

We now have the result:
Every functional defined on $L_q(0, 1)$ can be represented in the form (6). Conversely, if $\beta(t)$ is an arbitrary function belonging to $L_q(0, 1)$, then

$$\varphi(x) = \int_0^1 x(t)\,\beta(t)\,dt$$

is a linear functional defined on $L_p(0, 1)$ because $\varphi(x)$ is additive and according to Hölder's inequality, bounded.

Hence (6) represents the general form of the linear functionals defined on $L_p(0, 1)$ for an arbitrary $\alpha(t) \in L_q(0, 1)$.

The norm of such a functional can be easily found. First, (6) implies

$$|f(x)| = \left| \int_0^1 x(t)\,\alpha(t)\,dt \right| \leqq \left(\int_0^1 |x(t)|^p dt \right)^{\frac{1}{p}} \left(\int_0^1 |\alpha(t)|^q dt \right)^{\frac{1}{q}} = \|x\| \left(\int_0^1 |\alpha(t)|^q dt \right)^{\frac{1}{q}}.$$

Hence

$$\|f\| \leqq \left(\int_0^1 |\alpha(t)|^q dt \right)^{\frac{1}{q}}. \tag{7}$$

From (7) and (5) we conclude

$$\|f\| = \left(\int_0^1 |\alpha(t)|^q dt \right)^{\frac{1}{q}}.$$

Frequently, we consider the space $L(0,1)$ of the functions integrable in the sense of Lebesgue. Then $\|x\| = \int_0^1 |x(t)|\,dt$. The general form of the linear functionals defined on $L(0,1)$ then becomes $f(x) = \int_0^1 x(t)\,\alpha(t)\,dt$, where $\alpha(t)$ is bounded almost everywhere. For the norm we obtain

$$\|f\| = \inf_{\substack{E \subset [0,1] \\ m(E)=0}} \left\{ \sup_{t \in [0,1]-E} |\alpha(t)| \right\}.$$

GENERAL FORM OF LINEAR FUNCTIONALS ON c. Let $f(x)$ be a linear functional on c. We consider the elements $x_n = \{\xi_i^{(n)}\}$, where $\xi_i^{(n)} = \begin{cases} 1 \text{ if } i = n, \\ 0 \text{ if } i \neq n. \end{cases}$ Obviously, $x_n \in c$, because $\lim\limits_{i\to\infty} \xi^{(n)}$ exists and is equal to 0.

Let $x' = \{\xi_i^{(i)}\}$, $\xi_i^{(i)} = 1$ for all i; clearly $x' \in c$, too. Now we choose an arbitrary $x = \{\xi_1, \xi_2, \ldots\}$, for which $\lim\limits_i \xi_i = a$. The element

$$y_n = x - ax' - \sum_{k=1}^{n} (\xi_k - a)x_k$$

can also be written in the form

$$y_n = \{0, 0, \ldots, 0, \xi_{n+1} - a, \xi_{n+2} - a, \ldots\}.$$

Hence

$$\|y_n\| = \sup_{i \geqq n+1} |\xi_i - a| \to 0$$

as $n \to \infty$. Therefore

$$x - ax' = \lim_n \sum_{k=1}^{n} (\xi_k - a)\,x_k$$

or

$$x = ax' + \sum_{k=1}^{\infty} (\xi_k - a)\,x_k.$$

We put $f(x_k) = c_k, f(x') = c'$. The numbers c_k and c' are uniquely determined by the choice of f. Since f was assumed to be linear,

$$f(x) = \alpha f(x') + \sum_{k=1}^{\infty} (\xi_k - \alpha) f(x_k),$$

i.e.,

$$f(x) = \alpha c' + \sum_{k=1}^{\infty} (\xi_k - \alpha) c_k. \tag{8}$$

(8) holds for arbitrary $x \in c$. We note a few properties of the coefficients c' and c_k. To this end, we put

$$x_N = \{\xi_i^{(N)}\},$$

where $\xi_i^{(N)} = \begin{cases} \operatorname{sgn} c_i & \text{for } i \leqq N, \\ 0 & \text{for } i > N. \end{cases}$ Then, for these x_N

$$\alpha^{(N)} = \lim_i \xi_i^{(N)} = 0 \text{ and } f(x_N) = \sum_{k=1}^{N} |c_k|.$$

On the other hand,

$$f(x_N) \leqq \|f\| \|x_N\| = \|f\|$$

immediately implies

$$\sum_{k=1}^{N} |c_k| \leqq \|f\|.$$

As $N \to \infty$, we obtain

$$\sum_{k=1}^{\infty} |c_k| \leqq \|f\|. \tag{9}$$

The series formed from the absolute values of the c_k is therefore convergent. Hence (8) can be rewritten as

$$f(x) = \alpha c' + \sum_{k=1}^{\infty} \xi_k c_k - \alpha \sum_{k=1}^{\infty} c_k$$

or

$$f(x) = \alpha c_0 + \sum_{k=1}^{\infty} \xi_k c_k, \tag{10}$$

if we put $c' - \sum_{k=1}^{\infty} c_k = c_0$. So we have found:

Every linear functional defined on c is represented by an equality (10), *where the series* $\sum_{k=1}^{\infty} c_k$ *is absolutely convergent.*

Let $d_0, d_1, \dots, d_k, \dots$ be a sequence of numbers for which $\sum_{k=1}^{\infty} d_k$ is absolutely convergent. Then, obviously,

$$\varphi(x) = d_0 \alpha + \sum_{k=1}^{\infty} d_k \xi_k$$

is a linear functional defined on c. Trivially, $\varphi(x)$ is additive. The boundedness of $\varphi(x)$ follows from the inequality

$$|\varphi(x)| \leqq |d_0| |\alpha| + \sum_{k=1}^{\infty} |d_k| |\xi_k| \leqq \sup_k |\xi_k| \left(|d_0| + \sum_{k=1}^{\infty} |d_k| \right) = M \|x\|.$$

Thus the linear functionals defined on c have the form (10).

We will compute the norm of such a functional. (10) implies immediately

$$|f(x)| \leqq |c_0| |\alpha| + \sum_{k=1}^{\infty} |\xi_k| |c_k| \leqq \sup_k |\xi_k| \left(|c_0| + \sum_{k=1}^{\infty} |c_k| \right)$$

and
$$\|f\| \leqq |c_0| + \sum_{k=1}^{\infty} |c_k| \tag{11}$$

For the elements
$$x_N = \{\xi_i^{(N)}\}$$
with $\xi_i^{(N)} = \begin{cases} \operatorname{sgn} c_i \text{ for } i \leqq N, \\ \operatorname{sgn} c_0 \text{ for } i > N, \end{cases}$ we have

$$f(x_N) = |c_0| + \sum_{k=1}^{N} |c_k| + \sum_{k=N+1}^{\infty} c_k \operatorname{sgn} c_0$$

since $\alpha = \operatorname{sgn} c_0$. By

$$f(x_N) \leqq \|f\| \, \|x_N\| = \|f\|,$$
$$|c_0| + \sum_{k=1}^{N} |c_k| + \sum_{k=N+1}^{\infty} c_k \operatorname{sgn} c_0 \leqq \|f\|.$$

This inequality is valid for every N. Passing to the limit as $N \to \infty$, we obtain

$$|c_0| + \sum_{k=1}^{\infty} |c_k| \leqq \|f\|. \tag{12}$$

By comparison of (11) and (12), it follows that

$$\|f\| = |c_0| + \sum_{k=1}^{\infty} |c_k|.$$

GENERAL FORM OF LINEAR FUNCTIONALS ON l_p. Let $f(x)$ be a functional defined on $l_p, p > 1$. We consider in l_p the elements

$$x_n = \{\xi_i^{(n)}\}$$

with $\xi_i^{(n)} = \begin{cases} 1 \text{ for } i = n, \\ 0 \text{ for } i \neq n. \end{cases}$ If $x\ \{\xi_i\}$ is an arbitrary element of l_p, then we put

$$y_n = x - \sum_{k=1}^{n} \xi_k x_k.$$

Obviously, y_n can be rewritten as

$$y_n = \{0, 0, \ldots, 0, \xi_{n+1}, \xi_{n+2}, \ldots\}.$$

Hence
$$\|y_n\| = \left(\sum_{k=n+1}^{\infty} |\xi_k| \right)^{\frac{1}{p}} \to 0$$

as $n \to \infty$, because the series $\sum_{k=1}^{\infty} |\xi_k|^p$ converges. Consequently

$$x = \lim_n \sum_{k=1}^{n} \xi_k x_k = \sum_{k=1}^{\infty} \xi_k x_k.$$

This representation for x is also unique.

From the linearity of $f(x)$ we obtain

$$f(x) = \sum_{k=1}^{\infty} \xi_k f(x_k).$$

The numbers $f(x_k) = c_k$ are uniquely determined by f. Using them, we write the last equality as

$$f(x) = \sum_{k=1}^{\infty} \xi_k c_k. \tag{13}$$

Let $x_N = \{\xi_i^{(N)}\}$ with

$$\xi_i^{(N)} = \begin{cases} |c_i|^{q-1} \operatorname{sgn} c_i & \text{if } i \leqq N, \\ 0 & \text{if } i > N. \end{cases}$$

As before, again $\frac{1}{p}+\frac{1}{q}=1$ holds for q. Then

$$f(x_N) = \sum_{k=1}^{N} |c_k|^q$$

and on the other hand,

$$f(x_N) \leqq \|f\| \|x_N\| = \|f\| \left(\sum_{k=1}^{N} |c_k|^{(q-1)p} \right)^{\frac{1}{p}} = \|f\| \left(\sum_{k=1}^{N} |c_k|^q \right)^{\frac{1}{p}}.$$

Thus we obtain
$$\sum_{k=1}^{N} |c_k|^q \leqq \|f\| \left(\sum_{k=1}^{N} |c_k|^q \right)^{\frac{1}{p}},$$

which implies
$$\left(\sum_{k=1}^{N} |c_k|^q \right)^{\frac{1}{q}} \leqq \|f\|.$$

This inequality holds for arbitrary N. Therefore, as $n \to \infty$, we have

$$\left(\sum_{k=1}^{\infty} |c_k|^q \right)^{\frac{1}{q}} \leqq \|f\| \tag{14}$$

i.e.,
$$\{c_k\} \in l_q.$$

Conversely, if we take an arbitrary sequence $\{d_k\}, d_k \in l_q$, then

$$\varphi(x) = \sum_{k=1}^{\infty} \xi_k d_k$$

is a linear functional on l_p. Besides the trivial additivity, the boundedness of $\varphi(x)$ follows from Hölder's inequality.

Thus we have found:

Formula (13) *gives the general form of a linear functional defined on* l_p.

In addition, we will compute the norm of such a functional f. By means of Hölder's inequality, we obtain from (13)

$$|f(x)| = \left| \sum_{k=1}^{\infty} \xi_k c_k \right| \leqq \left(\sum_{k=1}^{\infty} |\xi_k|^p \right)^{\frac{1}{p}} \left(\sum_{k=1}^{\infty} |c_k|^q \right)^{\frac{1}{q}} = \left(\sum_{k=1}^{\infty} |c_k|^q \right)^{\frac{1}{q}} \|x\|$$

and consequently
$$\|f\| \leqq \left(\sum_{k=1}^{\infty} |c_k|^q \right)^{\frac{1}{q}}. \tag{15}$$

Using (14) and (15), we have $\|f\| = \left(\sum_{k=1}^{\infty} |c_k|^q \right)^{\frac{1}{q}}.$

COROLLARY. We consider the space l_2. Every linear functional defined on l_2 can be generally written in the form

$$f(x) = \sum_{k=1}^{\infty} \xi_k c_k$$

where $\sum_{k=1}^{\infty} c_k^2 < \infty$ and $\|f\| = \sqrt{\sum_{k=1}^{\infty} c_k^2}$. If we consider in particular the n-dimensional vector space E_n, then the general form of the linear functionals defined on this space is supplied by

$$f(x) = \sum_{k=1}^{n} \xi_k c_k.$$

Here $c_1, c_2, \ldots, c_n$ are arbitrary numbers and for the norm we have $\|f\| = \sqrt{\sum_{k=1}^{n} c_k^2}$.

Besides the spaces l_p we also consider in functional analysis the space l of number sequences $\{\xi_1, \xi_2, \ldots, \xi_n, \ldots\}$, for which $\sum_{i=1}^{\infty} |\xi_i| < \infty$. In l the norm is defined by $\|x\| = \sum_{i=1}^{\infty} |\xi_i|$. It can be shown that *every linear functional on l is of the form*

$$f(x) = \sum_{i=1}^{\infty} \xi_i c_i$$

where $\{c_i\}$ is a bounded sequence of real numbers. For the norm of such a functional we have

$$\|f\| = \sup_i |c_i|.$$

General Form of Linear Functionals on Hilbert Space. Let H be a Hilbert space and $f(x)$ an arbitrary linear functional defined on H. Let the set of zeros of this functional be designated by L, i.e., the totality of all $x \in H$ with $f(x) = 0$. Clearly L is a subspace of H. For, L is a linear system by virtue of the additivity of $f(x)$ and closed because $f(x)$ is continuous.

For an arbitrary element $x \in H$ not belonging to L, the projection of x on the subspace $H \dot{-} L$ will be designated by x_0. Then obviously $f(x_0) = \alpha \neq 0$. For $x_1 = \frac{x_0}{\alpha}$, $f(x_1) = 1$. Now, if x is an arbitrary element of H and $f(x) = \beta$, then

$$f(x) - \beta f(x_1) = 0$$

or

$$f(x - \beta x_1) = 0.$$

Therefore $x - \beta x_1 = z$ with $z \in L$ or

$$x = \beta x_1 + z.$$

This equality shows that H is the orthogonal sum of the subspace L and the one-dimensional subspace generated by x_1.

Since $x_1 \perp z$, we have

$$(x, x_1) = \beta \|x_1\|^2$$

or, since $\beta = f(x)$

$$f(x) = \left(x, \frac{x_1}{\|x_1\|^2}\right).$$

Designating $\frac{x_1}{\|x_1\|^2}$ by u,

$$f(x) = (x, u), \tag{16}$$

i.e., the representation of $f(x)$ as an inner product of the element x with a fixed element u.

u is determined uniquely by $f(x)$. If we had

$$f(x) = (x, u'),$$

then

$$(x, u - u') = 0,$$

for arbitrary $x \in H$, thus it follows that $u = u'$.

From (16), we obtain furthermore

$$|f(x)| = |(x, u)| \leqq \|u\| \|x\|,$$

i.e.,

$$\|f\| \leqq \|u\|.$$

On the other hand

$$f(u) = (u, u) = \|u\|^2.$$

Therefore $\|f\|$ cannot be less than $\|u\|$, consequently $\|f\| = \|u\|$.

Summarizing, we have: *Every linear functional $f(x)$ defined on a Hilbert space has the form*

$$f(x) = (x, u), \tag{16}$$

where u is uniquely determined by $f(x)$ and

$$\|f\| = \|u\|. \tag{17}$$

Conversely, for an arbitrary $u \in H$ there is defined by (16) a functional f and the norm of f satisfies the equality (17). Therefore, (16) is the general form of linear functionals on H.

From this there follow as above the representations of the linear functionals on the spaces l_2 and $L_2(0, 1)$.

§23. CONJUGATE SPACES AND ADJOINT OPERATORS

EXAMPLES OF CONJUGATE SPACES. As already shown above, the totality of all linear functionals $f(x)$ defined on a linear normed space E forms a Banach space $\bar{E}$. It is called *conjugate* to E. Using the results of §22, we can find the space $\bar{E}$ in some cases.

1. Let $E = C[0, 1]$. The set of functions $g(t)$ of bounded variation defined on $[0, 1]$ is obviously a linear system, if we define addition of two functions and multiplication of a function by a real number in the usual way. We define the norm of such a function by $\|g\| = \operatorname*{Var}_0^1 (g)$. If two functions of bounded variation are considered to be equal when they differ on their points of continuity by at most an additive constant, then the axioms for the norm are obviously fulfilled.

The normed linear space so obtained is that of the *space of functions of bounded variation.*

If we consider the space $\bar{E} = \overline{C[0, 1]}$ of all linear functionals defined on $C[0, 1]$, then to every linear functional $f \in \overline{C[0, 1]}$, there corresponds uniquely up to an additive constant a function $g(t)$ of bounded variation. Conversely, there is assigned to every function of bounded variation $g(t) + \text{const.}$ a functional $f \in \overline{C[0, 1]}$. Therefore, there exists a one-to-one correspondence between the set of all linear functionals of $\overline{C[0, 1]}$ and all elements of the space of functions of bounded variation. Clearly, under this association there corresponds to the sum $f_1(t) + f_2(t)$ the sum $g_1(t) + g_2(t)$ and to the functional $\lambda f(t)$ the function $\lambda g(t)$, if the functionals f_1, f_2 correspond to the functions $g_1(t), g_2(t)$. Therefore, this association between $\overline{C[0, 1]}$ and the space of all functions of bounded variation is an isomorphism. Since, in addition, $\|f\| = \operatorname*{Var}_0^1 (g) = \|g\|$, this correspondence is isometric.

From the standpoint of functional analysis, these two spaces are equal. Therefore, $\overline{C[0,1]}$ is the space of all functions of bounded variation.

The general form of the linear functional on $\overline{C[0,1]}$, and hence the space $\overline{\overline{C[0,1]}}$, too, were discovered by A. P. Artemenko and J. I. Grosberg [29], [34].

2. Let $E = L_p(0,1)$. Since to each functional $f \in \overline{L_p(0,1)}$ there corresponds a function $a(t) \in L_q(0,1)$ and conversely, there exists a one-to-one correspondence between $L_p(0,1)$ and $L_q(0,1)$. As above, we convince ourselves that this correspondence is isomorphic and isometric, i.e., $\overline{L_p(0,1)} = L_q(0,1)$ to within isomorphism and isometry. In particular, for $p = 2$, $L_2(0,1) = L_2(0,1)$. Hence $L_2(0,1)$ is called a *self-conjugate space.*

3. It can easily be seen that $\bar{l}_p = l_q$ and, in particular, that $\bar{l}_2 = l_2$.

4. Linear functionals on a Hilbert space H are generated by elements of the same space. Hence Hilbert spaces are self-conjugate.

By the same reasoning, n-dimensional Euclidean space is self-conjugate, too.

REGULAR SPACES. Let E be a normed linear space, $\bar{E}$ the conjugate space belonging to it. Since $\bar{E}$ is also linear and normed, we can arrive at $\bar{\bar{E}}, \bar{\bar{\bar{E}}}$, and so on.

$\bar{\bar{E}}$ is the space of linear functionals defined on $\bar{E}$. Now let $f \in \bar{E}$. f can be fixed, while x runs through all elements of E.

We can, however, fix x and vary f in $\bar{E}$. If, for instance,

$$f(x) = \int_0^1 x(t)\, dg(t),$$

then we have the first case if $g(t)$ is fixed and $x(t)$ varies; the second case if $x(t)$ is fixed and $g(t)$ varies.

If we fix x and vary f over all elements of $\bar{E}$, then there corresponds to each $f \in \bar{E}$ a real number. The expression $f(x)$, for fixed x and variable f, is therefore a functional F_x defined on $\bar{E}$. We therefore can write

$$f(x) = F_x(f).$$

Because, as we can easily see, F_x is even a linear functional, $F_x \in \bar{\bar{E}}$. For, we have

$$F_x(f_1 + f_2) = (f_1 + f_2)(x) = f_1(x) + f_2(x) = F_x(f_1) + F_x(f_2).$$

From
$$|F_x(f)| = |f(x)| \leqq \|x\| \, \|f\|$$
it follows in particular that
$$\|F_x\| \leqq \|x\|. \tag{1}$$

According to the first corollary to the Hahn-Banach theorem, there exists for every x, a linear functional f_0 such that $f_0(x) = \|x\|$ and $\|f_0\| = 1$. For such a functional, we have $|F_x(f_0)| = |f_0(x)| = \|x\|$ or $|F_x(f_0)| = \|f_0\| \, \|x\|$, which is the same thing. Therefore,

$$\|F_x\| \geqq \|x\|. \tag{2}$$

From (1) and (2), it follows that

$$\|F_x\| = \|x\|. \tag{3}$$

Clearly, also
$$F_{x_1+x_2}(f) = F_{x_1}(f) + F_{x_2}(f)$$
and
$$F_{\lambda x}(f) = \lambda F_x(f).$$

Therefore, there corresponds to every $x \in E$ a well-determined functional $F_x \in \bar{\bar{E}}$. This correspondence between the space E and the set $\{F_x\} \subseteq \bar{\bar{E}}$ is isomorphic and isometric [the bi-uniqueness between E and $\{F_x\}$ follows from (3)], i.e., $E \subseteq \bar{\bar{E}}$. If $E = \bar{\bar{E}}$, then the space E is called *regular*.[23]

Examples. 1. The n-dimensional Euclidean space is regular. In fact, if E is an n-dimensional Euclidean space, then $\bar{E}$ is also an n-dimensional Euclidean space, consequently $\bar{\bar{E}}$, too. If, however, an n-dimensional space is part of another one, then the two coincide. $E \subseteq \bar{\bar{E}}$ therefore implies $E = \bar{\bar{E}}$.

2. The space $L_p(0,1)$ is regular. For, we have

$$\overline{\overline{L_p(0,1)}} = \overline{(\overline{L_p(0,1)})} = \overline{L_q(0,1)} = L_p(0,1).$$

3. l_p is regular. This follows analogously to the above.

4. $C[0,1]$ is not regular. Assume $C[0,1]$ is regular. An arbitrary linear functional $F(f)$ defined on the space V of all functions of bounded variation then must be of the form $F_x(f) = f(x)$ if $x \in C[0,1]$ is chosen properly. Recalling the general form of the functionals defined on $C[0,1]$, we can write

$$F_x(f) = f(x) = \int_0^1 x(t)\,df(t) \tag{4}$$

for an arbitrary $F(f)$. We designate by $f(t)$ the function of bounded variation corresponding to the functional $f(x) \in \bar{C}[0,1]$. The functional

$$F_{x_0}(f) = f(t_0+0) - f(t_0-0)$$

assigns to every function $f(t)$ of bounded variation its jump at the point t_0. Obviously $F_{x_0}(f)$ is additive. From

$$|F_{x_0}(f)| = |f(t_0+0) - f(t_0-0)| \leqq \operatorname*{Var}_0^1(f) = \|f\|$$

there follows the boundedness of $F_{x_0}(f)$ and the fact that the norm of $F_{x_0}(f)$ cannot be greater than 1. In addition, $F_{x_0}(f) \not\equiv 0$. It suffices to consider $F_{x_0}(f_1)$ with

$$f_1(t) = \begin{cases} 0 & \text{for } 0 \leqq t < t_0, \\ 1 & \text{for } t_0 \leqq t \leqq 1. \end{cases}$$

Because of (4), there exists a continuous function $x_0(t)$ such that

$$F_{x_0}(f) = \int_0^1 x_0(t)\,df(t). \tag{5}$$

For

$$f_0(\mathrm{t}) = \int_0^t x_0(\tau)\,d\tau,$$

we have $F_{x_0}(f_0) = 0$, because $f_0(t)$ is continuous on $[0,1]$. On the other hand,

$$F_{x_0}(f_0) = \int_0^1 x_0(t)\,df_0(t) = \int_0^1 x_0(t)\,dt > 0,$$

since $x_0(t) \not\equiv 0$. Otherwise, $F_{x_0}(f) \equiv 0$. This is a contradiction.

Consequently, $C[0,1]$ cannot be regular.

A. I. Plessner has proved: *Either $E = \bar{\bar{E}}$ or all the spaces $E, \bar{E}, \bar{\bar{E}}, \bar{\bar{\bar{E}}}, \ldots$ are different.*

Adjoint Operators. Let the linear operator A map E into E_1 where E and E_1 are linear and normed. Furthermore, let $\varphi(y)$ be a linear functional defined on E_1.

---

[23] The term "reflexive" is also used by English speaking mathematicians. [Trans.]

Then $\varphi(y)$ is defined for $y = Ax$, where x is an arbitrary element in E. We have

$$\varphi(y) = \varphi(Ax) = f(x).$$

$f(x)$ represents a certain functional defined on E. It is obvious that $f(x)$ is linear. Hence, there corresponds to every $\phi \in \bar{E}_1$ a functional $f \in \bar{E}$.

The collection of all correspondences thus obtained forms a certain operator, defined on $\bar{E}_1$ and with range belonging to $\bar{E}$. This operator is designated by A^* and is called the *adjoint operator* of A. For the equality $\varphi(y) = f(x)$ we shall write

$$f = A^*\varphi.$$

Examples. 1. Let A be an operator $(E \to E)$, where E is the n-dimensional space. Then A is given by a matrix (a_{ij}) of order n and we can write $\eta_i = \sum_j a_{ij}\xi_j$ instead of

$$y = Ax$$

with $y = \{\eta_1, \eta_2, \ldots, \eta_n\}$, $x = \{\xi_1, \xi_2, \ldots, \xi_n\}$. For every linear functional $f \in \bar{E}$,

$$f = (f_1, f_2, \ldots, f_n), \quad f(x) = \sum_i f_i \xi_i$$

and therefore

$$f(Ax) = \sum_i f_i (Ax)_i = \sum_i \sum_j f_i a_{ij} \xi_i = \sum_i \sum_j a_{ij} f_i \xi_j = \sum_j \left(\sum_i a_{ij} f_i \right) \xi_j = \sum_j g_j \xi_j$$

with

$$g_j = \sum_i a_{ij} f_i.$$

The vector $g = (g_1, g_2, \ldots, g_n)$ is an element of $\bar{E}$. We obtain g from the vector $f = (f_1, f_2, \ldots, f_n)$ of the same space by the linear mapping $g = A^*f$, where A^* is the transposed matrix of A. Consequently, the transposed matrix corresponds in n-dimensional space to the adjoint operator.

2. We consider in $L_2(0, 1)$ the Fredholm operator

$$y = Kx(t) = \int_0^1 K(t, s)\, x(s)\, ds$$

with continuous kernel $K(s, t)$.

An arbitrary linear functional $f(y)$ on $L_2(0, 1)$ has the form

$$f(y) = \int_0^1 f(s)\, y(s)\, ds, \; f \in L_2(0, 1).$$

Therefore

$$f(Kx) = \int_0^1 f(s) \left(\int_0^1 K(s, t)\, x(t)\, dt \right) ds = \int_0^1 x(t) \left(\int_0^1 K(s, t) f(s)\, ds \right) dt$$

$$= \int_0^1 x(s) \left(\int_0^1 K(t, s) f(t)\, dt \right) ds = \int_0^1 x(s)\, g(s)\, ds$$

with

$$g(s) = \int_0^1 K(t, s) f(t)\, dt \quad (g = K^* f).$$

In this case there therefore corresponds an interchange of the variables in the kernel to arrive at the adjoint operator. The corresponding adjoint integral equation is the equation with the adjoint operator.

THEOREM 1. *The adjoint operator A^* of a linear operator A mapping the linear normed space E into the linear normed space E_1 is also a linear operator and* $\| A^* \| = \| A \|$.

First, it is clear that A^* is additive. Furthermore,

$$|f(x)| = |\varphi(Ax)| \leqq \|\varphi\| \|Ax\| \leqq \|\varphi\| \|A\| \|x\|.$$

This implies $\|f\| \leqq \|\varphi\| \|A\|$. Since $f = A^*\varphi$, we obtain

$$\|A^*\varphi\| \leqq \|A\| \|\varphi\|.$$

Consequently, A^* is a bounded operator for which

$$\|A^*\| \leqq \|A\|. \tag{6}$$

Let x_0 be any element of E. According to the first corollary to the Hahn-Banach theorem, there exists a functional $\varphi_0 \in \bar{E}_1$ with $\|\varphi\| = 1$ and $\varphi_0(Ax_0) = \|Ax_0\|$. We obtain from this,

$$\|A x_0\| = \varphi_0(A x_0) = f_0(x_0) \leqq \|f_0\| \|x_0\|$$
$$= \|A^* \varphi_0\| \|x_0\| \leqq \|A^*\| \|\varphi_0\| \|x_0\| = \|A^*\| \|x_0\|,$$

and hence

$$\|A\| \leqq \|A^*\|. \tag{7}$$

From (6) and (7) it follows that $\|A\| = \|A^*\|$.

INNER PRODUCT AND ORTHOGONAL ELEMENTS. Let $x \in E$ and $f \in \bar{E}$. The expression

$$(f, x) = f(x) \tag{8}$$

is a bilinear functional of the two variables x and f, i.e., it is linear in x and also in f.

If E is a Hilbert space, then $\bar{E} = E$, and (8) is the inner product of x and f [see formula (16) §22].

It is usual, also in the general case, if $\bar{E} \neq E$, to interpret (8) as an *inner product* of $f \in \bar{E}$ and $x \in E$.

If

$$(f, x) = 0,$$

then $f \in \bar{E}$ and $x \in E$ are called *orthogonal.*

THEOREM 2. *Let λ_0 be an eigenvalue of the linear operator $A \in (E \to E)$ and x_0 the corresponding eigen-element. Furthermore, let λ_1 be an eigenvalue of the adjoint operator A^* with the eigen-element f_0.*

Then the eigen-elements are orthogonal, if $\lambda_0 \neq \lambda_1$.

This theorem is a generalization of the theorem on the orthogonality of eigenfunctions of adjoint integral equations.

The relation between the operators A and A^* can be expressed by means of the inner product. We have

$$(f, Ax) = (A^*f, x)$$

for all $f \in \bar{E}$, $x \in E$. By hypothesis,

$$Ax_0 = \lambda_0 x_0, \; A^*f_0 = \lambda_1 f_0.$$

This and the above equality imply

$$\lambda_0(f_0, x_0) = (f_0, \lambda_0 x_0) = (f_0, Ax_0) = (A^*f_0, x_0) = \lambda_1(f_0, x_0)$$

or

$$(\lambda_1 - \lambda_0)(f_0, x_0) = 0.$$

Because, by hypothesis, $\lambda_1 - \lambda_0 \neq 0$,

$$(f_0, x_0) = 0.$$

BIORTHOGONAL SEQUENCES. The sequences $\{x_n\}, x_n \in E$ and $\{f_n\}$, $f_n \in \bar{E}$, are called *biorthogonal* if for all n and m

$$(f_n, x_m) = \delta_{nm} \tag{9}$$

If E is self conjugate, for instance a Hilbert space, then both sequences belong to E. If $\{x_n\}$ and $\{f_n\}$ coincide, then the biorthogonality becomes ordinary orthogonality.

Let $\{x_n\}$ and $\{f_n\}$ be biorthogonal, and let $x \in E$ be representable as a series

$$x = \sum_{n=1}^{\infty} c_n x_n. \tag{10}$$

We have

$$(f_m, x) = \lim_k \left(f_m, \sum_{i=1}^{k} c_i x_i \right) = \lim_k \sum_{i=1}^{k} c_i (f_m, x_i).$$

For $k \geqq m$, $\sum_{i=1}^{k} c_i (f_m, x_i) = c_m$ because of (9), for in this sum all terms except $c_m (f_m, x_m) = c_m$ vanish. This implies $(f_m, x) = c_m$ and we obtain for (10)

$$x = \sum_{n=1}^{\infty} (f_n, x) x_n. \tag{11}$$

If

$$f = \sum_{n=1}^{\infty} d_n f_n, \tag{12}$$

then analogously

$$d_n = (f, x_n).$$

We call (11) and (12) *Fourier series* of the biorthogonal sequences.

P. L. Chebyshev and A. A. Markov first investigated non-trivial examples of biorthogonal sequences in connection with problems of interpolation.[24]

CONJUGATE SPACES OF COMPLEX LINEAR SPACES. All concepts introduced in this paragraph can be generalized for the case of a complex linear space $\mathscr{E}$. Again we call the totality of all linear functionals on $\mathscr{E}$ the conjugate space $\bar{\mathscr{E}}$ of $\mathscr{E}$.

As just before we call the number $f(x)$ the inner product (x, f) with $x \in \mathscr{E}$ and $f \in \bar{\mathscr{E}}$. In order to preserve the properties of this inner product also for the complex Hilbert space, (x, f) must be a linear functional of x and conjugate-linear in f:

$$(x, \lambda f) = \bar{\lambda}(x, f).$$

Also multiplication by a complex number λ in $\bar{\mathscr{E}}$ is thus defined: λf is the linear functional f_λ on $\mathscr{E}$ with

$$f_\lambda(x) = \bar{\lambda} f(x).$$

The concept of the adjoint operator A^* of an operator A of $(\mathscr{E} \to \mathscr{E}_1)$ is defined as follows: A^* is that operator of $(\bar{\mathscr{E}}_1 \to \bar{\mathscr{E}})$ for which

$$(Ax, f) = (x, A^* f)$$

holds for all $x \in \mathscr{E}$ and all $f \in \bar{\mathscr{E}}$.

All properties of the adjoint operator generalize immediately to the complex case except for one modification. The theorem of the orthogonality of the eigen-elements x_0 and f_0 of the operators A, A^*, respectively, holds for

$$Ax_0 = \lambda_0 x_0, \quad A^* f_0 = \lambda_1 f_0,$$

if $\lambda_0 \neq \bar{\lambda}_1$.

---

[24] Markov, A. A., *Collected Works*, 1948, pp. 212–230.

§24. WEAK CONVERGENCE OF SEQUENCES OF FUNCTIONALS

WEAK CONVERGENCE. Let E be normed linear space. A sequence $\{f_n\}$ of functionals of $\bar{E}$ is called *weakly convergent* to the functional $f_0 \in \bar{E}$, if the sequence $f_n(x) \to f_0(x)$ for every $x \in E$.

If a sequence of linear functionals converges weakly to f_0, then it can be easily seen that all of its subsequences converge weakly to f_0. A sequence cannot converge weakly to two different limits.

Furthermore, there holds

THEOREM 1. *If a sequence $\{f_n\}$ of linear functionals converges weakly to the functional f_0, then the norms of the f_n are bounded.*

This theorem is a special case of the Banach theorem on the boundedness of the norm of an everywhere convergent sequence of linear operators (viz. §19).

THEOREM 2. *A sequence $\{f_n\}$ of linear functionals converges weakly to a linear functional f_0 if and only if*

1) *The sequence $\{\|f_n\|\}$ is bounded, and*

2) *$f_n(x) \to f_0(x)$ for an arbitrary x from the set G for which the linear combinations of its elements are everywhere dense in E.*

Necessity. The necessity of the first condition follows from Theorem 1, that of the second one is evident.

Sufficiency. Let $M = \sup_n \|f_n\|$ for arbitrary $x \in E$. Choose an x_0 which is a linear combination of elements of G such that $\|x - x_0\| < \frac{\varepsilon}{4M}$. We estimate

$$
\begin{aligned}
|f_n(x) - f_0(x)| &\leqq |f_n(x) - f_n(x_0)| + |f_n(x_0) - f_0(x_0)| + |f_0(x_0) - f_0(x)| \\
&\leqq |f_n(x_0) - f_0(x_0)| + (\|f_n\| + \|f_0\|)\,\|x - x_0\| < |f_n(x_0) - f_0(x_0)| + \frac{\varepsilon}{2}.
\end{aligned}
$$

According to the second condition, $f_n(x_0) \to f_0(x_0)$, there exists an n_0 such that

$$|f_n(x_0) - f_0(x_0)| < \frac{\varepsilon}{2}$$

for all $n \geqq n_0$.

Therefore, for $n \geqq n_0$

$$|f_n(x) - f_0(x)| < \varepsilon$$

and because x is an arbitrary element of E, we have proved the weak convergence of the sequence $\{f_n\}$ to the functional f_0.

WEAK COMPACTNESS. The following theorem is important for applications in functional analysis.

THEOREM 3. *If the space E is separable, then every sphere in the conjugate space $\bar{E}$ is weakly compact*, i.e., *from an arbitrary sequence $\{f_n\}$ of linear funcitonals with bounded norms, we can choose a subsequence which converges weakly to the functional f_0.*

Proof. It was shown earlier that $\bar{E}$ is complete in the sense of weak convergence of operators. It is therefore sufficient to show that from every sequence $\{f_n\}$ of linear functionals with bounded norms, there can be selected a weakly convergent subsequence. Without loss of generality, we assume that $\|f_n\| \leqq 1$. Let $x_1, x_2, \ldots, x_n, \ldots$ be a countable set which is everywhere dense in E. Since

$$|f_n(x_1)| \leqq \|f_n\| \|x_1\| \leqq \|x_1\|,$$

then $\{f_n(x_1)\}$ is a bounded sequence of numbers. We can select from it a convergent subsequence

$$f_{n_1^{(1)}}(x_1), f_{n_2^{(1)}}(x_1), \ldots, f_{n_k^{(1)}}(x_1), \ldots$$

We now investigate the sequence $\{f_{n_k^{(1)}}\}$.

Because

$$|f_{n_k^{(1)}}(x_2)| \leqq \|x_2\|$$

$\{f_{n_k^{(1)}}(x_2)\}$ is also a bounded sequence of numbers. From this, there can again be selected a convergent subsequence

$$f_{n_1^{(2)}}(x_2), f_{n_2^{(2)}}(x_2), \ldots, f_{n_k^{(2)}}(x_2), \ldots$$

Continuing this procedure, we construct the sequences, $\{f_{n_k^{(3)}}(x_3)\}, \{f_{n_k^{(4)}}(x_4)\}, \ldots$. It is essential that each succeeding subsequence is selected from the preceding one, and the subsequence $\{f_{n_k^{(m)}}\}$ converges for all $x_i, i = 1, 2, \ldots, m$.

We now construct the "diagonal sequence"

$$f_{n_1^{(1)}}, f_{n_2^{(2)}}, \ldots, f_{n_k^{(k)}}, \ldots \qquad (\omega)$$

It can easily be seen that this subsequence converges for every x_m from the denumerable everywhere dense set considered. In fact, it is sufficient to remark that $f_{n_m^{(m)}}, f_{n_{m+1}^{(m+1)}}, \ldots$ is a subsequence of $\{f_{n_k^{(m)}}\}$ which converges for x_m, according to the construction. Therefore, all sequences $\{f_{n_k^{(k)}}\}$ converge for all x_m.

Now let $x \in E$ be arbitrary. If $\varepsilon > 0$, we can find an x_k such that

$$\|x - x_k\| < \frac{\varepsilon}{3}.$$

Then, for arbitrary $f \in \bar{E}$,

$$|f(x_k) - f(x)| = |f(x_k - x)| \leqq \|f\| \|x_k - x\| < \|f\| \frac{\varepsilon}{3}.$$

Especially because $\|f_{n_m^{(m)}}\| \leqq 1$,

$$|f_{n_m^{(m)}}(x_k) - f_{n_m^{(m)}}(x)| < \frac{\varepsilon}{3}$$

for every $n_m^{(m)}$.

Because the sequence (ω) converges for every x_n,

$$|f_{n_m^{(m)}}(x_k) - f_{n_p^{(p)}}(x_k)| < \frac{\varepsilon}{3}$$

for a sufficiently large m and p and this implies

$$|f_{n_m^{(m)}}(x) - f_{n_p^{(p)}}(x)| \leqq |f_{n_m^{(m)}}(x) - f_{n_m^{(m)}}(x_k)|$$
$$+ |f_{n_m^{(m)}}(x_k) - f_{n_p^{(p)}}(x_k)| + |f_{n_p^{(p)}}(x_k) - f_{n_p^{(p)}}(x)| < \frac{\varepsilon}{3} + \frac{\varepsilon}{3} + \frac{\varepsilon}{3} = \varepsilon.$$

This, however, proves the convergence of the sequence $\{f_{n_k^{(k)}}(x)\}$, i.e., the weak convergence of the subsequence (ω) of $\{f_n\}$.

COROLLARY. *In the spaces l_p and $L_p(0,1)$, every sphere is weakly compact.* This follows from $l_p=\bar{l}_q$, $L_p(0,1)=\overline{L_q(0,1)}$ and the separability of l_p and $L_p(0,1)$.

Remark. One can easily prove that in $\bar{E}$ every sphere is weakly closed and, consequently, weakly compact in itself.

HELLY'S PRINCIPLE OF CHOICE. We consider the space V of functions $g(t), 0 \leqq t \leqq 1$ of bounded variation. Because V is conjugate to the separable Banach space $C[0,1]$, every sphere in V is weakly compact in itself.

THEOREM 4 (E. Helly's Principle of Choice). *Let $M=\{g(t)\}$, $0\leqq t\leqq 1$, be an infinite set of functions of bounded variation which are continuous on the right.*

Let
$$|g(t)| \leqq K \text{ and } \operatorname*{Var}_0^1 (g) \leqq K$$
for every $g(t)\in M$.

Then there exists a sequence $\{g_n(t)\} \subseteq M$ which converges everywhere on $[0,1]$ *to a function $g_\omega(t)$ of bounded variation.*

As is well-known, every function of bounded variation can be represented as a difference of two non-decreasing functions. Because of this it is sufficient to prove the theorem for the case where M consists of non-decreasing functions. From the weakly compact set M, we can select a sequence $\{g_n(t)\}$ such that the functionals belonging to it converge weakly to the functional belonging to the function $g_0(t)$ of bounded variation:

$\int_0^1 f(\tau)\,dg_n(\tau) \to \int_0^1 f(\tau)\,dg_0(\tau)$ for every continuous function f. $g_0(t)$ has at most a denumerable number of discontinuities $\{t_j\}$.

Let $g_0(t)$ be continuous at the point t. We investigate the continuous functions $\varphi_1(\tau), \varphi_2(\tau), \ldots, \varphi_m(\tau), \ldots$ where $\varphi_m(\tau)=1$ for $0\leqq\tau\leqq t$, $\varphi(\tau)=0$ for $t+\dfrac{1}{m}\leqq \tau\leqq 1$, $\varphi_m(\tau)$ linear on the interval $\left(t, t+\dfrac{1}{m}\right)$.

Then
$$\int_0^1 \varphi_m(\tau)\,dg_n(\tau) \to \int_0^1 \varphi_m(\tau)\,dg_0(\tau) \text{ for } n\to\infty. \tag{1}$$
Furthermore, from the continuity of $g_0(\tau)$ at $\tau=t$ and the boundedness of $\varphi_m(\tau)$ it follows moreover that
$$\int_t^{t+\frac{1}{m}} \varphi_m(\tau)\,dg_0(\tau) \to 0 \text{ as } m\to\infty.$$
Hence
$$\int_0^1 \varphi_m(\tau)\,dg_0(\tau) = \int_0^1 \varphi_m(\tau)\,dg_0(\tau) + \int_t^{t+\frac{1}{m}} \varphi_m(\tau)\,dg_0(\tau) \to g_0(t)-g_0(0). \tag{2}$$

Because $g_n(\tau)$ is a non-decreasing function and $\varphi_m(\tau) \geqq 0$, we finally obtain

$$\int_0^1 \varphi_m(\tau)\, dg_n(\tau) \geqq \int_0^t \varphi_m(\tau)\, dg_n(\tau) = g_n(t) - g_n(0). \tag{3}$$

It follows from (1), (2), (3) that:

$$\overline{\lim_n}\, [g_n(t) - g_n(0)] \leqq g_0(t) - g_0(0).$$

If we now substitute the functions $\varphi_m\left(\tau + \frac{1}{m}\right)$ for $\varphi_m(\tau)$, then we obtain

$$\underline{\lim_n}\, [g_n(t) - g_n(0)] \geqq g_0(t) - g_0(0).$$

The two inequalities imply

$$\lim_n\, [g_n(t) - g_n(0)] = g_0(t) - g_0(0). \tag{4}$$

By hypothesis, $|g_n(0)| \leqq K$, $|g_n(t_j)| \leqq K$, $(n = 1, 2, \ldots)$, there exists a subsequence $\{g_{n_i}(t)\}$ of $\{g_n(t)\}$ such that $g_{n_i}(0) \to g_0$, $g_{n_i}(t_j) \to g_j$ as $n_i \to \infty$. This, together with (4) imply that

$$g_{n_i}(t) \to g_0(t) - g_0(0) + g_0 \text{ as } n_i \to \infty, \tag{5}$$

for every $t \in \{t_j\}$. The function $g_\omega(t_j) = g_j$, $g_\omega(t) = g_0(t) - g_0(0) + g_0$ for $t \in \{t_j\}$ is of bounded variation, and for every $t \in [0, 1]$,

$$g_{n_i}(t) \to g_\omega(t)$$

q.e.d.

APPLICATION TO QUADRATURE FORMULAS [9]. In the space $C[0, 1]$, we consider the functional

$$f(x) = \int_0^1 x(t)\, d\sigma(t)$$

with non-decreasing $\sigma(t)$ and the sequence of functionals

$$f_n(x) = \sum_{k=1}^{k_n} c_k^{(n)} x(t_k^{(n)}), \quad n = 1, 2, 3, \ldots$$

The $c_k^{(n)}$ are chosen such that $f(x)$ and $f_n(x)$ coincide for all polynomials of degree less than or equal to n:

$$f(x) = f_n(x), \text{ if } x = \sum_{l=o}^{n} a_l t^l.$$

The f_n are used for the approximate evaluation of the functional f. The relation

$$f(x) \approx f_n(x)$$

which becomes an equality for all polynomials of degree less than or equal to n, is called a *quadrature formula*.

We consider a sequence of quadrature formulas

$$f(x) \approx f_n(x).$$

There now arises the question whether the sequence $f_n(x)$ converges to the value

$f(x)$ for every $x(t) \in C[0,1]$ as $n \to \infty$. In other words: Does the sequence $\{f_n\}$ of functionals converge weakly to the functional f?

THEOREM 5. *The quadrature formulas converge,* i.e., *for every continuous function* $x(t)$

$$\lim_n \sum_{k=1}^{kn} c_k^{(n)} x(t_k^{(n)}) = \int_0^1 x(t)\, d\sigma(t)$$

if and only if

$$\sum_{k=1}^{kn} | c_k^{(n)} | \leqq M = \text{const.}$$

holds for every n.

Indeed, according to the definition of the functionals f_n,

$$f_m(p_n) = f(p_n) \text{ for } m \geqq n,$$

for all polynomials $p_n(t)$ of degree n. Moreover, it is obvious that

$$\|f_n\| \leqq \sum_{k=1}^{kn} |c_k^{(n)}| \leqq M.$$

Consequently, the sequence $\{f_n\}$ converges for the set of all polynomials which lie everywhere dense in $C[0,1]$ to the functional f, and the norms are bounded. Now, however, the theorem to be proved follows immediately from Theorem 2 of this paragraph.

THEOREM 6 (V. A. STEKLOV). *If all coefficients* $c_k^{(n)}$ *of the quadrature formulas are positive, then the sequence of quadrature formulas* $f(x) \approx f_n(x)$ *converges for every continuous function* $x(t)$.

In fact, for an arbitrary n and for $x_0(t) \equiv 1$,

$$f_n(x_0) = f(x_0).$$

Therefore

$$\sum_{k=1}^{kn} | c_k^{(n)}| = \sum_{k-1}^{kn} c_k^{(n)} = \int_0^1 d\sigma = \sigma(1) - \sigma(0),$$

and the hypotheses of the preceding theorem are fulfilled.

§25. WEAK CONVERGENCE OF SEQUENCES OF ELEMENTS

ELEMENTARY THEOREMS. We introduce the concept of weak convergence of elements of normed linear spaces.

Let E be a normed linear space, $\{x_n\}$ a sequence of elements of E and $x_0 \in E$. If for all functionals $f \in \bar{E}$ the sequence $f(x_n) \to f(x_0)$ as $n \to \infty$, then we say that $\{x_n\}$ *converges weakly* to x_0, and we write:

$$x_n \overset{wk}{\to} x_0.$$

x_0 is a *weak limit* of the sequence $\{x_n\}$.

A sequence cannot converge weakly to two different limits.

We assume that $x_n \overset{wk}{\to} x_0$ and $x_n \overset{wk}{\to} y_0$, i.e., for an arbitrary linear f

$$f(x_n) \to f(x_0) \text{ and } f(x_n) \to f(y_0).$$

Consequently $f(x_0) = f(y_0)$ or

$$f(x_0 - y_0) = 0.$$

If we now choose an f_0 with $\|f_0\| = 1$ and $f_0(x_0 - y_0) = \|x_0 - y_0\|$, then we obtain $\|x_0 - y_0\| = 0$, i.e., $x_0 = y_0$.

It can also be easily seen that every arbitrary subsequence $\{x_{n_k}\}$ converges weakly to x_0 if $x_n \overset{wk}{\rightarrow} x_0$.

As before, the norm convergence will be called strong convergence ($x_n \rightarrow x$).

It is, moreover, obvious that from the strong convergence of $\{x_n\}$ to x_0, there follows the weak convergence to the same limit.

The converse is not always true. A sequence can converge weakly to a certain element without converging strongly to this element.

For instance, consider in $L_2(0, 1)$, the sequence $x_n(t) = \sin n\pi t$. For every linear functional f

$$f(x_n) = \int_0^1 \sin n\pi t \, \alpha(t) \, dt$$

with $\alpha(t)$ square integrable. Obviously, $f(x_n)$ is the nth Fourier coefficient of $\alpha(t)$ with respect to $\sin n\pi t$. Consequently, $f(x_n) \rightarrow 0$ as $n \rightarrow \infty$. This implies, however, that $x_n \overset{wk}{\rightarrow} \theta$. On the other hand, it can be easily seen that x_n cannot converge strongly to the zero element; for

$$\| x_n - \theta \|^2 = \int_0^1 \sin^2 n\pi t \, dt = \frac{1}{2}.$$

There holds, however,

THEOREM 1. *In a finite-dimensional space, strong and weak convergences coincide.*

It is sufficient to prove that in a finite-dimensional space, weak convergence implies strong convergence to the same element. Let E be a finite-dimensional space and $\{x_n\}$ a sequence with $x_n \overset{wk}{\rightarrow} x_0$. Because E is finite-dimensional, there exists a finite system of linearly independent elements $e_1, e_2, \ldots, e_k$. Every $x \in E$ can be represented in the form $x = \xi_1 e_1 + \xi_2 e_2 + \ldots + \xi_k e_k$ with real $\xi_j, j = 1, \ldots, k$. Hence

$$x_n = \xi_1^{(n)} e_1 + \xi_2^{(n)} e_2 + \ldots + \xi_k^{(n)} e_k,$$
$$x_0 = \xi_1^{(n)} e_1 + \xi_2^{(n)} e_2 + \ldots + \xi_k^{(n)} e_k.$$

We now consider functionals $f_i \in \bar{E}$ for which $f_i(e_i) = 1, f_i(e_k) = 0$, for $k \neq i$. Therefore

$$f_i(x_n) = \xi_i^{(n)} \text{ and } f_i(x_0) = \xi_i^{(0)}.$$

Since, however, for every linear f, the sequence $f(x_n) \rightarrow f(x_0)$, also $f_i(x_n) \rightarrow f_i(x_0)$, i.e.,

$$\xi_i^{(n)} \rightarrow \xi_i^{(0)} \text{ for } i = 1, 2, \ldots, k.$$

Let the greatest of the numbers $\|e_i\|$ $(i = 1, 2, \ldots, k)$ be M. For a given $\varepsilon > 0$ there exists an n_0 such that

$$| \xi_i^{(n)} - \xi_i^{(0)} | < \frac{\varepsilon}{Mk}$$

for all $i = 1, 2, \ldots, k$ and $n \geqq n_0$. There also holds the estimation

$$\|x_n - x_0\| = \left\| \sum_{i=1}^{k} (\xi_i^{(n)} - \xi_i^{(0)}) e_i \right\| \leqq \sum_{i=1}^{k} |\xi_i^{(n)} - \xi_i^{(0)}| \, \|e_i\| < \varepsilon.$$

Consequently, $x_n \to x_0$ (strongly).

THEOREM 2. *If $\{x_n\}$ converges weakly to x_0, then there exists a sequence $\left\{\sum_{i=1}^{m} \lambda_i x_i\right\}$ of linear combinations which converges strongly to x_0.*

x_0 belongs to the closed linear manifold $\bar{L}$ generated by $x_1, x_2, \ldots, x_n, \ldots$.

Let us assume that x_0 does not belong to $\bar{L}$. Then there exists according to the second corollary to the Hahn-Banach theorem (p. 95) a linear functional $f \in \bar{E}$ with $f(x_0) = 1$ and $f(x_n) = 0$ for $n \neq 0$. This means, however, that $f(x_n)$ does not converge to $f(x_0)$ and this contradicts the hypothesis that $x_n \overset{wk}{\to} x_0$.

THEOREM 3. *Let A be an operator defined on a normed linear space E. Let its range be in the normed linear space E_1.*

If the sequence $\{x_n\}, x_n \in E$, converges weakly to $x_0 \in E$, then $\{Ax_n\}$, $Ax_n \in E_1$ converges weakly to $Ax_0 \in E_1$.

We choose an arbitrary functional $\varphi \in \bar{E}_1$. Then $\varphi(Ax_n) = f(x_n)$ with $f \in \bar{E}$. Similarly $\varphi(Ax_0) = f(x_0)$. Because $x_n \overset{wk}{\to} x_0, f(x_n) \to f(x_0)$ or $\varphi(Ax_n) \to \varphi(Ax_0)$. Since φ is an arbitrary functional in $\bar{E}_1, Ax_n \overset{wk}{\to} Ax_0$. Every linear operator A is not only strongly but weakly continuous.

THEOREM 4. *If a sequence $\{x_n\}$ converges weakly to x_0, then the norms of the elements are bounded.*

We consider the $x_n (n = 1, 2, \ldots)$ as elements of the space $\bar{\bar{E}}$. Then weak convergence of $\{x_n\}$ to x_0 means that the sequence $\{x_n\}$ of functionals, $x_n \in \bar{\bar{E}}$, converges to $x_0 \in \bar{\bar{E}}$ for all $f \in \bar{E}$. Then, however, the sequence $\{\|x_n\|\}$ is bounded according to the Banach theorem, q.e.d.

Remark. If x_0 is the weak limit of the sequence $\{x_n\}$, then $\|x_0\| \leqq \overline{\lim_n} \|x_n\|$. The existence of the above limit follows from the theorem just proved. Let us now assume that $\|x_0\| > \overline{\lim_n} \|x_n\|$. Then there exists a number C with $\|x_0\| > C > \overline{\lim_n} \|x_n\|$. Consequently, we can find an n_0 such that

$$\|x_0\| > C > \|x_n\|$$

for $n \geqq n_0$. Now we construct a linear functional f_0 in such a way that $\|f_0\| = 1$ and $f_0(x_0) = \|x_0\|$. Then $f_0(x_n) \leqq \|f_0\| \|x_n\| = \|x_n\| < C$ for all $n \geqq n_0$. Consequently, $f_0(x_n)$ does not converge to $f_0(x_0)$ and this contradicts the hypothesis that $x_n \overset{wk}{\to} x_0$.

There are cases possible in which $\|x_0\| < \overline{\lim_n} \|x_n\|$ can be realized, as the following example shows.

In the space $L_2(0, 1)$, we consider the functions

$$x_n(t) = \sqrt{2} \sin n\pi t.$$

Then $\|x_n\| = 1$ and hence $\lim_n \|x_n\| = 1$. On the other hand, we have for an arbitrary linear functional f

$$f(x_n)=\sqrt{2}\int_0^1 \alpha(t)\sin n\pi t\,dt=\sqrt{2}\,c_n$$

where the c_n are the Fourier coefficients of $\alpha(t)\in L_2(0,1)$. Consequently, the sequence $f(x_n)\to 0$ for every linear f and as $n\to\infty$, i.e., $c_n \overset{wk}{\to} \theta$. Hence $x_0=\theta$ and $\|x_0\|=0<1=\lim\limits_n \|x_n\|$.

THEOREM 5. *The sequence* $\{x_n\}$ *converges weakly to* x_0 *if and only if*

1) *The sequence* $\{\|x_n\|\}$ *is bounded and*

2) $f(x_n)\to f(x_0)$ *for every* f *from the set* Γ *of linear functionals whose linear combinations are everywhere dense in* $\bar{E}$.

This theorem is a special case of Theorem 2 in §24. For, weak convergence of a sequence $\{x_n\}$, $x_n\in E$, to $x_0\in E$ is equivalent to weak convergence of the same sequence if the x_n and x_0 are considered as linear functionals defined on $\bar{E}$.

WEAK CONVERGENCE IN SPECIAL SPACES.

THEOREM 6. *A sequence* $\{x_n\}, x_n=\{\xi_i^{(n)}\}, \xi_i^{(n)}\in l_p$, *converges weakly to* $x_0=\{\xi_i^{(0)}\}$, $\xi_i^{(0)}\in l_p$, *if and only if*

1) *The sequence* $\{\|x_n\|\}$ *of the norms is bounded and*

2) $\xi_i^{(n)}\to\xi_i^{(0)}$ as $n\to\infty$ *and for all* i *(in general, however, not uniformly).*

For the proof, we note that the linear combinations of the $f_i=\{0,0,\ldots,0,1,0,\ldots\}$ are everywhere dense in $l_q=\bar{l}_p$. According to the general Theorem 5, it is necessary and sufficient for $x_n\overset{wk}{\to}x_0$ that the first condition be fulfilled and that for every fi, the sequence ${}_i(x_n)=\xi_i^{(n)}\to\xi_i^{(0)}=f_i(x^0)$.

Therefore, in l_p weak convergence is equivalent to coordinate-wise convergence if the norms of the terms of the sequence are bounded.

THEOREM 7. *In order for a sequence* $\{x_n(t)\}, x_n(t)\in L_p(0,1)$ *to converge weakly to* $x_0(t)\in L_p(0,1)$, *it is necessary and sufficient that*

1) *The sequence* $\{\|x_n\|\}$ *be bounded and*

2) $\int_0^\tau x_n(t)\,dt\to\int_0^\tau x_0(t)\,dt$ *for arbitrary* $\tau\in[0,1]$.

Condition 1) coincides with that of Theorem 5. We investigate 2. To do this, we put

$$\alpha_\tau(t)=\begin{cases}1 \text{ for } 0\leqq t\leqq\tau,\\ 0 \text{ for } \tau<t\leqq 1.\end{cases}$$

Then the linear combinations of the functions $\alpha_\tau(t)$, i.e., the sums

$$\sum_{i=1}^{n} c_i[\alpha_{\tau_i}(t)-\alpha_{\tau_{i-1}}(t)]$$

with $0=\tau_0<\tau_1<\ldots<\tau_{n-1}<\tau_n=1$ lie everywhere dense in $L_q(0,1)=\overline{L_p(0,1)}$. In order for $x_n(t)\overset{wk}{\to}x_0$, it is therefore necessary and sufficient that the first condition be fulfilled and that

$$\int_0^1 x_n(t)\,\alpha_\tau(t)\,dt\to\int_0^1 x_0(t)\,\alpha_\tau(t)\,dt$$

or
$$\int_0^{\tau} x_n(t)\,dt \to \int_0^{\tau} x_0(t)\,dt$$
as $n \to \infty$ and for every $\tau \in [0, 1]$.

Remark. Let E be a space with weakly compact unit sphere $S_1(\|x\| \leqq 1)$. We consider an $f(x)$ which is not necessarily linear but weakly continuous, i.e.,
$$f(x_n) \to f(x_0)$$
as $x_n \overset{wk}{\to} x_0$. Then $f(x)$ is bounded on S_1 and assumes the least upper and greatest lower bound. The proof of this assertion is the same as that of the corresponding theorem for continuous functionals which are defined on a compact set of a metric space.

§26. UNIVERSALITY OF C[0, 1]

The Soviet mathematician P. S. Uryson proved in 1925 that there exist 'universal' spaces. A separable metric space E is called *universal* if there can be found for every separable metric space an isometric subspace of E. Later, the Polish mathematicians S. Banach and S. Mazur proved that $C[0, 1]$ is universal.

The proof of the theorem of Banach and Mazur is based on the weak compactness of conjugate spaces.

LEMMA. *Let N be a closed subspace of the set s for whose elements* $x = \{\xi_1, \xi_2, \ldots, \xi_n, \ldots\}$, $|\xi_i| \leqq 1$ for $i = 1, 2, \ldots$. *Then N is the continuous image of a closed subset P of the closed interval* $[0, 1]$

Proof. We choose an arbitrary point $x = \{\xi_j\}$ of the set N. If the coordinate ξ_i of x is non-negative, then we write it in the binary system in the form
$$0.\ \beta_{i1}\beta_{i2}\beta_{i3}\ldots,$$
where the β_{ik} are 0 or 1. If ξ_i is negative, then we put $\xi_i = -1 + \eta_i$ with non-negative η_i and we have for ξ_i
$$\bar{1}.\ \nu_{i1}\nu_{i2}\nu_{i3}\ldots,$$
where the ν_{ik} are the binary digits of the number η_i. Therefore every coordinate of the point x is written in the form
$$\xi_i = \bar{\alpha}_{i0}.\ \alpha_{i1}\alpha_{i2}\alpha_{i3}\ldots \tag{1}$$
with $\alpha_{ik} = 0$ or 1.

We now associate with every point x the number
$$y = 0.\ \alpha_{10}\alpha_{20}\alpha_{11}\alpha_{30}\alpha_{21}\alpha_{12}\alpha_{03}\ldots \tag{2}$$
of the interval $[0, 1]$ and interpret it as a representation in the ternary system, where, however, only the digits 0 and 1 appear. Conversely, we also associate with each $y \in [0, 1]$ which is written in the ternary system in the form (2) using only the digits 0 and 1, the point $x \in s$ with the coordinates ξ_i. The ξ_i are defined by
$$\left.\begin{array}{l} \xi_1 = \bar{\alpha}_{10}.\ \alpha_{11}\,\alpha_{12}\,\alpha_{13}\ldots, \\ \xi_2 = \bar{\alpha}_{20}.\ \alpha_{21}\,\alpha_{22}\,\alpha_{23}\ldots, \\ \xi_3 = \bar{\alpha}_{30}.\ \alpha_{31}\,\alpha_{32}\,\alpha_{33}\ldots, \\ \ldots\ldots\ldots\ldots \end{array}\right\} \tag{3}$$

Now we consider the set P of the $y \in [0, 1]$ whose ternary representation contains only the digits 0 and 1 and the $x \in N$ corresponding to them in s. For convenience, we choose for the $y \in P$ which are rational in the ternary system that representation in which all the digits from a certain place on are equal to 0. Then, every $y \in P$ has a unique ternary representation (2). Therefore the mapping

$$x = \varphi(y),\ y \in P,\ x \in N$$

by means of (3) is uniquely defined too.

Now we prove the continuity of this mapping. Let $y_n \to y$, where y_n and y contain in the ternary system only zeros and ones. The number of coinciding digits in these representations of the y_n and y increases without bound as $n \to \infty$. We now investigate the corresponding points

$$x_n = \varphi(y_n) \text{ and } x = \varphi(y).$$

According to the construction of the x_n and x there also increases the number of the first digits which coincide for the coordinates $\xi_i^{(n)}$ and ξ_i of the points x_n and x, respectively, without bound. Consequently, $\xi_i^{(n)} \to \xi_i$ as $n \to \infty$ and $i = 1, 2, \ldots$. According to the definition of convergence in s, this means that $x_n \to x$ and this proves the continuity of the mapping $x = \varphi(y)$.

In addition, we shall prove that the set P is closed. Let the sequence $\{y_n\}$, $y_n \in P$, converge to y. The y_n and y have ternary expansions consisting only of the digits 0 and 1. Because of the continuity of the operator φ, $x_n \to x$, where $x_n = \varphi(y_n)$ and $x = \varphi(y)$. According to the definition with $y_n \in P$, $x_n \in N$ and because N is closed, $x \in N$ also. Then, however, $y \in P$ also and the lemma is proved.

THEOREM 1. *Every separable Banach space E is isometrically isomorphic to a subspace of* $C[0, 1]$.

Proof. We designate the sphere $\|f\| \leqq 1$ in $\bar{E}$ by S and use the weak convergence in S. Then, according to the remark on p. 119, §24, S is a set compact in itself. Let the elements of the sequence $\{a_n\}$ be a denumerable set everywhere dense in the unit sphere $\|x\| \leqq 1$ of the space E. It is sufficient for the weak convergence of a sequence $\{f^{(k)}\}$ of functionals of S to the functional $f \in S$ that

$$f^{(k)}(a_n) \to f(a_n),\ n = 1, 2, 3, \ldots$$

as $k \to \infty$. If we assign to each $f \in S$ an element of the space s with coordinates $\{f(a_i)\}$, then the convergence of a sequence of functionals in the set S becomes equivalent to the convergence of the sequence of the associated elements to the element in s associated with the limit. Since, furthermore,

$$|f(a_i)| \leqq \|f\|\,\|a_i\| \leqq 1,$$

the elements of the space s associated with the elements $f \in S$ form a set N satisfying the hypothesis of the lemma proved above.[25] Thus N, and consequently S, is a continuous image of a closed subset P of $[0, 1]$, and to every $t \in P$ there corresponds an $f_t \in S$. The collection of all f_t coincides with S and $f_{t_n} \overset{wk}{\to} f_t$ as $t_n \to t$.

We select an $x \in E$. According to the definition of weak convergence of functionals,

$$f_{t_n}(x) \to f_t(x) \text{ as } t_n \to t.$$

---

[25] That N is closed follows from the fact that S is compact in itself.

Therefore, if x is fixed, $f_t(x)$ is a continuous function of $t \in P$. We will designate it by $\varphi_x(t)$:

$$f_t(t) = \varphi_x(t). \tag{4}$$

We extend the function $\varphi_x(t)$ defined on P linearly and continously onto the intervals adjacent to P. Thus $\varphi_x(t)$ is continuous on the closed interval [0, 1], i.e., it belongs to $C[0,1]$. By definition of the norm in $C[0,1]$,

$$\|\varphi_x\|_C = \max_{t \in [0,1]} |\varphi_x(t)|.$$

By the linearity of $\varphi_x(t)$ on the intervals adjacent to P, however, the maximum of $\varphi_x(t)$ on [0, 1] coincides with the maximum of $\varphi_x(t)$ on P. Therefore

$$\|\varphi_x\|_C = \max_{t \in P} |\varphi_x(t)|.$$

On the other hand, for $t \in P$, according to (4),

$$|\varphi_x(t)| = |f_t(x)| \leqq \|f_t\| \|x\| \leqq \|x\|_E$$

and consequently

$$\max_{t \in P} |\varphi_x(t)| \leqq \|x\|_E. \tag{5}$$

Moreover, for a given x, we can construct an F with norm 1 for which

$$|F(x)| = \|x\|_E.$$

Since $F \in S$, there exists a $t_0 \in P$ with

$$f_{t_0} = F.$$

Consequently,

$$|f_{t_0}(x)| = \|x\|_E,$$

i.e.,

$$|\varphi_x(t_0)| = \|x\|_E$$

and therefore

$$\max_{t \in P} |\varphi_x(t)| \geqq \|x\|_E. \tag{6}$$

From (5) and (6) it follows that

$$\|\varphi_x\|_C = \max_{t \in P} |\varphi_x(t)| = \|x\|_E. \tag{7}$$

Let $x \in E$ be associated with $\varphi_x(t)$ and $y \in E$ with $\varphi_y(t)$. It is apparent from the construction of the functions that $x+y$ and λx correspond to the functions $\varphi_x(t)+\varphi_y(t)$ and $\lambda\varphi_x(t)$, respectively. We consequently have an isomorphic mapping of E onto a subspace of $C[0,1]$. The isomorphism implies that to the element $x-y$ there is assigned the function $\varphi_x(t)-\varphi_y(t)$, and we therefore obtain by (7), the relation

$$\|x-y\|_E = \|\varphi_x - \varphi_y\|_C,$$

i.e., the mapping of E onto a subspace of $C[0,1]$ is not only isomorphic but also isometric. This completes the proof of the theorem.

THEOREM 2 (FRÉCHET). *Every separable metric space X is isometric to a subspace of a separable Banach space.*

Proof. Let $M = \{x_0, x_1, x_2, \ldots, x_n, \ldots\}$ be a countable set everywhere dense in X. Let there correspond to every $x \in X$ a point $y = \{\eta_i\}$ of the space m with

$$\eta_i = \rho(x, x_i) - \rho(x_0, x_i), \quad i = 1, 2, 3, \ldots .$$

By the triangle inequality,

$$|\eta_i| = |\rho(x, x_i) - \rho(x_0, x_i)| \leqq \rho(x, x_0),$$

and therefore the $\{\eta_i\}$ form a bounded sequence, i.e., y is a point of the space m. Under our mapping, let there correspond to the elements $x \in X$ and $x' \in X$ the elements $y = \{\eta_i\}$ and $y' = \{\eta_i'\}$.

$$\|y - y'\| = \sup_i |\eta_i - \eta_i'|. = \sup_i |[\rho(x, x_i) - \rho(x_0, x_i)] - [\rho(x', x_i) - \rho(x_0, x_i)]|$$

$$= \sup_i |\rho(x, x_i) - \rho(x', x_i)| \leqq \rho(x, x'). \tag{8}$$

Now let ε be an arbitrary positive number smaller than $\rho(x, x')$. There exists an x_n of our countable everywhere dense set M such that $\rho(x, x_n) \leqq \frac{\varepsilon}{2}$. Consequently

$$\rho(x', x_n) \geqq \rho(x', x) - \rho(x_n, x) \geqq \rho(x', x) - \frac{\varepsilon}{2} > 0$$

and therefore

$$|\eta_n - \eta_n'| = |\rho(x_n, x) - \rho(x_n, x')| \geqq \rho(x', x_n) - \frac{\varepsilon}{2}$$

$$\geqq \rho(x', x) - \frac{\varepsilon}{2} - \frac{\varepsilon}{2} = \rho(x, x') - \varepsilon.$$

This implies

$$\|y - y'\| \geqq \rho(x, x') - \varepsilon. \tag{9}$$

Since $\varepsilon > 0$ was arbitrary, (9) implies

$$\|y - y'\| \geqq \rho(x, x'), \tag{10}$$

and it results from (8) and (10) that

$$\|y - y'\| = \rho(x, x'). \tag{11}$$

Thus the distance of two points x and x' in X is equal to the distance of the corresponding points y and y' in m. Hence X is isometric with a certain $L \subset M$. With E being the subspace of m generated by the elements of the set L and since m is separable, E is a separable Banach space. X is isometric with a subspace of it and this completes the proof.

THEOREM 3 (BANACH–MAZUR). *Every separable metric space is isometric with a subspace of* $C[0, 1]$.

The proof of this theorem follows immediately from the theorems 1 and 2.

In conclusion, we will mention another property of the universality of $C[0, 1]$. In connection with several questions regarding the problem of moments and the theory of linear integral equations, M. G. Krein established the cone-theory in Banach spaces [10,], [36].

In a space E, a closed convex set $K \subset E$ is called a *cone* if for every $x \neq 0$, $x \in K$ and $\lambda \geqq 0$, it follows that $\lambda x \in K$, and for $\lambda < 0$, $\lambda x \notin K$, and if for $x, y \in K$, $x + y \in K$ also. This set K is called *normal* if for any two elements x, $y \in K$ with $\|x\| = \|y\| = 1$, there holds $\|x + y\| > \delta$, where δ is a fixed non-negative number. Thus, for instance, the collection of non-negative functions in $C[0, 1]$ is a normal cone. There holds the following theorem of M. G. Krein:

Let K be a normal cone of a separable space E. There exists a linear one-to-one mapping of E into a subspace of $C[0, 1]$ *under which the elements of K and only those are mapped into the non-negative functions.*

If E is not separable, then there holds an analogous theorem. We substitute the space C of the functions continuous on a certain bicompact set for $C[0, 1]$.

CHAPTER IV

COMPLETELY CONTINUOUS OPERATORS

§27. COMPLETELY CONTINUOUS OPERATORS

DEFINITION. An operator A defined on a normed linear space E whose range is in the normed linear space E_1 is called *completely continuous* if it maps every bounded set of E into a compact set in E_1.

We saw above that every linear operator A transforms a bounded set into a bounded one. Moreover, one can easily see that every linear operator A takes a compact set into a compact one. Let M be a compact set in E and M_1 the image of this set in E_1 under the operator A (i.e., $M_1 = \{Ax\}$, where x runs through the set M). We choose an arbitrary sequence $\{y_n\}, y_n \in M_1$. Then let $x_n \in M$ be a pre-image of y_n. We now investigate the sequence $\{x_n\}, x_n \in M$. Because M is compact, we can select from the x_n, a subsequence $\{x_{n_i}\}$ converging to $x_0 \in E$. By the continuity of A, the sequence $\{y_{n_i}\}$ tends to $y_0 = Ax_0, y_0 \in E_1$. In this manner, we can select from an arbitrary sequence $\{y_n\}, y_n \in M$, a convergent subsequence, i.e., M_1 is compact.

In general, complete continuity is stronger than simple continuity, which we require for every linear operator. For instance, the identity operator is not completely continuous in an infinite dimensional space E because it maps the unit sphere into itself, and this is not compact.

Example. Let $E = E_1 = C[0, 1]$ and

$$y(s) = \int_0^1 K(s, t)\, x(t)\, dt = Ax,$$

where $K(s, t)$ is a kernel continuous on the square $0 \leqq s, t \leqq 1$. We shall prove that the operator A is completely continuous. Let $\{x(s)\}$ be a bounded set of functions of $C[0, 1]$, $\|x\| \leqq K$. Obviously, the function

$$y(s) = \int_0^1 K(s, t)\, x(t)\, dt$$

is uniformly bounded, where $x(t)$ is from the set mentioned. For, if $M = \max_{s,t} |K(s, t)|$ then $|y(s)| \leqq MK$. The functions $y(s)$ are also equicontinuous: For an arbitrary $\varepsilon > 0$, there exists, because of the uniform continuity of the kernel $K(s, t)$ a δ such that

$$|K(s_1, t) - K(s_2, t)| < \frac{\varepsilon}{K}$$

for $|s_1 - s_2| < \delta$ and every $t \in [0, 1]$. Then

$$|y(s_1) - y(s_2)| \leqq \int_0^1 |K(s_1, t) - K(s_2, t)|\, |x(t)|\, dt < \varepsilon$$

for all functions $y(s)$ under consideration, if $|s_1 - s_2| < \delta$, q.e.d. By the theorem of Arzelà, the set $\{y(s)\}$ with the metric of the space $C[0, 1]$ is compact, and this proves the complete continuity of the operator A.

THEOREM 1. *A completely continuous operator A maps a weakly convergent sequence into a strongly convergent one.*

Let the sequence $\{x_n\}$ converge weakly to x_0 and $Ax_n = y_n$, $Ax_0 = y_0$. We assume that y_n does not converge strongly to y_0. Then there exists for an $\varepsilon_0 > 0$, an increasing sequence of indices $n_1, n_2, \ldots, n_k, \ldots$ such that $\| y_{n_k} - y_0 \| \geqq \varepsilon_0$. We consider the sequence $\{x_{n_k}\}$. By hypothesis, the sequence $\{x_n\}$ converges weakly and is therefore bounded. Thus $\{x_{n_k}\}$ is also a bounded sequence. Since A is completely continuous, $\{y_{n_k}\}$ is compact and therefore contains a subsequence $\{y_{n_{k_j}}\}$, converging strongly to a certain y_0'. A fortiori, $y_{n_{k_j}} \xrightarrow{wk} y_0'$, i.e.,

$$Ax_{n_{k_j}} \xrightarrow{wk} y_0'. \tag{1}$$

Since $x_{n_{k_j}} \xrightarrow{wk} x_0$, and according to Theorem 3, §25,

$$Ax_{n_{k_j}} \xrightarrow{wk} Ax_0 = y_0. \tag{2}$$

From (1) and (2), there follows $y_0' = y_0$. Thus, on one hand $y_{n_{kj}} \to y_0$ and, on the other hand, $\| y_{n_{k_j}} - y_0 \| \geqq \varepsilon_0$. This contradiction implies the assertion.

THEOREM 2. *If a sequence $\{A_n\}$ of completely continuous operators mapping E into the complete space E_1 converges uniformly to the operator A,* i.e., *if* $\| A_n - A \| \to 0$, *then A is also a completely continuous operator.*

Proof. It has to be shown that A maps every bounded set of E into a compact set in E_1. We consider an arbitrary bounded sequence $\{x_n\}, x_n \in E$. Since A_1 is completely continuous, $\{A_1 x_n\}$ is a compact set. Therefore, there exists a subsequence $\{x_n^{(1)}\} \subset \{x_n\}$ such that $\{A_1 x_n^{(1)}\}$ converges. Since A_2 is completely continuous, the sequence $\{A_2 x_n^{(1)}\}$ is compact. Therefore, there exists a subsequence $\{x_n^{(2)}\} \subset \{x_n^{(1)}\}$ such that $\{A_1 x_n^{(2)}\}$ and $\{A_2 x_n^{(2)}\}$ converge.

We obtain the sequences $\{x_n^{(k)}\}$ by continuation of this process. Then the sequences $\{A_i x_n^{(k)}\}$, $i = 1, 2, \ldots, k$, are convergent. Considering the diagonal sequence

$$x_1^{(1)}, x_2^{(2)}, \ldots, x_n^{(n)}, \ldots,$$

we easily see that the sequences $\{A_k x_n^{(n)}\}$ converge for arbitrary k. We now prove that $\{Ax_n^{(n)}\}$ is also convergent. We estimate:

$$\begin{aligned} \|Ax_n^{(n)} - Ax_m^{(m)}\| &\leqq \|Ax_n^{(n)} - A_k x_n^{(n)}\| + \|A_k x_n^{(n)} - A_k x_m^{(m)}\| + \|A_k x_m^{(m)} - Ax_m^{(m)}\| \\ &\leqq \|A - A_k\| (\|x_n^{(n)}\| + \|x_m^{(m)}\|) + \|A_k x_n^{(n)} - A_k x_m^{(m)}\|. \end{aligned}$$

For a given $\varepsilon > 0$, we choose k so large that $\| A - A \| < \dfrac{\varepsilon}{4M}$ where $M = \sup\limits_n \|x_n\|$. Then we determine n_0 so that

$$\| A_k x_n^{(n)} - A_k x_m^{(m)} \| < \frac{\varepsilon}{2}$$

for $n, m \geqq n_0$ and thus obtain

$$\| Ax_n^{(n)} - Ax_m^{(m)} \| < \varepsilon$$

for $n, m \geqq n_0$. Hence we have selected a subsequence $\{x_n^{(n)}\}$ of $\{x_n\}$ such that $\{Ax_n^{(n)}\}$ converges in itself, and consequently to $y_0 \in E_1$. Since $\{x_n\}$ was an arbitrary bounded sequence, the complete continuity of the operator A is established.

Example. If $E = E_1 = L_2(0, 1)$, then the operator

$$Ax = y(s) = \int_0^1 K(s,t)\,x(t)\,dt \text{ where } \int_0^1\int_0^1 K^2(s,t)\,ds\,dt < \infty$$

is completely continuous.

We first assume that $K(s, t)$ is a continuous kernel. Let G be a bounded set of $L_2(0, 1)$ and let

$$\int_0^1 x^2(t)\,dt \leqq R,$$

for all $x(t) \in G$. Now we consider the set of functions

$$y(s) = \int_0^1 K(s,t)\,x(t)\,dt, \quad x(t) \in G$$

and we shall prove that the $y(s)$ are uniformly bounded and equicontinuous. This implies the compactness of the set $y(s)$ in the sense of uniform convergence and, a fortiori, in the sense of convergence in the mean.

We have

$$|y(s)| = \left|\int_0^1 K(s,t)\,x(t)\,dt\right| \leqq \left(\int_0^1 |K(s,t)|^2\,dt\right)^{\frac{1}{2}} \left(\int_0^1 x^2(t)\,dt\right)^{\frac{1}{2}} \leqq MR^{\frac{1}{2}}$$

where $M = \max|K(s,t)|$. Consequently, the functions $y(s)$ are uniformly bounded. Moreover,

$$|y(s_1) - y(s_2)| \leqq \left(\int_0^1 [K(s_1,t) - K(s_2,t)]^2 dt\right)^{\frac{1}{2}} \left(\int_0^1 x^2(t)\,dt\right)^{\frac{1}{2}} < \varepsilon$$

for $|s_1 - s_2| < \delta$, where δ is chosen in such a manner that

$$|K(s_1,t) - K(s_2,t)| < \frac{\varepsilon}{\sqrt{R}}$$

for $|s_1 - s_2| < \delta$. The estimate $|y(s_1) - y(s_2)| < \varepsilon$ depends on neither the location of s_1, s_2 in $[0, 1]$ nor the choice of $y(s)$. Consequently, the $y(s)$ are equicontinuous.

Therefore, the operator A is completely continuous in the case of a continuous kernel. Now let $K(s, t)$ be an arbitrary square integrable kernel. We select a sequence $\{K_n(s,t)\}$ of continuous kernels which converges in the mean to $K(s, t)$, i.e., a sequence such that

$$\int_0^1\int_0^1 [K(s,t) - K_n(s,t)]^2\,ds\,dt \to 0 \text{ as } n \to \infty.$$

We put

$$\int_0^1 K_n(s,t)\,x(t)\,dt = A_n x.$$

Now
$$\|Ax - A_n x\| = \left(\int_0^1 \left[\int_0^1 K(s,t)\,x(t)\,dt - \int_0^1 K_n(s,t)\,x(t)\,dt\right]^2 ds\right)^{\frac{1}{2}}$$
$$= \left(\int_0^1 \left[\int_0^1 (K(s,t) - K_n(s,t))\,x(t)\,dt\right]^2 ds\right)^{\frac{1}{2}}$$
$$\leqq \left(\int_0^1 \left[\int_0^1 (K(s,t) - K_n(s,t))^2 dt \int_0^1 x^2(t)\,dt\right] ds\right)^{\frac{1}{2}}$$
$$= \left(\int_0^1\int_0^1 (K(s,t) - K_n(s,t))^2\,ds\,dt\right)^{\frac{1}{2}} \|x\|.$$

Consequently

$$\|A - A_n\| \leqq \left(\int_0^1\int_0^1 (K(s,t) - K_n(s,t))^2\,ds\,dt\right)^{\frac{1}{2}},$$

which implies that $\|A - A_n\| \to 0$ as $n \to \infty$. Since all the A_n are completely continuous, A is also completely continuous by Theorem 2.

THEOREM 3. *The range of a completely continuous operator is separable.*

Proof. Let G_n be a continuous image of the sphere $\|x\| \leqq n$. Since A is completely continuous, G_n is compact and therefore a separable set (viz p. 38). Let T_n be a denumerable set everywhere dense in G_n. Because $G = \sum_{n=1}^{\infty} G_n$ is the range of A, $\sum_{n=1}^{\infty} T_n$ becomes a denumerable set everywhere dense in G, q.e.d.

THEOREM 4. *If A is a completely continuous operator mapping E into $\bar{E}_1$, then the adjoint operator A^* mapping $\bar{E}_1$ into $\bar{E}$ is completely continuous.*

Proof. Let $\{\varphi_n\}$ be a bounded sequence of functionals from $\bar{E}_1$, $\|\varphi_n\| \leqq M (n = 1, 2, \ldots)$ and let T be a denumerable set everywhere dense in the range G of A. By means of the diagonal process, we can select a subsequence $\{\varphi n_i\}$ which converges for all $y \in T$. Because the norms of the $\{\varphi_{n_i}\}$ are bounded, this subsequence will converge even for all $y \in G$ (see Theorem 2, §24). We put

$$\lim_{i\to\infty} \varphi_{n_i}(Ax) = \lim_{i\to\infty} f_{n_i}(x) \to f(x).$$

Now we show that

$$\|f_{n_i} - f\| \to 0$$

as $i \to \infty$.

Let $x_i \in E$ with

$$\|x_i\| = 1 \text{ and } |f(x_i) - f_{n_i}(x_i)| \geqq \tfrac{1}{2} \|f - f_{n_i}\|. \tag{3}$$

If $\|f_{n_i} - f\|$ did not tend to 0, there would exist an $\eta > 0$ such that $\|f - f_{n_i}\| > \eta$ for $i = 1, 2, \ldots$. Then we would obtain from (3), after having put $\varphi_{n_i} = \varphi_i'$ for short,

$$|\varphi'(Ax_i) - \lim_j \varphi_j'(Ax_i)| > \frac{\eta}{2}. \tag{4}$$

By virtue of $\|x_i\| = 1$, there exists a sequence of indices k_i, such that $\lim_{i\to\infty} Ax_{k_i} = y_0$ exists. Then there exists for every $\varepsilon > 0$ an N such that the inequalities

$$\|y_0 - Ax_{k_i}\| < \varepsilon \text{ and } |\varphi_{k_i}'(y_0) - \lim_j \varphi_{k_j}'(y_0)| < \varepsilon$$

are fulfilled for every $i \geqq N$ and consequently

$$|\varphi_{k_j}'(Ax_{k_i}) - \lim_j \varphi_{k_j}'(Ax_{k_i})| \leqq |\varphi_{k_i}'(Ax_{k_i} - y_0)| + |\varphi_{k_i}(y_0) - \lim_j \varphi_{k_j}'(y_0)|$$
$$+ |\lim_j \varphi_{k_j}'(y_0 - Ax_{k_i})| \leqq M\varepsilon + \varepsilon + M\varepsilon.$$

However, by (4), this is impossible for sufficiently small ε.

The convergence $\|f_{n_i} - f\| \to 0$ follows from this contradiction. Because $f_{n_i} = A^*\varphi_{n_i}$, $\|A^*\varphi_{n_i} - f\| \to 0$, i.e., there was selected a convergent subsequence of $\{A^*\varphi_n\}$. This proves the theorem.

§28. BANACH SPACES WITH BASIS

DEFINITION. Let E be an infinite dimensional Banach space. The sequence $e_1, e_2, \ldots, e_n, \ldots$ from E is called a *basis* for E, if every $x \in E$ is representable uniquely in the form $x = \sum_{i=1}^{\infty} \xi_i e_i$ with real ξ_i $(i = 1, 2, \ldots)$. The uniqueness of the representation is obviously equivalent to the requirement that $\sum_{i=1}^{\infty} \xi_i e_i = \theta$ if and only if $\xi_i = 0$ for all i.

Examples. 1. Let $E = l_p$. Then the elements

$$e_1 = \{1, 0, 0, \ldots\},$$
$$e_2 = \{0, 1, 0, \ldots\},$$
$$\cdots\cdots\cdots$$

form a basis for l_p. For, when representing linear functionals in l_p we saw that for every $x \in l_p$, $x = \sum_{i=1}^{\infty} \xi_i e_i$ uniquely.

2. Let $E = C[0, 1]$. We consider the sequence

$$t,\ 1 - t;\ u_{00}(t);\ u_{10}(t);\ u_{11}(t);\ u_{20}(t),\ u_{21}(t),\ u_{22}(t),\ u_{23}(t);\ \ldots, \tag{1}$$

of elements of $C[0, 1]$, where the $u_{kl}(t)$ $(k = 1, 2, 3, \ldots;\ 0 \leqq l < 2^k)$ are defined in the following manner: For t in

$$\left(\frac{l}{2^k}, \frac{l+1}{2^k}\right),$$

$u_{kl}(t)$ has the form of an isosceles triangle with altitude 1 (in Fig. 2, the function $u_{22}(t)$ is drawn), outside this interval $u_{kl}(t) = 0$. Every function $x(t) \in C[0, 1]$ is representable by a series

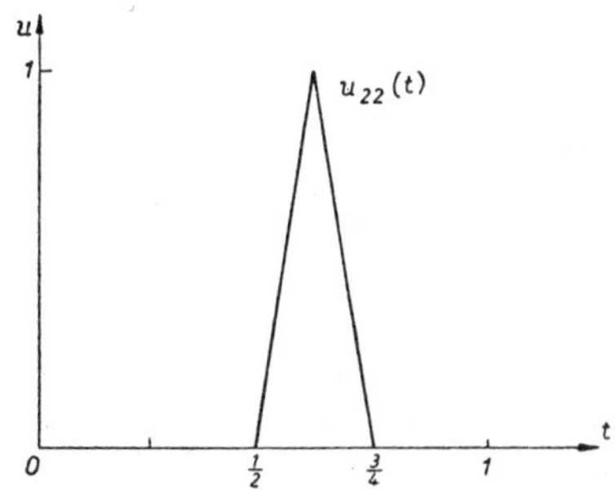

Fig. 2

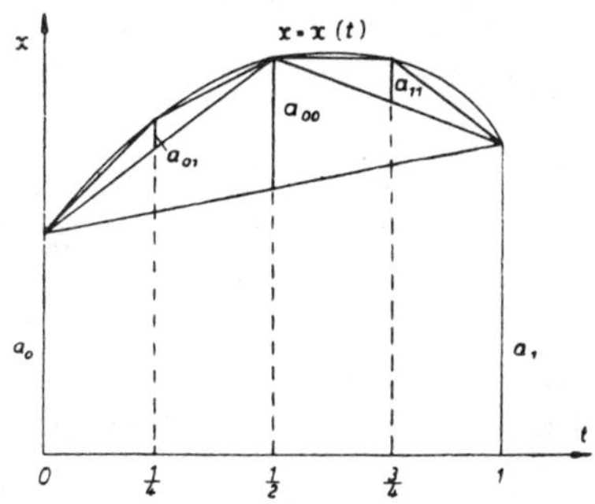

Fig. 3

$$x(t) = a_0(t) + a_1(1 - t) + \sum_{k=0}^{\infty} \sum_{l=0}^{2k-1} a_{kl} u_{kl}(t) \tag{2}$$

with $a_0 = x(1)$, $a_1 = x(0)$. The coefficients a_{kl} can be determined uniquely by the geometrical construction given in Fig. 3. The image of the partial sum

$$a_0 t + a_1(1 - t) + \sum_{k=0}^{s-1} \sum_{l=0}^{2k-1} a_{kl} u_{kl}(t)$$

of the series (2) is obviously a polygon with $2^s + 1$ vertices which lie on the curve $x = x(t)$ with equidistant abscissas. The collection of functions (1) forms a basis for $C[0, 1]$.

If E has a basis, then E is obviously separable. In a space with basis, the set of linear combinations $\sum_{i=1}^{n} r_i e_i$ with rational coefficients r_i is a denumerable everywhere dense set. There have been constructed bases for all known separable Banach spaces. Their existence for an arbitrary separable space of type B, however, has not been proved yet.

Let E be a Banach space with the basis $e_1, e_2, \ldots, e_n, \ldots$. We consider the linear system E_1 whose elements are all number sequences $y = \{\eta_1, \eta_2, \ldots, \eta_n, \ldots\}$ for which the series $\sum_{i=1}^{\infty} \eta_i e_i$ converges.

We introduce a norm in E_1 by

$$\| y \| = \sup_n \left\| \sum_{i=1}^{n} \eta_i e_i \right\|.$$

Now we shall prove that E_1 is a space of type B.

One shows immediately that the axioms for the norm are fulfilled.

Now let there be given a convergent sequence

$$\{y_k\}, y_k = \{\eta_1^{(k)}, \eta_2^{(k)}, \ldots\} \in E_1.$$

Then, for every $\varepsilon > 0$,

$$\| y_m - y_k \| = \sup \left\| \sum_{i=1}^{n} (\eta_i^{(m)} - \eta_i^{(k)}) e_i \right\| < \varepsilon \text{ for } m, k \geqq m_0(\varepsilon)$$

and consequently

$$\left\| \sum_{i=1}^{n} (\eta_i^{(m)} - \eta_i^{(k)}) e_i \right\| < \varepsilon \tag{3}$$

for $m, k \geqq m_0$ and arbitrary n. Thus

$$\| (\eta_n^{(m)} - \eta_n^{(k)}) e_n \| \leqq \left\| \sum_{i=1}^{n} (\eta_i^{(m)} - \eta_i^{(k)}) e_i \right\| + \left\| \sum_{i=1}^{n-1} (\eta_i^{(m)} - \eta_i^{(k)}) e_i \right\| < 2\varepsilon$$

and therefore for every n,

$$| \eta_n^{(m)} - \eta_n^{(k)} | < \frac{2\varepsilon}{\| e_n \|}, \text{ for } m, k \geqq m_0.$$

Hence the number sequence $\{\eta_n^{(m)}\}, m = 1, 2, \ldots$ converges to an $\eta_n^{(0)}$ and this is true for every n.

In inequality (3), we now pass to the limit as $k \to \infty$ and we obtain for $m \geqq m_0$ and every n,

$$\left\| \sum_{i=1}^{n} (\eta_i^{(m)} - \eta_i^{(0)}) e_i \right\| \leqq \varepsilon. \tag{4}$$

Putting

$$s_n^{(m)} = \sum_{i=1}^{n} \eta_i^{(m)} e_i, \quad s_n^{(0)} = \sum_{i=1}^{n} \eta_i^{(0)} e_i$$

and taking into account inequality (4), we obtain

$$\| s_{n+p}^{(0)} - s_n^{(0)} \| \leqq \| s_{n+p}^{(m)} - s_n^{(m)} \| + 2\varepsilon$$

for $m \geqq m_0$ and arbitrary n and p.

Let $\omega > 0$ be given. We first choose ε so that $2\varepsilon < \frac{\omega}{2}$. Then, there is determined an n_0 for a fixed $m \geqq m_0$ so that

$$\| s_{n+p}^{(m)} - s_n^{(m)} \| < \frac{\omega}{2}$$

for every $n \geqq n_0$ and every $p > 0$ (this is always possible because of the convergence of the series $\sum_{i=1}^{\infty} \eta_i^{(m)} e_i$). Then

$$\| s_{n+p}^{(0)} - s_n^{(0)} \| < \omega$$

for $n \geqq n_0$ and arbitrary $p > 0$, i.e., the series

$$\sum_{i=1}^{\infty} \eta_i^{(0)} e_i$$

converges, and consequently $y_0 = \{\eta_1^{(0)}, \eta_2^{(0)}, \ldots, \eta_n^{(0)}, \ldots\} \in E_1$. Besides this, (4) implies that for $m \geqq m_0$,

$$\sup_n \left\| \sum_{i=1}^{n} (\eta_i^{(m)} - \eta_i^{(0)}) e_i \right\| \leqq \varepsilon$$

i.e.,

$$\| y_0 - y_m \| < \varepsilon$$

for $m \geqq m_0$. This proves the completeness of E_1, hence E_1 is a Banach space.

Obviously, there is associated with every $x = \sum_{i=1}^{\infty} \xi_i e_i \in E$ exactly one $y_x = \{\xi_i\} \in E_1$. Conversely, there corresponds to every $y = \{\eta_i\} \in E_1$ exactly one $x_y \in E$:

$$x_y = \sum_{i=1}^{\infty} \eta_i e_i.$$

In this manner, we have defined an operator A_0 mapping E_1 onto E biuniquely, i.e., $x = A_0 y$. The additivity of A_0 can easily be seen. A_0 is also bounded, for

$$\| A_0 y \| = \| x \| = \left\| \sum_{i=1}^{\infty} \eta_i e_i \right\| \leqq \sup_n \left\| \sum_{i=1}^{n} \eta_i e_i \right\| = \| y \|.$$

Thus we have a linear operator A_0 mapping E_1 onto E biuniquely. By the Banach theorem, there exists an inverse operator $y = A_0^{-1} x$ which is also linear.

Let

$$x = \sum_{i=1}^{\infty} \xi_i e_i$$

be an arbitrary element of E. We define a functional f_k by putting $f_k(x) = \xi_k$. Obviously, f_k is additive. Furthermore,

$$|f_k(x)| = |\xi_k| = \frac{|\xi_k| \, \|e_k\|}{\|e_k\|} = \frac{\| \xi_k e_k \|}{\| e_k \|} = \frac{\left\| \sum_{i=1}^{k} \xi_i e_i - \sum_{i=1}^{k-1} \xi_i e_i \right\|}{\| e_k \|}$$

$$\leqq 2 \sup_n \left\| \sum_{i=1}^{n} \xi_i e_i \right\| \frac{1}{\| e_k \|} = \frac{2 \| y \|}{\| e_k \|} = \frac{2 \| A_0^{-1} x \|}{\| e_k \|} \leqq \frac{2 \, \| A_0^{-1} \|}{\| e_k \|} \| x \|.$$

This implies the boundedness of f_k, for

$$\| f_k \| \leqq 2 \frac{\| A_0^{-1} \|}{\| e_k \|}.$$

Constructing an f_k for every k, we obtain an infinite sequence of functionals $f_1, f_2, \ldots, f_n, \ldots$ of $\bar{E}$. Now, every $x \in E$ can be written in the form

$$x = \sum_{i=1}^{\infty} f_i(x)\, e_i.$$

In particular, we put $x = e_j$. Then

$$\xi_i = \begin{cases} 1 \text{ if } i = j, \\ 0 \text{ if } i \neq j, \end{cases} \tag{5}$$

$$f_i(e_j) = \begin{cases} 1 \text{ if } i = j, \\ 0 \text{ if } i \neq j. \end{cases}$$

We thus recognize that the elements $\{e_i\}$ and the functionals $\{f_i\}$ are biorthogonal sequences.

Now let F be an arbitrary functional of $\bar{E}$. Because

$$x = \sum_{i=1}^{\infty} f_i(x)\, e_i = \lim_n \sum_{i=1}^{n} f_i(x)\, e_i,$$

$$F(x) = \lim_n \sum_{i=1}^{n} F[f_i(x)\, e_i] = \lim_n \sum_{i=1}^{n} f_i(x)\, F(e_i) = \sum_{i=1}^{\infty} f_i(x)\, F(e_i).$$

We put $F(e_i) = c_i$ and obtain for every linear $F \in \bar{E}$,

$$F(x) = \sum_{i=1}^{\infty} c_i f_i(x)$$

or

$$F = \sum_{i=1}^{\infty} c_i f_i. \tag{6}$$

Evidently the representation (6) is unique. (6) is convergent for every $x \in E$, i.e., the sequence of partial sums is weakly convergent, q.e.d.

Let x be an arbitrary element of E. Then

$$x = \sum_{i=1}^{\infty} \xi_i e_i = \sum_{i=1}^{n} \xi_i e_i + \sum_{i=n+1}^{\infty} \xi_i e_i = S_n x + R_n x.$$

S_n and R_n are clearly linear operators for a fixed n. Their additivity is obvious and the boundedness follows from the inequalities

$$\| S_n x \| \leqq \sup_n \left\| \sum_{i=1}^{n} \xi_i e_i \right\| = \| A_0^{-1} x \| \leqq \| A_0^{-1} \| \| x \|$$

and

$$\| R_n x \| \leqq 2 \| A_0^{-1} \| \| x \|.$$

CRITERION FOR COMPACTNESS

THEOREM. *The set M of a Banach space E with basis is compact if and only if M is bounded and if there exists for every $\varepsilon > 0$, an n_0 such that $\| R_n x \| < \varepsilon$ for $n \geqq n_0$ and every $x \in M$.*

Necessity. The boundedness of M follows from Corollary 3, Theorem 3, §9. We prove the validity of the second condition.

To do this, we take an $\varepsilon' > 0$ and construct a finite ε'-net $[x_1, x_2, \dots, x_k]$ for M. For an arbitrary $x \in M$, there is an x_i which belongs to the ε'-net so that $\| x - x_i \| < \varepsilon'$. Thus

$$\| R_n x \| = \| x - S_n x \| \leqq \| x - x_i \| + \| x_i - S_n x \|$$
$$\leqq \| x - x_i \| + \| S_n x_i - S_n x \| + \| R_n x_i \| \leqq (1 + \| A_0^{-1} \|) \| x - x_i \| + \| R_n x_i \|$$
$$< (1 + \| A_0^{-1} \|) \varepsilon' + \| R_n x_i \|.$$

For every fixed $x, R_n x \to \theta$ as $n \to \infty$. Therefore, there exists an n_0 such that $\| R_n x_i \| < \varepsilon'$ for $n \geqq n_0 (i = 1, 2, \ldots, k)$. Hence

$$\| R_n x \| < (2 + \| A_0^{-1} \|) \varepsilon',$$

for $n \geqq n_0$. Now taking $\varepsilon' = \dfrac{\varepsilon}{2 + \| A_0^{-1} \|}$, we obtain the required inequality because n_0 does not depend on the choice of $x \in M$.

Sufficiency. We shall show that the hypotheses imply the existence of a finite ε-net for M. For a given ε, we choose an n_0 so that $\| R_{n_0} x \| < \dfrac{\varepsilon}{2}$ for all $x \in M$. Then we investigate the set M_{n_0} consisting of the elements $S_{n_0} x$ with $x \in M$. M_{n_0} can be considered as a set defined by $e_1, e_2, \ldots, e_{n_0}$ and lying in the n_0-dimensional space $E_{n_0} \subset E$. According to

$$\| S_{n_0} x \| \leqq \| A_0^{-1} \| \| x \|$$

and the boundedness of M, M_{n_0} is bounded and therefore compact. Therefore, there exists in E_{n_0} a finite $\dfrac{\varepsilon}{2}$-net for M_{n_0}. This net is obviously an ε-net for M, and so we have accomplished what we set out to do.

Example. The fundamental cube U in l_2 (see p. 35) is compact. Indeed, for an arbitrary $x \in U, \| R_n x \| \leqq \left(\sum_{k=n+1}^{\infty} \dfrac{1}{k^2} \right)^{\frac{1}{2}}$ which implies $\| R_n x \| \to 0$ uniformly in x.

Let A be an operator mapping E into E and let $x \in E$. Then

$$x = \lim_n x_n, \text{ if } x_n = \sum_{i=1}^{n} \xi_i e_i.$$

Consequently,

$$y = Ax = A(\lim_n x_n) = \lim_n Ax_n = \lim_n \sum_{i=1}^{n} \xi_i Ae_i.$$

Since Ae_i is again in E,

$$Ae_i = \sum_{k=1}^{\infty} a_{ki} e_k.$$

Therefore

$$y = Ax = \lim_n \sum_{i=1}^{n} \xi_i \left(\sum_{k=1}^{\infty} a_{ki} e_k \right) \tag{7}$$

and

$$y = \sum_{k=1}^{\infty} \eta_k e_k \tag{8}$$

because $y \in E$. Let $\{f_i\}$ be a sequence biorthogonal to $\{e_i\}$. We obtain from (7) and (8):

$$\eta_m = f_m(y) = f_m\left(\lim_n \sum_{i=1}^{n} \xi_i \left(\sum_{k=1}^{\infty} a_{ki} e_k \right) \right) = \lim_n f_m\left(\sum_{i=1}^{n} \xi_i \left(\sum_{k=1}^{\infty} a_{ki} e_k \right) \right) \tag{9}$$
$$= \lim_n \sum_{i=1}^{n} \xi_i f_m\left(\sum_{k=1}^{\infty} a_{ki} e_k \right) = \lim_n \sum_{i=1}^{n} \xi_i a_{mi} = \sum_{i=1}^{\infty} a_{mi} \xi_i.$$

Equality (9) shows that A determines uniquely an infinite matrix $\{a_{mi}\}$ and moreover, the components of $y = Ax$ are defined uniquely by those of x.

Now we consider the adjoint operator A^* which maps $\bar{E}$ into itself.

Let $A^*f = \varphi$, i.e., for every $x \in E$,

$$f(Ax) = \varphi(x).$$

Furthermore

$$f = \sum_{i=1}^{\infty} c_i f_i$$

and

$$\varphi = \sum_{i=1}^{\infty} d_i f_i.$$

In addition,

$$f(Ax) = f\left(\sum_{k=1}^{\infty}\left(\sum_{i=1}^{\infty} a_{ki}\xi_i\right) e_k\right) = \lim_n f\left(\sum_{k=1}^{n}\left(\sum_{i=1}^{\infty} a_{ki}\xi_i\right) e_k\right)$$

$$= \lim_n \sum_{k=1}^{n}\left(\sum_{i=1}^{\infty} a_{ki}\,\xi_i\right) f(e_k) = \lim_n \sum_{k=1}^{n}\left(\sum_{i=1}^{\infty} a_{ki}\xi_i\right) c_k$$

$$= \lim_n \sum_{i=1}^{\infty}\left(\sum_{k=1}^{n} a_{ki}c_k\right) \xi_i.$$

On the other hand

$$f(Ax) = \varphi(x) = \sum_{i=1}^{\infty} d_i f_i(x) = \sum_{i=1}^{\infty} d_i \xi_i.$$

It follows that

$$\sum_{i=1}^{\infty} d_i\xi_i = \lim_n \sum_{i=1}^{\infty}\left(\sum_{k=1}^{n} a_{ki}c_k\right) \xi_i. \tag{10}$$

Now let $x = e_m$, i.e., $\xi_m = 1$ and $\xi_i = 0$ for $i \neq m$. Then (10) implies

$$d_m = \lim_n \sum_{k=1}^{n} a_{km}c_k = \sum_{k=1}^{\infty} a_{km}c_k. \tag{11}$$

This equality shows that the transposed matrix corresponds to the adjoint operator. Such a representation of operators with their adjoints holds, for example, in l_2.

From the matric representation of operators, we readily obtain

1) $(A + B)^* = A^* + B^*$,
2) $(AB)^* = B^*A^*$ and
3) $(A^{-1})^* = (A^*)^{-1}$ *if* A^{-1} exists.

§29. LINEAR OPERATOR EQUATIONS WITH COMPLETELY CONTINUOUS OPERATORS IN A BANACH SPACE WITH BASIS

As F. Riess has shown, the theory of Fredholm integral equations carries over to operator equations with completely continuous operators. We restrict ourselves to equations in spaces with basis.

DECOMPOSITION OF A COMPLETELY CONTINUOUS OPERATOR. We consider the equation

$$Ax - x = y, \tag{1}$$

where x, y are elements of a Banach space E with basis and A a completely continuous operator mapping E into itself. Simultaneously, we consider the equation

$$A^*f - f = \varphi \tag{2}$$

with $f, \varphi \in \bar{E}$ and A^* the adjoint of A. As was shown earlier, A^* is completely continuous and $\| A^* \| = \| A \|$. Let S be the unit sphere of E and M the totality of all elements $y = Ax$ with $x \in S$. The complete continuity of A implies the compactness of M. We take an ε with $0 < \varepsilon < 1$. Then, according to the theorem of §28 (see p. 136), there exists an n such that $\| R_n y \| < \varepsilon$ for all $y \in M$. If n is fixed, we obtain

$$Ax = y = S_n y + R_n y = S_n(Ax) + R_n(Ax) = A_1 x + A_2 x,$$

where A_1 and A_2 are obviously linear operators. Putting $x = \sum_{i=1}^{\infty} \xi_i e_i, y = \sum_{i=1}^{\infty} \eta_i e_i$,

$$A_1 x = S_n(Ax) = \sum_{k=1}^{n} \eta_k e_k = \sum_{k=1}^{n} \left(\sum_{i=1}^{\infty} a_{ki} \xi_i \right) e_k.$$

We see from this that A_1 is finite dimensional, i.e., for every x, the element $A_1 x$ belongs to a finite dimensional subspace defined by the basis elements $e_1, e_2, \ldots, e_n$. Furthermore, $\sup_{x \in S} \| A_2 x \| = \sup_{y \in M} \| R_n y \| \leqq \varepsilon$. Therefore

$$\| A_2 \| \leqq \varepsilon.$$

Thus we have decomposed the operator A into a sum of two operators one of which is finite dimensional and the norm of the other is not greater than an arbitrarily small ε. We also say the completely continuous operators are almost finite dimensional.

We pass from the equality $A = A_1 + A_2$ to that with the adjoint operators: $A^* = A_1^* + A_2^*$. Here $\| A_2^* \| = \| A_2 \| \leqq \varepsilon$. Because the norms of A_2 and A_2^* were less than 1, there exist, according to Theorem 2 §20, the operators $(I - A_2)^{-1}$ and $(I - A_2^*)^{-1}$. By means of these, the equations (1) and (2) can be transformed into a simpler form.

REDUCTION OF OPERATOR EQUATIONS TO SYSTEMS OF LINEAR EQUATIONS. Having rewritten equation (1) in the form

$$A_1 x - (I - A_2) x = y \tag{3}$$

we introduce an element x_1 by

$$x_1 = (I - A_2) x$$

Then $x = (I - A_2)^{-1} x_1$. The correspondence between x and x_1 is biunique.

Equation (3) becomes

$$A_1 (I - A_2)^{-1} x_1 - x_1 = y. \tag{4}$$

Writing (2) in the form

$$A_1^* f - (I - A_2^*) f = \varphi \tag{5}$$

and applying the operator $(I - A_2^*)^{-1}$, we obtain

$$(I - A_2^*)^{-1} A_1^* f - f = (I - A_2^*)^{-1} \varphi. \tag{6}$$

We introduce the notation

$$A_1 (I - A_2)^{-1} = B.$$

Obviously B is a finite dimensional operator mapping E into a subspace defined by the basis elements $e_1, e_2, \ldots, e_n$.

Furthermore,

$$(I - A_2^*)^{-1} A_1^* = [(I - A_2)^*]^{-1} A_1^* = [(I - A_2)^{-1}]^* A_1^* = [A_1 (I - A_2)^{-1}]^* = B^*.$$

Finally designating $(I - A_2)^{-1}\varphi$ by φ_1, we arrive at

$$Bx_1 - x_1 = y, \tag{7}$$

$$B^*f - f = \varphi_1, \tag{8}$$

which is equivalent to (1) and (2). These new equations have, however, the advantage that the operator B is finite dimensional.

Let B correspond to the matrix (a_{ik}). Since B is finite dimensional, $a_{ik} = 0$ for $i > n$. Therefore, to every $x = \sum_{k=1}^{\infty} \xi_k e_k$, there is assigned a

$$z = \sum_{i=1}^{n} \zeta_i e_i$$

by means of B where $\zeta_i = \sum_{k=1}^{\infty} a_{ik}\,\xi_k \,(i = 1, 2, \ldots, n)$. To the operator B^* there corresponds the transposed matrix (a_{ki}).

We rewrite (7) by use of the matrix representation of B and the basis representation of the elements. Putting $x_1 = \sum_{i=1}^{\infty} \xi_i e_i$, $y = \sum_{i=1}^{\infty} \eta_i e_i$,

$$\sum_{i=1}^{n}\left(\sum_{k=1}^{\infty} a_{ik}\,\xi_k\right) e_i - \sum_{i=1}^{\infty} \xi_i e_i = \sum_{i=1}^{\infty} \eta_i e_i.$$

This equality holds only if the coefficients of the basis elements e_i are equal on both sides. Therefore

$$\sum_{k=1}^{\infty} a_{ik}\,\xi_k - \xi_i = \eta_i \text{ for } i = 1, 2, \ldots, n;$$

$$-\xi_i = \eta_i \text{ for } i = n+1, n+2, \ldots .$$

Thus we have found the ξ_i for $i > n$. In order to find the remaining ξ_i, we write the first system of equations in the form

$$\sum_{k=1}^{n} a_{ik}\,\xi_k + \sum_{k=n+1}^{\infty} a_{ik}\,\xi_k - \xi_i = \eta_i.$$

Replacing ξ_k by $-\eta_k$ in the second sum, we obtain

$$\sum_{k=1}^{n} a_{ik}\,\xi_k - \xi_i = \eta_i + \sum_{k=n+1}^{\infty} a_{ik}\eta_k \text{ for } i = 1, 2, \ldots, n. \tag{9}$$

Thus we have a system of n linear equations in n unknowns to determine the coefficients $\xi_1, \xi_2, \ldots, \xi_n$. Since the remaining coefficients of the development are known, (7) leads essentially to the system (9) of equations.

GENERALIZATION OF THE FREDHOLM THEOREM. Using the matrix of B^*, the equation (8) can be rewritten in the form

$$\sum_{i=1}^{\infty}\left(\sum_{k=1}^{\infty} a_{ki}\,c_k\right) f_i - \sum_{i=1}^{\infty} c_i f_i = \sum_{i=1}^{\infty} d_i f_i$$

where $\{f_i\}$ is a basis for $\bar{E}$, $f = \sum_{i=1}^{\infty} c_i f_i$ and $\varphi_1 = \sum_{i=1}^{\infty} d_i f_i$.

Again we arrive at

$$\sum_{k=1}^{n} a_{ki} c_k - c_i = d_i \text{ for } i = 1, 2, \ldots, n, n+1, n+2, \ldots .$$

We separate this system into

$$\sum_{k=1}^{n} a_{ki}c_k - c_i = d_i \text{ for } i = 1, 2, \ldots, n \tag{10}$$

and

$$\sum_{k=1}^{n} a_{ki}c_k - c_i = d_i \text{ for } i = n+1, n+2, \ldots . \tag{11}$$

One first solves (10) and then determines the remaining coefficients from (11). Thus equation (8) leads to (10).

So we have reduced solving the operator equations

$$Ax - x = y, \tag{1}$$

$$A^*f - f = \varphi \tag{2}$$

to solving the system of linear algebraic equations

$$\sum_{k=1}^{n} a_{ik}\xi_k - \xi_i = \eta_i + \sum_{k=n+1}^{\infty} a_{ik}\eta_k, \tag{9}$$

$$\sum_{k=1}^{n} a_{ki}c_k - c_i = d_i, \tag{10}$$

$$(i = 1, 2, \ldots, n) .$$

The systems (9) and (10) are transposes of each other. Therefore, if one of them is solvable for an arbitrary right side, so is the other and both solutions are determined uniquely. The homogeneous systems have only the trivial solution.

These considerations imply the

THEOREM. *If one of the equations* (1) *or* (2) *is solvable for an arbitrary right side, then this is true for the other, too. The solvability of these equations for arbitrary right sides implies the uniqueness and, in particular, the existence of the trivial solution only, for the homogeneous equations. If* $Ax - x = \theta$ *has a non-trivial solution, then* $A^*f - f = \bar{\theta}$ *(where* $\bar{\theta}$ *is the zero element in* $\bar{E}$*) also has a non-trivial solution. The number of linearly independent solutions is the same for both equations. In this case, the inhomogeneous equation* $Ax - x = y$ *has a solution only if* y *is orthogonal to all solutions* f *of the conjugate homogeneous equation* $A^*f - f = \bar{\theta}$.

In the same way, we can investigate equations depending on a parameter:

$$Ax - \lambda x = y,$$

$$A^*f - \lambda f = \varphi.$$

Remark. In a complex space, λ is to be replaced by $\bar{\lambda}$ in the second equation.

In conclusion, we prove that

For an arbitrary Banach space,

$$Ax = x, \tag{*}$$

with A *a completely continuous operator, has only a finite number of linearly independent solutions.* (This is a generalization of the third Fredholm theorem.)

We designate the totality of all solutions of (*) by E_1. Obviously E_1 is a subspace of E. A does not change the elements of E_1 and maps an arbitrary bounded set $S_0 \subset E_1$ into itself. On the other hand, A being a completely continuous operator, transforms S_0 into a compact set. It follows from the Theorem of §16, (p. 69) that E_1 is finite dimensional, q.e.d.

CHAPTER V

ELEMENTS OF SPECTRAL THEORY OF SELF ADJOINT OPERATORS ON HILBERT SPACES

§30. SELF ADJOINT OPERATORS

ADJOINT OPERATORS ON HILBERT SPACES. If we consider linear operators defined on a Hilbert space, then, because of the self conjugacy of this space and the existence of the scalar product, we can investigate a class of operators: The Hermitian or self adjoint ones. These operators are, for many purposes, more than arbitrary linear operators on Banach spaces. They play an important role in analysis and theoretical physics and there exists an extensive literature concerning them.

Let H be a Hilbert space and let A be a linear operator defined on H, whose range is in H. We consider a linear functional

$$f_y(x) = (Ax, y). \tag{1}$$

$f_y(x)$ has, being a linear functional on a Hilbert space, the form

$$f_y(x) = (x, y^*),$$

where $y^* \in H$ is uniquely defined by f_y. When changing y, f_y is obviously changed also, and therefore y^*, too; we thus obtain the operator

$$y^* = A^*y,$$

defined on H and with range in H. This A^* is connected with A by

$$(Ax, y) = (x, A^*y) \tag{2}$$

and is the *adjoint* operator to A (see p. 114). As can easily be seen, A^* is uniquely defined by (2). If, for all x and y,

$$(Ax, y) = (x, A^*y) \text{ and } (Ax, y) = (x, A_1^*y)$$

then

$$A^*y = A_1^*y$$

for all y. This means, however, that

$$A^* = A_1^*.$$

According to Theorem 1, §23, A^* is a bounded operator with

$$\| A^* \| = \| A \|. \tag{3}$$

We designate the operator adjoint to A^* by A^{**}.

According to (2), for all $x, y \in H$

$$(A^*x, y) = \overline{(y, A^*x)} = \overline{(Ay, x)} = (x, Ay).$$

This implies $A^{**} = A$ and analogously $(A^{**})^* = A^{***} = A^*, A^{****} = A$, etc. It follows readily that

$$(A_1 + A_2)^* = A_1^* + A_2^*,$$

$$(\lambda A)^* = \bar{\lambda} A^*,$$
$$(A_1 A_2)^* = A_2^* A_1^*.$$

As has been shown at the end of §28, the first and the third formulas are also valid for adjoint operators on arbitrary normed linear spaces.

SELF ADJOINT OPERATORS. A bounded linear operator A is called a *bounded self adjoint* (or *Hermitian*) operator if $A^* = A$.

Examples. 1. In an n-dimensional complex space which can be considered as a finite-dimensional analog of a Hilbert space, the linear operators can be interpreted as matrices (a_{ik}) with complex numbers a_{ik}. The operator adjoint to (a_{ik}) is $(\bar{a}_{ki})$. A self adjoint operator is a Hermitian matrix, i.e., a matrix with $a_{ik} = \bar{a}_{ki}$.

If (a_{ik}) is real, then self adjointness means symmetry.

2. As follows from the results of §28, an operator A on a Hilbert space is represented by an infinite matrix. Analogously to the n-dimensional case, the adjoint operator is of the form $(\bar{a}_{ki})$. Self adjointness of A again means $a_{ik} = \bar{a}_{ki}$.

3. For a Fredholm operator in L_2 with the kernel $K(s, t)$, the adjoint operator has the kernel $\overline{K(t, s)}$. Self adjointness is identical with

$$K(s, t) = \overline{K(t, s)}.$$

In case of a real kernel, we have symmetry.

In $L_2(0, 1)$ we consider the operator A assigning to each $x(t) \in L_2(0, 1)$ the function $Ax = tx(t) \in L_2(0, 1)$. It is easy to verify that this operator is self-adjoint.

According to the preceding, if A is self-adjoint and λ real, λA is also self-adjoint. If A_1 and A_2 are self-adjoint, then $A_1 A_2$ is self-adjoint if and only if A_1 and A_2 are permutable. Finally, if $A_n \to A$ in the sense of convergence in norm and if all A_n are self-adjoint operators, then A is also self-adjoint.

We interpret (Ax, y) with A self-adjoint as a functional of x and y and write $A(x, y)$ for it. Then

$$A(\lambda x_1 + \mu x_2, y) = \lambda A(x_1, y) + \mu A(x_2, y),$$
$$A(x, y) = \overline{A(y, x)},$$

as can easily be seen. Such a functional is called a *bilinear Hermitian form.* This form is bounded, i.e.,

$$|A(x, y)| \leqq C_A \|x\| \|y\|,$$

where C_A is a constant ($C_A = \|A\|$).

In this manner every self-adjoint operator A generates a bounded bilinear Hermitian form

$$A(x, y) = (Ax, y) = (x, Ay).$$

Conversely, if a bounded bilinear Hermitian form $A(x, y)$ is given, then it defines a self-adjoint A satisfying the equality

$$(Ax, y) = A(x, y).$$

For, if we fix y in $A(x, y)$, we obtain a linear functional in x. It follows that

$$A(x, y) = (x, y^*),$$

where y^* is uniquely determined. Thus we arrive at an operator B, defined by the equation

$$By = y^*$$

in such a way that

$$A(x, y) = (x, By) = (B^*x, y).$$

We put $B^* = A$ and obtain

$$(Ax, y) = A(x, y).$$

Obviously, A is an additive operator. We convince ourselves immediately of the boundedness of A, because

$$|(Ax, y)| = |A(x, y)| \leqq C_A \|x\| \|y\|.$$

Putting $y = Ax$ and dividing by $\|Ax\|$, we have

$$\|Ax\| \leqq C_A \|x\|.$$

Finally, we see that the operator A is uniquely defined.

We shall show that $A^* = A$.

Since the considered form is Hermitian, $A(x, x)$ is real for all x. Therefore, in particular, $A(x+y, x+y)$ and $A(x+iy, x+iy)$ are real. Consequently,

$$(A(x+y), x+y) = (x+y, A(x+y)).$$

and

$$(A(x+iy), x+iy) = (x+iy, A(x+iy)).$$

We decompose the inner product and obtain

$$(Ax, y) + (Ay, x) = (x, Ay) + (y, Ax),$$

$$(Ax, y) - (Ay, x) = (x, Ay) - (y, Ax).$$

Addition yields

$$(Ax, y) = (x, Ay)$$

for arbitrary x and y, i.e., A is self-adjoint.

QUADRATIC FORMS. In a bilinear Hermitian form $A(x, y)$ we put $x = y$. We obtain a quadratic form $A(x, x)$ which assumes real values for all x. We have

$$A(\alpha x + \beta y, \alpha x + \beta y) = \alpha\bar{\alpha} A(x, x) + \alpha\bar{\beta} A(x, y) + \bar{\alpha}\beta A(y, x) + \beta\bar{\beta} A(y, y).$$

$A(x, x)$ is called a *quadratic Hermitian form.* For every bilinear Hermitian form $A(x, y)$, we can form the corresponding Hermitian form $A(x, x)$. The converse also holds: By a quadratic Hermitian form $A(x, x)$, there is uniquely determined a bilinear Hermitian form $A(x, y)$. It is defined by the equality

$$A(x, y) = \frac{1}{4} \{[A(x_1, x_1) - A(x_2, x_2)] + i[A(x_3, x_3) - A(x_4, x_4)]\}$$

with

$$x_1 = x + y, x_2 = x - y$$

and

$$x_3 = x + iy, x_4 = x - iy.$$

If $A(x, x)$ is a bounded quadratic Hermitian form, i.e., if

$$|A(x, x)| \leqq C_A \|x\|^2,$$

then the corresponding bilinear Hermitian form is also bounded, as is readily seen. The converse is evident.

Thus we have shown: Every self adjoint operator A generates a unique bounded quadratic Hermitian form $A(x, x) = (Ax, x)$. Conversely, every bounded quadratic

Hermitian form $A(x,x)$ generates a bilinear Hermitian form, and consequently a self adjoint operator A such that $(Ax,x)=A(x,x)$.

Let $m=\inf\limits_{\|x\|=1}(Ax,x)$ and $M=\sup\limits_{\|x\|=1}(Ax,x)$. We call m and M the *greatest lower* and the *least upper* bounds of the self adjoint operator A.

We now show that

$$\|A\|=\max\{|m|,|M|\}=\sup_{\|x\|=1}|(Ax,x)|.$$

For, if $\|x\|=1$, then

$$|(Ax,x)|\leqq\|Ax\|\,\|x\|\leqq\|A\|\,\|x\|^2=\|A\|$$

and consequently

$$C_A=\sup_{\|x\|=1}|(Ax,x)|\leqq\|A\|. \tag{4}$$

On the other hand, we have for arbitrary $y\in H$

$$(Ay,y)\leqq C_A\|y\|^2.$$

If $z\in H$ and is different from the zero element, then we put

$$\lambda=\left[\frac{\|Az\|}{\|z\|}\right]^{\frac{1}{2}}\quad\text{and } u=\frac{1}{\lambda}Az.$$

We find

$$\|Az\|^2=(A\lambda z,u)=\frac{1}{4}\{(A(\lambda z+u),\lambda z+u)-(A(\lambda z-u),\lambda x-u)\}$$

$$\leqq\frac{1}{4}C_A\{\|\lambda z+u\|^2+\|\lambda z-u\|^2\}=\frac{1}{2}C_A\{\|\lambda z\|^2+\|u\|^2\}$$

$$=\frac{1}{2}C_A\left\{\lambda^2\|z\|^2+\frac{1}{\lambda^2}\|Az\|^2\right\}=C_A\|z\|\,\|Az\|.$$

Hence

$$\|Az\|\leqq C_A\|z\|$$

and consequently

$$\|A\|\leqq C_A=\sup_{\|x\|=1}|(Ax,x)|. \tag{5}$$

From the inequalities (4) and (5), we obtain the asserted equality.

From what we have shown, it follows in particular that $A=B$ if the self adjoint operators A and B satisfy the equality

$$(Ax,x)=(Bx,x)$$

for all $x\in H$.

§31. THE SPECTRUM OF SELF ADJOINT OPERATORS

We consider a family of operators $A_\lambda=A-\lambda I$ with the complex parameter λ.

EIGENVALUES. According to the earlier definition (see p. 92), λ is called an eigenvalue of the operator A if there exists a solution $x\neq\theta$ for the equation

$$Ax=\lambda x. \tag{1}$$

The element x is called an *eigenelement*. By E_λ we designate the subspace generated by all solutions of (1) with the eigenvalue λ.

We mention the following properties of the eigenvalues of a self adjoint operator A.

1) *If x is an eigenelement of the eigenvalue λ, then*

$$(Ax, x) = \lambda \| x \|^2. \tag{2}$$

This follows from (1) by inner multiplication by x. If, in particular, for the eigenelement $x, \| x \| = 1$, then

$$(Ax, x) = \lambda. \tag{3}$$

This implies:

2) *The eigenvalues of a self adjoint operator A are real numbers.*

3) *Eigenelements which belong to different eigenvalues are othogenal.*

This is a consequence of Theorem 2, §23, and the equality $A = A^*$.

4) Let G_λ be the range of the operator A_λ, i.e., the collection of all $y \in H$ such that

$$y = A_\lambda x.$$

The subspace $\bar{G}_\lambda$ is the orthogonal complement of E_λ, i.e.,

$$H = \bar{G}_\lambda + E_\lambda. \tag{4}$$

For, if $y \in G_\lambda$ and $x \in E_\lambda$, then

$$y = A_\lambda x', \ A_\lambda x = \theta,$$

which implies

$$(y, x) = (A_\lambda x', x) = (x', A_\lambda x) = (x', \theta) = 0.$$

If $y \in \bar{G}_\lambda$, then $y = \lim_n y_n$ with $y_n \in G_\lambda$ and since, according to the preceding $(y_n, x) = 0$ for all $x \in E_\lambda$,

$$(y, x) = \lim_n (y_n, x) = 0.$$

Conversely, let x be an element such that

$$(y, x) = 0$$

for all $y \in G_\lambda$. For an arbitrary $x_1 \in H, A_\lambda x_1 \in G_\lambda$ and therefore

$$(A_\lambda x_1, x) = 0 \text{ and } (x_1, A_\lambda x) = 0.$$

Since x_1 was arbitrary, we obtain from this $A_\lambda x = \theta$, i.e., x is in E_λ, q.e.d.

COROLLARY. *If λ is not an eigenvalue of the self adjoint operator A, then* (4) *becomes*

$$\bar{G}_\lambda = H.$$

With E we designate the direct sum of all the E_λ, or what amounts to the same, the closed linear subspaces spanned by all eigenelements.

In E we can select from the elements of E_λ a dense (finite or infinite) orthogonal system of eigenelements. Since the eigenelements from different E_λ are orthogonal, we obtain an orthogonal system $\{x_n\}$ of eigenelements dense in E.

SPECTRUM. If λ is a point of the spectrum, then the bounded linear operator $R_\lambda = A_\lambda^{-1} = (A - \lambda I)^{-1}$ cannot exist.

As we know, all eigenvalues belong to the spectrum. As we shall see later, however, not every point of the spectrum is an eigenvalue.

For the existence of a bounded operator A_λ^{-1}, it is necessary and sufficient that H be mapped biuniquely onto itself, i.e., that

$$G = \bar{G}_\lambda = H.$$

THEOREM 1. *The point λ is a regular value of the self adjoint operator A if and only if there exists a positive constant C such that*

$$\| A_\lambda x \| = \| Ax - \lambda x \| \geqq C \| x \| \tag{5}$$

for all $x \in E$.

Proof. If there exists a bounded operator A_λ^{-1} with $\| A_\lambda^{-1} \| = d$, then $y = A_\lambda x$ implies

$$\| x \| = \| A_\lambda^{-1} y \| \leqq d \| A_\lambda x \|.$$

Therefore, $\| A_\lambda x \| \geqq \dfrac{1}{d} \| x \|$, and the necessity is proved.

Conversely, let (5) be valid. Then the equation $A_\lambda x = \theta$ holds only for $x = \theta$. Hence λ is not an eigenvalue, and by the corollary mentioned on p. 146

$$\bar{G}_\lambda = H.$$

Now we prove that $\bar{G}_\lambda$ is closed, i.e., $G_\lambda = \bar{G}_\lambda = H$. Let $y = \lim_n y_n$ with $y_n \in G_\lambda$ and we have to show that $y \in G_\lambda$. By definition of G_λ, $y_n = Ax_n$. Furthermore,

$$\| y_{n+p} - y_n \| = \| A(x_{n+p} - x_n) \| \; C > \| x_{n+p} - x_n \|. \tag{6}$$

The sequence $\{y_n\}$ converges to y and is therefore a fundamental sequence. By (6), $\{x_n\}$ is also a fundamental sequence. Since H is a complete space, there exists

$$x = \lim_{n \to \infty} x_n.$$

Here

$$Ax = \lim_{n \to \infty} Ax_n = \lim_{n \to \infty} y_n = y.$$

Consequently, $y \in G_\lambda$. The above remark implies the assertion.

COROLLARY. *λ belongs to the spectrum of a self adjoint operator A if and only if there exists a sequence $\{x_n\}$ such that*

$$\| Ax_n - \lambda x_n \| \leqq C_n \| x_n \|, \; C_n \to 0 \text{ as } n \to \infty. \tag{7}$$

In (7), we can put $\| x_n \| = 1$. We then obtain:

If

$$\| x_n \| = 1, \textit{ then } \| Ax_n - \lambda x_n \| \to 0. \tag{8}$$

THEOREM 2. *The complex numbers $\lambda = \alpha + \beta i$ with $\beta \neq 0$ are regular values of the self adjoint operator A.*

If

$$y = A_\lambda x = Ax - \lambda x$$

then

$$(y, x) = (Ax, x) - \lambda(x, x),$$

$$(x, y) = \overline{(y, x)} = (Ax, x) - \bar{\lambda}(x, x).$$

This implies

$$(x, y) - (y, x) = (\lambda - \bar{\lambda})(x, x) = 2\beta i \| x \|^2$$

or

$$2 |\beta| \| x \|^2 = |(x, y) - (y, x)| \leqq |(x, y)| + |(y, x)| \leqq 2 \| y \| \| x \|$$

and hence $\| y \| \geqq |\beta| \| x \|$, $\| A_\lambda x \| \geqq |\beta| \| x \|$, respectively. (9)

Theorem 1 and (9) imply the assertion.

THEOREM 3. *The spectrum of a self adjoint operator A lies entirely in the interval $[m, M]$ of the real axis, where $M = \sup_{\|x\| = 1} (Ax, x)$ and $m = \inf_{\|x\| = 1} (Ax, x)$.*

The last theorems imply that the spectrum can lie only on the real axis. We shall show now that the real λ exterior to $[m, M]$ are regular.

Let, for instance, $\lambda > M$, therefore $\lambda = M + d, d > 0$.

We have

$$(A_\lambda x, x) = (Ax, x) - \lambda(x, x) \leqq M(x, x) - \lambda(x, x) = -d\|x\|^2,$$

implying

$$|(A_\lambda x, x)| \geqq d\|x\|^2. \tag{10}$$

On the other hand, $|(A_\lambda x, x)| \leqq \|A_\lambda x\| \|x\|$. Consequently,

$$\|A_\lambda x\| \geqq d\|x\|,$$

and thus the regularity of λ is demonstrated.

THEOREM 4. *M and m are points of the spectrum.*

Let us remark that if we replace A by A_λ, then the spectrum is shifted by λ to the left and m and M become $m - \lambda$ and $M - \lambda$. Therefore we can assume without loss of generality that $M \geqq m \geqq 0$. Then $M = \|A\|$ (see p. 145). We prove that M is a point of the spectrum.

By definition of M, there exists a sequence $\{x_n\}$ with $\|x_n\| = 1$ such that $(Ax_n, x_n) = M - \delta_n, \delta_n \geqq 0, \delta_n \to 0$ as $n \to \infty$.

Moreover,

$$\|Ax_n\| \leqq \|A\| \|x_n\| = \|A\| = M.$$

Therefore

$$\begin{aligned}\|Ax_n - Mx_n\|^2 &= (Ax_n - Mx_n, Ax_n - Mx_n)\\ &= \|Ax_n\|^2 - 2M(Ax_n, x_n) + M^2\|x_n\|^2\\ &\leqq M^2 - 2M(M - \delta_n) + M^2 = 2M\delta_n\end{aligned}$$

or

$$\|Ax_n - Mx_n\| \leqq \sqrt{2M\delta_n}.$$

Hence

$$\|Ax_n - Mx_n\| \to 0, \|x_n\| = 1,$$

which, together with the corollary to Theorem 1, implies that M belongs to the spectrum.

COROLLARY. *Every self adjoint operator has a non-empty spectrum.*[26]

Examples. 1. If A is the identity operator I, then its spectrum consists of the single eigenvalue 1 for which the space of eigenelements $E_1 = H$. For $\lambda \neq 1$, $R_\lambda = \dfrac{1}{\lambda - 1} I$ is a bounded operator.

2. We define the operator A of $(L_2(0, 1) \to L_2(0, 1))$ by

$$Ay(t) = ty(t) \quad (0 \leqq t \leqq 1).$$

Obviously, $m \geqq 0$ and $M \leqq 1$. We show that the interval $[0, 1]$ is the spectrum of A. This implies $m = 0$ and $M = 1$.

Indeed, let $0 \leqq \lambda \leqq 1$ and $\varepsilon > 0$. We consider those intervals $[\lambda, \lambda + \varepsilon]$, $[\lambda - \varepsilon, \lambda]$ which lie in $[0, 1]$. Let

$$y_\varepsilon(t) = \begin{cases} \dfrac{1}{\sqrt{\varepsilon}} & \text{for } t \in [\lambda, \lambda + \varepsilon],\\ 0 & \text{for } t \notin [\lambda, \lambda + \varepsilon].\end{cases}$$

---

[26] In the theory of normed rings, it is shown that an arbitrary bounded operator defined on a Banach space has a non-empty spectrum [6].

Since

$$\int_0^1 y_\varepsilon^2\,dt = \int_\lambda^{\lambda+\varepsilon} \frac{1}{\varepsilon}\,dt = 1,$$

$$y_\varepsilon \in L_2(0,1) \text{ and } \|y_\varepsilon\| = 1.$$

Furthermore $$A_\lambda y_\varepsilon(t) = (t-\lambda)\,y_\varepsilon(t),$$

from which $$\|A_\lambda y_\varepsilon(t)\|^2 = \frac{1}{\varepsilon}\int_\lambda^{\lambda+\varepsilon} (t-\lambda)^2 dt = \frac{\varepsilon^2}{3}.$$

As $\varepsilon \to 0$, $\|A_\lambda y_\varepsilon\| \to 0$. Consequently, λ with $0 \leqq \lambda \leqq 1$ is a point of the spectrum.

A does not have an eigenvalue, for

$$A_\lambda y(t) = (t-\lambda)\,y(t),$$

and if $A_\lambda y(t) = \theta$, then $(t-\lambda)\,y(t)$ vanishes on $[0, 1]$ almost everywhere and therefore $y(t) = 0$ almost everywhere, i.e., $y(t) = \theta$.

INVARIANT SPACES. The subspace $H_1 \subset H$ is called an *invariant space of the self adjoint operator* A, if for $x \in H_1$, also

$$Ax \in H_1.$$

Examples of invariant spaces are G_λ and E_λ. We will give several properties of such spaces.

1. The invariance of H_1 implies the invariance of its orthogonal complement H_2. H_2 is the totality of all y such that $(y, x) = 0$ for all $x \in H_1$. But if $x \in H_1$, $Ax \in H_1$, too. Therefore, for $y \in H_2, (y, Ax) = 0$ or $(Ay, x) = 0$. Consequently, Ay is orthogonal to every $x \in H_1$, i.e., $Ay \in H_2$, q.e.d.

In the invariant space H_1, A defines an operator A_{H_1} in $(H_1 \to H_1)$, and for $x \in H_1, A_{H_1}x = Ax$. A_{H_1} is a self adjoint operator.

2. If the invariant spaces H_1 and H_2 are orthogonal complements of each other, then the spectrum of A is the set-theoretic join of the spectra of A_{H_1} and A_{H_2}.

Let λ be a point of the spectrum of A_{H_1} (or A_{H_2}). Then there exists a sequence $\{x_n\}, x_n \in H_1$ (correspondingly, H_2) in such a way that $\|x_n\| = 1$ and $\|A_{H_1}x_n\| \to 0$. Since $\|A_\lambda x_n\| = \|A_{H_1}x_n\| \to 0$, λ belongs to the spectrum of A.

Now let λ belong to the spectrum of neither A_{H_1} nor A_{H_2}. Then there exists a positive number c such that for arbitrary $x_1 \in H_1$ and $x_2 \in H_2$,

$$\|A_\lambda x_1\| = \|A_{H_1\lambda}x_1\| \geqq c\,\|x_1\| \text{ and } \|A_{H_2\lambda}x_2\| \geqq c\,\|x_2\|.$$

But every $x \in H$ is of the form

$$x = x_1 + x_2 \text{ with } x_1 \in H_1,\ x_2 \in H_2, \text{ and } \|x\|^2 = \|x_1\|^2 + \|x_2\|^2.$$

This implies

$$\|A_\lambda x\| = \|A_\lambda x_1 + A_\lambda x_2\| = (\|Ax_1\|^2 + \|Ax_2\|^2)^{\frac{1}{2}}$$
$$\geqq c(\|x_1\|^2 + \|x_2\|^2)^{\frac{1}{2}} = c\,\|x\|.$$

Hence λ is not a point of the spectrum of A.

POINT SPECTRA AND CONTINUOUS SPECTRA. Let E be the linear closed subspace spanned by all eigenelements of a self adjoint operator A and let G be its orthogonal complement. As we know, H is representable as the orthogonal sum of the spaces E and G. E is an invariant space of A; consequently, the spectrum of A is the set-theoretic sum of the spectra of A_E and A_G. The spectrum of A_E is called *point spectrum* and that of A_G *continuous spectrum* of the operator A. If $E=H$, then A has a *pure point spectrum.* This appears with completely continuous operators, as we shall see below. If the operator has no eigenelement, then the spectrum of the operator is *purely continuous.* The operator A in Example 2 is of that sort.

§32. OPERATORS WITH PURE POINT SPECTRUM

Let the self adjoint operator A have a pure point spectrum. Then $E=H$ and consequently there exists a complete orthonormal system $\{x_n\}$ of eigenelements with

$$Ax_n = \lambda_n x_n, \tag{1}$$

where the λ_n are eigenvalues.

Each x is representable in the form of a Fourier series

$$x = \sum_n c_n x_n \text{ with } c_n = (x, x_n). \tag{2}$$

By P_n we designate the operator defined by

$$P_n x = (x, x_n) x_n = c_n x_n.$$

[P_n is a projection operator on the straight line tx_n, $-\infty < t < \infty$. (See §33).]

The series (2) can also be written in the form

$$x = Ix = \sum_n P_n x$$

or in operator form

$$I = \sum_n P_n. \tag{3}$$

It follows easily that

$$\left.\begin{aligned} P_n P_m &= 0 \text{ for } n \neq m, \\ P_p^2 &= P_n. \end{aligned}\right\} \tag{4}$$

By (1) and (2),

$$Ax = \sum_n \lambda_n c_n x_n = \sum_n \lambda_n P_n x. \tag{5}$$

(Since $|\lambda_n| \leqq \|A\|$ and $\sum_n c_n^2$ converges, $\sum_n (\lambda_n c_n)^2$ converges.)

For (5), we also write

$$A = \sum_n \lambda_n P_n. \tag{6}$$

Thus from (5) and (2) follows:

$$(Ax, x) = \sum_n \lambda_n c_n^2. \tag{7}$$

Therefore, we have reduced the quadratic form (Ax, x) to a sum of squares. It is possible to write (7) in the form

$$(Ax, x) = \sum_n \lambda_n (P_n x, x). \tag{8}$$

In fact $(P_n x, x) = (c_n x_n, x) = c_n (x_n, x) = c_n^2$.

Now let λ not be contained in the closed subspace generated by $\{\lambda_n\}$. Then there exists a constant $d > 0$ such that $|\lambda - \lambda_n| > d$. We have

$$A_\lambda x = (A - \lambda I)x = \sum_n (\lambda_n - \lambda) P_n x.$$

From this and (4) we obtain easily

$$R_\lambda x = A_\lambda^{-1} x = \sum_n \frac{1}{\lambda_n - \lambda} P_n x. \tag{9}$$

Because $P_n x = c_n x_n$,

$$R_\lambda x = \sum_n \frac{c_n}{\lambda - \lambda_n} x_n,$$

and since

$$\left| \frac{c_n}{\lambda_n - \lambda} \right| \leqq \frac{|c_n|}{d},$$

it follows that

$$\| R_\lambda x \| \leqq \frac{1}{d} \left(\sum_n c_n^2 \right)^{\frac{1}{2}} = \frac{1}{d} \| x \|$$

and

$$\| R_\lambda \| \leqq \frac{1}{d}.$$

Since λ therefore does not belong to the spectrum, we can rewrite (9) in the form

$$R_\lambda = \sum_n \frac{1}{\lambda_n - \lambda} P_n. \tag{10}$$

These formulas are analogous to those for quadratic forms and for symmetrical (Hermitian) matrices in the n-dimensional case. They differ from them only by the infinite summation.

Hilbert first investigated in his work [33] the general theory of self adjoint operators and of the corresponding forms (Ax, x), which he interpreted as limits of quadratic forms in n variables as $n \to \infty$. If n increases without bound, the finite sums can become infinite sums as well as integral expressions (these were treated in §§36 and 37). The pure point spectrum is especially simple because of its complete analogy with the finite-dimensional case. In this work by Hilbert the class of completely continuous operators with a pure point spectrum was investigated.

THEOREM 1. *Every non-zero point of the spectrum of a self adjoint completely continuous operator A is an eigenvalue of A.*

If $\lambda \neq 0$ is a point of the spectrum of A, then there exists a sequence $\{x_n\}$, $x_n \in H$, such that

$$\| x_n \| = 1, \| A x_n - \lambda x_n \| \to 0.$$

Putting $A x_n - \lambda x_n = y_n$, $\| y_n \| \to 0$ and

$$x_n = \frac{1}{\lambda}(A x_n - y_n).$$

A maps the sequence $\{x_n\}$ on a compact sequence $\{A x_n\}$. Therefore, there exists a convergent subsequence $\{A x_{n_k}\}$. With it, there also converges the subsequence

$$x_{n_k} = \frac{1}{\lambda}(A x_{n_k} - y_{n_k}). \tag{11}$$

If $x_{n_k} \to x$, then $A x_{n_k} \to A x$ and $y_{n_k} \to \theta$. Therefore, (11) implies

$$x = \frac{1}{\lambda} Ax \text{ or } Ax = \lambda x$$

with $\|x\| = \lim_k \|x_{n_k}\| = 1$.

Hence x is an eigenelement and λ an eigenvalue of A.

COROLLARY 1. *Every self adjoint completely continuous operator different from 0 has at least one eigenvalue.*

This assertion follows immediately from Theorem 1 and the corollary to Theorem 4, §31.

COROLLARY 2. *Every invariant space H_1 of a self adjoint completely continuous operator A contains an eigenelement.*

With A, $A_{H_1} \in (H_1 \to H_1)$ is also completely continuous. By Corollary 1 this operator has an eigenvalue λ_1. Consequently, there exists in H_1 an eigenelement of the operator A_{H_1} and, therefore, also of A.

COROLLARY 3. *A self adjoint completely continuous operator has a pure point spectrum.*

The invariant space G which is orthogonal to all eigenvectors is empty. If G were not empty, then by Corollary 2, G would have to contain an eigenelement and this contradicts the definition of G.

THEOREM 2. *The set $\{\lambda_n\}$ of eigenvalues of a self adjoint completely continuous operator A can have only one limit point $\lambda = 0$.*

If there would exist a sequence $\{\lambda_n\}$ of eigenvalues with $|\lambda_n| \geqq c > 0$, then because of the orthogonality for eigenelements x_n with $\|x_n\| = 1$,

$$\|Ax_n - Ax_m\|^2 = \|\lambda_n x_n - \lambda_m x_m\|^2 = \lambda_n^2 + \lambda_m^2 \geqq 2c^2 \text{ if } m \neq n.$$

Then, however, in contradiction to the complete continuity of A, the set $\{Ax_n\}$ would not be compact.

§33. PROJECTION OPERATORS

We now introduce the concept of a *projection* operator.

Let $L \subset H$ and let each $x \in H$ be uniquely representable in the form

$$x = y + z$$

where $y \in L$ and $z \perp L$. We put $Px = y$ and obtain an operator which is defined on all of H and whose range is the subspace L. This operator is called a *projection operator* and designated by P_L. We shall prove that *P_L is a self adjoint operator with $\|P_L\| = 1$ and $P_L^2 = P_L$.*

P_L is an additive operator. For, if

$$x_1 = y_1 + z_1 \text{ and } x_2 = y_2 + z_2$$

with $y_1, y_2 \in L$ and $z_1, z_2 \perp L$, then

$$x_1 + x_2 = (y_1 + y_2) + (z_1 + z_2)$$

with $y_1 + y_2 \in L$ and $z_1 + z_2 \perp L$. This implies

$$P_L(x_1 + x_2) = y_1 + y_2 = P_L x_1 + P_L x_2.$$

Furthermore, because of the orthogonality of y and z,

$$\|x\|^2 = \|y+z\|^2 = (y+z, y+z) = (y,y) + (z,z) = \|y\|^2 + \|z\|^2.$$

Consequently, $\|y\| \leqq \|x\|$,

i.e., $\|P_L x\| \leqq \|x\|$

for every x. Hence $\|P_L\| \leqq 1$.

For $x \in L, P_L x = x$ implies $\|P_L x\| = \|x\|$,

therefore $\|P_L\| = 1$.

We show that P_L is self adjoint. Let $x_1, x_2 \in H$ and let y_1, y_2 be their projections on L. We have

$$(P_L x_1, x_2) = (y_1, x_2) = (y_1, y_2)$$

and analogously $(x_1, P_L x_2) = (x_1, y_2) = (y_1, y_2)$.

It follows that $(P_L x_1, x_2) = (x_1, P_L x_2)$.

Since, for all $x \in H$, we have $P_L x \in L$, it follows that $P_L^2 x = P_L(P_L x) = P_L x$ for every $x \in H$, i.e.,

$$P_L^2 = P_L,$$

and the theorem is proved.

Now we prove the converse: *Every self adjoint operator P with $P^2 = P$ is a projection operator on a subspace L.*

We consider the set L_P of elements of the form $y = Px$ where x runs through all of H. By the additivity and homogeneity of P, L_P is a linear manifold. It is easy to show that L_P is closed. For, if $y_n \to y$ with $y_n \in L_P$, then $y_n = Px_n$ with $x_n \in H$ and consequently $Py_n = P^2 x_n = Px_n = y_n$. By the continuity of $P, y_n \to y$ implies the convergence of $Py_n \to Py$. Since $Py_n = y_n, y_n \to Py$ and $y = Py$, i.e., $y \in L_P$. Finally, $x - Px \perp Px$ by the self adjointness of P and because of $P^2 = P$, for

$$(x - Px, Px) = (Px - P^2 x, x) = 0.$$

It follows now from the definition of L_P that P is a projection operator on this subspace, q.e.d.

In addition, we shall point out several simple properties of projection operators. Two projection operators P_1, P_2 are called *orthogonal* if $P_1 P_2 = 0$. This is equivalent to $P_2 P_1 = 0$. Analogously, for $P_1 P_2 = 0, (P_1 P_2)^* = P_2^* P_1^* = P_2 P_1 = 0$, and conversely.

The projection operators P_1 and P_2 are orthogonal if and only if the corresponding subspaces L_1 and L_2 are orthogonal.

If $P_1 P_2 = 0$, then for $x_1 \in L_1$, $x_2 \in L_2$.

$$(x_1, x_2) = (P_1 x_1, P_2 x_2) = (P_2 P_1 x_1, x_2) = (\theta, x_2) = 0.$$

Let $L_1 \perp L_2$, then it follows for all $x \in H, P_2 x \in L_2$ and therefore $P_1 P_2 x = \theta$, i.e., $P_1 P_2 = 0$.

LEMMA 1. *The sum of two projection operators P_{L_1}, P_{L_2} is a projection operator if and only if these operators are orthogonal. If $P_{L_1} P_{L_2} = 0$, then*

$$P_{L_1} + P_{L_2} = P_{L_1 \dot{+} L_2}.$$

Necessity. Let

$$P = P_{L_1} + P_{L_2}$$

be a projection operator. Then

$$(P_{L_1} + P_{L_2})^2 = P_{L_1} + P_{L_2},$$

which implies

$$P_{L_1}P_{L_2} + P_{L_2}P_{L_1} = 0.$$

We multiply by P_{L_1} on the left and obtain

$$P_{L_1}P_{L_2} + P_{L_1}P_{L_2}P_{L_1} = 0.$$

If we now multiply by P_{L_1} on the right, then

$$P_{L_1}P_{L_2}P_{L_1} = 0.$$

By the preceding equality, $P_{L_1}P_{L_2} = 0$ and consequently $P_{L_2}P_{L_1} = 0$.

Sufficiency. Let

$$P_{L_1}P_{L_2} = P_{L_2}P_{L_1} = 0.$$

Then

$$(P_{L_1} + P_{L_2})^2 = P_{L_1} + P_{L_2}.$$

Consequently, $P_{L_1} + P_{L_2}$ is a projection operator.

By the hypothesis $P_{L_1}P_{L_2} = 0$, the subspaces L_1 and L_2 are orthogonal. If $x \in H$, then

$$Px = P_{L_1}x + P_{L_2}x = x_1 + x_2 \in L_1 \dotplus L_2,$$

where $(x - Px, Px) = (Px - P^2x, x) = 0$. This means, however, that P is a projection operator on $L_1 \dotplus L_2$. This proves the lemma.

LEMMA 2. *The product of two projection operators P_{L_1}, P_{L_2} is a projection operator if and only if P_{L_1} and P_{L_2} are permutable. If this condition is fulfilled, then $P_{L_1}P_{L_2} = P_{L_1 \cdot L_2}$.*

Necessity. Because $P = P_{L_1}P_{L_2}$ is self adjoint,

$$P_{L_1}P_{L_2} = (P_{L_1}P_{L_2})^* = P^*_{L_2}P^*_{L_1} = P_{L_2}P_{L_1},$$

and the permutability is proved.

Sufficiency. If $P_{L_1}P_{L_2} = P_{L_2}P_{L_1}$, then $P = P_{L_1}P_{L_2}$ is self adjoint. In addition,

$$(P_{L_1}P_{L_2})^2 = P_{L_1}P_{L_2}P_{L_1}P_{L_2} = P^2_{L_1}P^2_{L_2} = P_{L_1}P_{L_2},$$

and, consequently, P is a projection operator.

Let $x \in H$ be arbitrary. Then

$$Px = P_{L_1}P_{L_2}x = P_{L_2}P_{L_1}x$$

lies in L_1 and L_2, i.e., in $L_1 \cdot L_2$. Now let $y \in L_1 \cdot L_2$. Then

$$Py = P_{L_1}(P_{L_2}y) = P_{L_1}y = y.$$

Therefore P is a projection operator on $L_1 \cdot L_2$, and this proves the lemma.

The projection operator P_{L_2} is called a *part* of the projection operator P_{L_1} if $P_{L_1}P_{L_2} = P_{L_2}$. Passing to the adjoint operator, we convince ourselves that this definition is equivalent to $P_{L_2}P_{L_1} = P_{L_2}$. It follows immediately from the definition that P_{L_2} is a part of P_{L_1} if and only if L_2 is a subspace of L_1.

A projection operator P_{L_2} is a part of a projection operator P_{L_1} if and only if the inequality $\|P_{L_2}x\| \leqq \|P_{L_1}x\|$ is fulfilled for all $x \in H$.

It follows from $P_{L_2}P_{L_1}x = P_{L_2}x$ that $\|P_{L_2}x\| \leqq \|P_{L_2}\| \|P_{L_1}x\| = \|P_{L_1}x\|$.

Conversely, if this condition is fulfilled for all $x \in L_2$, then

$$\|P_{L_1}x\| \geqq \|P_{L_2}x\| = \|x\|.$$

$$\|P_{L_1}x\| \leqq \|x\|$$

implies

$$\|P_{L_1}x\| = \|x\|.$$

Therefore $x \in L_1$. Hence, for all $x \in H, P_{L_2}x \in L_1$ and this means that $P_{L_1}P_{L_2}x = P_{L_2}x$, i.e., $P_{L_1}P_{L_2} = P_{L_2}$, q.e.d.

LEMMA 3. *The difference $P_1 - P_2$ of two projection operators is a projection operator if and only if P_2 is a part of P_1. If this condition is fulfilled, then $L_{P_1-P_2}$ is the orthogonal complement of L_{P_2} in L_{P_1}.*

Necessity. If $P_1 - P_2$ is a projection operator, then

$$E - (P_1 - P_2) = (E - P_1) + P_2$$

is a projection operator, too.[27] Then, according to Lemma 1,

$$(E - P_1)P_2 = 0,$$

i.e., $P_1P_2 = P_2$.

Sufficiency. Let P_2 be a part of P_1. Then $E - P_1$ is orthogonal to P_2 and by Lemma 1, $(E - P_1) + P_2$ is a projection operator and so is, consequently, $P_1 - P_2$. Finally, by the hypothesis $P_1P_2 = P_2$, it follows that $P_1 - P_2$ and P_2 are orthogonal. Then, again by Lemma 1,

$$L_{P_1} = L_{P_1 - P_2} + L_{P_2},$$

q.e.d.

§34. POSITIVE OPERATORS

A self adjoint operator A different from the zero operator is called *positive*, $A > 0$, if

$$(Ax, x) \geqq 0$$

for all $x \in H$. If $A - B > 0$, then the self adjoint operator A is called *greater* than the self adjoint operator B. We also say that B is *smaller* than A. It is easy to verify that $A^2 > 0$, for every self adjoint operator different from 0 and that AA^* and A^*A, for every linear operator $A \neq 0$, are positive.

THEOREM. *The product of two positive self adjoint permutable operators A and B is positive.*

Proof. We put

$$A_1 = \frac{A}{\|A\|}, A_2 = A_1 - A_1^2, \ldots, A_{n+1} = A_n - A_n^2, \ldots$$

and show that

$$0 \leqq A_n \leqq E \tag{1}$$

for every n. For $n = 1$, this is obvious. Let (1) be valid for $n = k$. Then

$$(A_k^2(E - A_k)x, x) = ((E - A_k)A_kx, A_kx) \geqq 0,$$

i.e.,

$$A_k^2(E - A_k) \geqq 0$$

and analogously,

$$A_k(E - A_k)^2 \geqq 0.$$

[27] For the rest of this chapter, we shall designate the identity operator by E.

Therefore, $$A_{k+1} = A_k^2(E - A_k) + A_k(E - A_k)^2 \geqq 0$$
and $$E - A_{k+1} = (E - A_k) + A_k^2 \geqq 0.$$
Thus (1) is valid for $n = k + 1$.

Moreover,
$$A_1 = A_1^2 + A_2 = A_1^2 + A_2^2 + A_3 = \ldots = A_1^2 + A_2^2 + \ldots + A_n^2 + A_{n+1},$$
which implies
$$\sum_{k=1}^{n} A_k^2 = A_1 - A_{n+1} \leqq A_1.$$
Therefore,
$$\sum_{k=1}^{n} (A_k x, A_k x) \leqq (A_1 x, x).$$
Consequently, the series
$$\sum_{k=1}^{\infty} \| A_k x \|^2$$
converges and hence $\| A_k x \| \to 0$ as $k \to \infty$. Thus
$$\Big(\sum_{k=1}^{n} A_k^2\Big)x = A_1 x - A_{n+1}x \to A_1 x.$$
Since B is obviously permutable with all A_k, we obtain
$$(ABx, x) = \| A \| (BA_1 x, x) = \| A \| \lim_n \sum_{k=1}^{n} (BA_k^2 x, x)$$
$$= \| A \| \lim_n \sum_{k=1}^{n} (BA_k x, A_k x) \geqq 0.$$

From what we have proved it follows easily that *if $\{A_n\}$ is a monotonic non-decreasing sequence of self adjoint mutually permutable operators, if for all n, $A_n B = BA_n$, and if*
$$A_1 \leqq A_2 \leqq \ldots \leqq A_n \leqq \ldots \leqq B,$$
then the A_n converge to A and $A \leqq B$. (An analogous assertion holds for monotonically decreasing sequences.)

We consider the self adjoint operators $C_n = B - A_n$. These are positive, permutable, and form a monotonically decreasing sequence. Thus, for $m < n$, the operators
$$(C_m - C_n)C_m \text{ and } C_n(C_m - C_n)$$
are also positive, which implies
$$(C_m^2 x, x) \geqq (C_m C_n x, x) \geqq (C_n^2 x, x).$$

A monotonically decreasing non-negative number sequence $\{(C^2 x, x)\}$ has a limit. From the above, the sequence $\{(C_m C_n x, x)\}$ converges to the same limit as $n, m \to \infty$. Therefore, as $n, m \to \infty$,
$$\|C_m x - C_n x\|^2 = ((C_m - C_n)^2 x, x) = (C_m^2 x, x) - 2(C_m C_n x, x) + (C_n^2 x, x) \to 0.$$
Hence, the sequence $\{C_n x\}$ and consequently, also $\{A_n x\}$ converge to Ax for arbitrary x, i.e., $Ax = \lim_n A_n x$. Obviously, A is a self adjoint operator satisfying the inequality $A \leqq B$, q.e.d.

§35. SQUARE ROOTS OF POSITIVE OPERATORS

A self adjoint operator B is called *square root* of the positive operator A if $B^2 = A$. We shall prove that *the positive square root B of an arbitrary positive self adjoint operator A exists and is unique. It is permutable with each operator permutable with A.*

Proof. We can assume without loss of generality that $A \leqq E$ and we put $B_0 = 0$ and

$$B_{n+1} = B_n + \tfrac{1}{2}(A - B_n^2). \tag{1}$$

All B_n are self adjoint, positive and permutable with every operator permutable with A. In particular, $B_n B_m = B_m B_n$. We convince ourselves immediately of the validity of the equalities

$$E - B_{n+1} = \tfrac{1}{2}(E - B_n)^2 + \tfrac{1}{2}(E - A) \tag{2}$$

and

$$B_{n+1} - B = \tfrac{1}{2}[(E - B_{n-1}) + (E - B_n)](B_n - B_{n-1}). \tag{3}$$

(2) implies $B_n \leqq E$ for all n. We verify easily that $B_n \leqq B_{n+1}$. For $n = 0$, this is obvious from the inequality

$$B_1 = \tfrac{1}{2}A \geqq 0 = B_0.$$

Equality (3) shows furthermore that $B_{n+1} - B_n \geqq 0$, if $B_n - B_{n-1} \geqq 0$. Consequently, $B_n \leqq B_{n+1}$ for all n. In this manner, we have obtained a bounded monotonically increasing sequence $\{B_n\}$ of self adjoint operators.

This sequence converges, according to the preceding statements, to a positive B. Passing to the limit, (1) implies

$$B = B + \tfrac{1}{2}(A - B^2),$$

i.e.,

$$B^2 = A.$$

The existence of a positive square root B of an operator A is proved. Finally, B is permutable with every operator permutable with A because the B_n have this property.

Let B_1 be another positive square root of A. Since B_1 permutes with A, $B_1 B = B B_1$. If $x \in H$ and $y = (B - B_1)x$, then

$$\begin{aligned}(By, y) + (B_1 y, y) &= ((B + B_1)y, y)\\ &= ((B + B_1)(B - B_1)x, y) = ((B^2 - B_1^2)x, y) = 0.\end{aligned}$$

Because B and B_1 are positive, it follows that $(By, y) = (B_1 y, y) = 0$. Due to the fact that roots are positive, $B = C^2$, where C is a self adjoint operator. Now $\| Cy \|^2 = (C^2 y, y) = (By, y) = 0$, therefore $Cy = 0$. Consequently, $By = C(Cy) = 0$ and analogously $B_1 y = 0$. Thus

$$\|B_1 x - Bx\|^2 = ((B - B_1)^2 x, x) = ((B - B_1)y, x) = 0,$$

i.e.,

$$Bx = B_1 x$$

for arbitrary $x \in H$, and this proves the uniqueness of the square root.

Example. In $L_2(0, 1)$, the operator defined by

$$Af(t) = tf(t)$$

has the square root B where

$$Bf(t) = \sqrt{t}f(t).$$

LEMMA 1. *Let A and B be self adjoint permutable operators and let $A^2 = B^2$. Then there exists a projection operator P with the properties:*

1) *Every bounded linear operator C permutable with $A - B$ permutes with P.*
2) *$Ax = 0$ implies $Px = x$, and*
3) *$A = (2P - E)B$.*

We consider the set of all $x \in H$ for which $(A - B)x = 0$. This set forms a subspace $L \subset H$. Let P be a projection operator on L. If C is permutable with $A - B$ and if $y \in L$, then Cy also belongs to L. For, $(A - B)Cy = C(A - B)y = 0$. Therefore $CPx \in L$ for all $x \in H$ and hence $PCPx = CPx$, i.e., $PCP = CP$. Analogously, $C^*P = PC^*P$. This implies $PC = (C^*P)^* = (PC^*P)^* = PCP$. Consequently, $CP = PC$ and this proves 1).

Let $Ax = 0$. Then

$$\| Bx \|^2 = (Bx, Bx) = (B^2x, x) = (A^2x, x) = \| Ax \|^2 = 0,$$

i.e., $Bx = 0$. Hence
$$(A - B)x = 0,$$
consequently
$$Px = x$$
and this proves 2) also.

Finally,

$$(A - B)(A + B) = A^2 - B^2 = 0.$$

Therefore
$$(A + B)x \in L$$
for all x and hence
$$P(A + B)x = (A + B)x,$$
i.e.,
$$P(A + B) = A + B.$$
In addition, we have $P(A - B) = 0$, and thus

$$P(A + B) - P(A - B) = A + B$$

and
$$A = (2P - E)B.$$
So the lemma is proved.

§36. SPECTRAL DECOMPOSITION OF A SELF ADJOINT OPERATOR

DECOMPOSITION OF THE IDENTITY. We generalize formulas (3), (6), and (10) of §32 for arbitrary self adjoint operators.

THEOREM 1. *For every self adjoint operator A, there exists a projection operator E_+ with the properties:*

1) *An arbitrary bounded linear operator C permutable with A is permutable with E_+.*
2) *$AE_+ \geqq 0, A(E - E_+) \leqq 0$, and*
3) *If $Ax = 0$, then $E_+x = x$.*

Let E_+ be a projection operator, projecting all of H on the null subspace of the operator $A - B$, where B is the positive square root of A^2. From Lemma 1 of the preceding § it follows that 1) and 3) hold. By virtue of this lemma

$$A = (2E_+ - E)B,$$

therefore

$$AE_+ = BE_+ \geqq 0, \quad A(E - E_+) = -(E - E_+)B \leqq 0$$

and this proves 2).

We note that the equality $A=(2E_+ - E)B$ implies $BE_+ = \frac{1}{2}(A+B)$ and hence

$$AE_+ = \tfrac{1}{2}(A+B),$$
$$A(E-E_+) = \tfrac{1}{2}(A-B).$$

The operator AE_+ is designated by A_+ and is called the *positive part of A*. For $A(E-E_+)$ we write A_- and A_- is called the *negative part of A*. Then

$$A = A_+ + A_-.$$

Examples. 1. Let A be a symmetric matrix of nth order with eigenvalues $\lambda_1, \lambda_2, \ldots, \lambda_n$, where $\lambda_1, \lambda_2, \ldots, \lambda_k$ are smaller than 0 and $\lambda_{k+1}, \lambda_{k+2}, \ldots, \lambda_n$ are greater than 0. Hence (see p. 63)

$$A = U(\lambda_1, \lambda_2, \ldots, \lambda_k, \lambda_{k+1}, \ldots, \lambda_n)\, U^{-1}.$$

Then

$$A_+ = U(0, \ldots, 0, \lambda_{k+1}, \ldots, \lambda_n)\, U^{-1} \text{ and } A_- = U(\lambda_1, \ldots, \lambda_k, 0, \ldots, 0)\, U^{-1}.$$

2. Let the operator A in $L_2(0, 1)$ be defined by the equation

$$Af(t) = tf(t).$$

Then

$$A_+ f(t) = \frac{t+|t|}{2} f(t), \quad A_- f(t) = \frac{t-|t|}{2} f(t).$$

THEOREM 2. *Every self adjoint operator A generates a family $\{E_\lambda\}$ of projection operators, depending on a real parameter λ, $-\infty<\lambda<\infty$ and satisfying the following conditions:*

1) *$AC=CA$ implies $E_\lambda C = CE_\lambda$ for all λ,*
2) *$E_\lambda \leqq E_\mu$ if $\lambda<\mu$,*
3) *E_λ is continuous on the left*, i.e., $E_{\lambda-0}=E_\lambda$,
4) *$E_\lambda = 0$ for $-\infty<\lambda\leqq m$ and $E_\lambda = E$ for $M<\lambda<\infty$, where m and M are the greatest lower and least upper bounds of A, respectively.*

The family $\{E_\lambda\}$ is called a *decomposition of the identity* generated by A. We first give examples of such decompositions.

1. Let A be a symmetric matrix of nth order

$$A = U(\lambda_1, \ldots, \lambda_n)U^{-1}$$

with $\lambda_1<\lambda_2<\lambda_3<\ldots<\lambda_n$ and let e_i be eigenvectors corresponding to the eigenvalues λ_i. For $\lambda_i<\lambda\leqq\lambda_{i+1}$, let there be designated by E_λ the operator which projects on the i-dimensional subspace generated by the vectors $e_1, e_2, \ldots, e_i$. Let $E_\lambda=0$ for $\lambda\leqq\lambda_1$ and let $E_\lambda = E$ for $\lambda>\lambda_n$.

The analog holds also for the case considered in §32.

2. In $L_2(-1, 1)$ let A be defined by the equation

$$Af(t) = tf(t).$$

Then

$$E_\lambda f(t) = \varphi(t)$$

where $\varphi(t)=0$ for $t\geqq\lambda$, $\varphi(t)=f(t)$ for $t<\lambda$. For $\lambda\leqq-1$, obviously $E_\lambda=0$ and for $\lambda>1, E_\lambda=E$.

We now present the proof of Theorem 2.

Let λ be a positive real number and $A_\lambda = A-\lambda E$. We designate by E_λ the projection operator $E-E_+(\lambda)$. $E_+(\lambda)$ is a projection operator constructed for A_λ according to Theorem 1.

Condition 1) is obviously fulfilled. This implies, in particular, that E_λ and E_μ are permutable for arbitrary λ and μ.

We consider 2) and investigate the projection operator

$$P = E_\lambda(E - E_\mu)$$

for $\lambda < \mu$.

We have $$E_\lambda P = E_\lambda^2(E - E_\mu) = E_\lambda(E - E_\mu) = P \qquad (\alpha)$$

and analogously $$(E - E_\mu)P = P. \qquad (\beta)$$

By definition of E_λ,

$$(A - \lambda E)\, E_\lambda \leqq 0, \qquad (\gamma)$$

$$(A - \mu E)(E - E_\mu) \geqq 0. \qquad (\delta)$$

We put $Px = y$ for an arbitrary $x \in H$. (α) and (β) imply

$$E_\lambda y = E_\lambda P x = P x = y$$

and analogously $$(E - E_\mu) y = y.$$

It now follows from (γ) and (δ) that

$$((A - \lambda E)\, E_\lambda y, y) = ((A - \lambda E)\, y, y) \leqq 0,$$

$$((A - \mu E)(E - E_\mu)\, y, y) = ((A - \mu E)\, y, y) \geqq 0.$$

From these two inequalities we obtain

$$((\mu - \lambda)\, y, y) \leqq 0$$

or $$(\mu - \lambda)\, \| y \|^2 \leqq 0.$$

Because, by hypothesis, $\mu - \lambda > 0$, this implies that $y = Px = 0$, and since $x \in H$ was arbitrary, $P = 0$, i.e.,

$$E_\lambda(E - E_\mu) = E_\lambda - E_\lambda E_\mu = 0$$

and 2) is proved.

We consider the half open interval $\Delta = [\lambda, \mu)$ of the number line. For the projection operator $E(\Delta) = E_\mu - E_\lambda$,

$$E_\mu E(\Delta) = E(\Delta)$$

and $$(E - E_\lambda) E(\Delta) = E(\Delta).$$

Therefore $$(A - \mu E)\, E(\Delta) = (A - \mu E) E_\mu E(\Delta) \leqq 0,$$

$$(A - \lambda E)\, E(\Delta) = (A - \lambda E)(E - E_\lambda)\, E(\Delta) \geqq 0$$

and consequently $$\lambda E(\Delta) \leqq A E(\Delta) \leqq \mu E(\Delta). \qquad (1)$$

(We use the fact that we can manipulate operator inequalities, obeying the usual sign rules.)

We now prove 3). For arbitrary $x \in H$, the expression $(E_\lambda x, x)$ is a non-decreasing function of λ. Therefore, $\lim\limits_{\substack{\lambda \to \mu \\ \lambda < \mu}} (E_\lambda x, x)$ exists. We obtain from this the convergence of

$$\| E_\nu x - E_\lambda x \|^2 = ((E_\nu - E_\lambda)\, x, x) = (E_\nu x, x) - (E_\lambda x, x) \to 0$$

for $\lambda < \nu < \mu, \lambda \to \mu$ and $\nu \to \mu$. Consequently,

$$\lim_{\substack{\lambda \to \mu \\ \lambda < \mu}} E_\lambda x = x' = E_{\mu - 0} x$$

exists for all $x \in H$. It is easy to see that $E_{\mu - 0}$ is a projection operator. We prove that

$$E_{\mu-0} = E_\mu.$$

Let $$E(\Delta_0) = E_\mu - E_{\mu-0}.$$

Then $$E(\Delta) = E_\mu - E_\lambda \to E(\Delta_0) \text{ as } \lambda \to \mu, \lambda < \mu$$

in the sense of strong convergence of operators. Now, passing to the limit in the inequality (1), which is obviously possible, we obtain

$$\mu E(\Delta_0) = AE(\Delta_0).$$

Let $x \in H$ be arbitrary and $y = E(\Delta_0)x$. By the preceding equality,

$$(A - \mu E)y = 0,$$

from which we obtain $E_\mu y = 0$ according to 3) of Theorem 1. Passing to the limit,

$$E_\mu E(\Delta_0) = E(\Delta_0)$$

and consequently $$E(\Delta_0)x = E_\mu E(\Delta_0)x = E_\mu y = 0.$$

Since $x \in H$ was arbitrary, $E(\Delta_0) = 0$ and this proves 3).[28]

4) may be proved without difficulty. Let $\lambda < m$ and $E_\lambda \neq 0$. Then there exists an x such that $E_\lambda x \neq 0$. We put $E_\lambda x = y$ and have $E_\lambda y = y$, where $\|y\| = 1$ can be assumed. Then,

$$(Ay, y) - \lambda = (Ay, y) - \lambda(y, y) = ((A - \lambda E)y, y) = ((A - \lambda E)E_\lambda y, y) \leqq 0,$$

i.e., $$(Ay, y) \leqq \lambda < m,$$

which contradicts the definition of the number m. Consequently, $E_\lambda = 0$ for $\lambda < m$ and by left continuity, $E_m = 0$ also. In a similar way, we show that $E_\lambda = E$ for $\lambda > M$.

SPECTRAL DECOMPOSITION OF A SELF ADJOINT OPERATOR.

THEOREM 3. *For every positive* ε, *the following holds*

$$A = \int_m^{M+\varepsilon} \lambda dE_\lambda, \tag{2}$$

where the integral is to be interpreted as limit of integral sums in the sense of uniform convergence in the space of the operators.

Proof. Let the half-open interval $[m, M + \varepsilon)$ with $\varepsilon > 0$ be decomposed into the half-open intervals $\Delta_1, \Delta_2, \ldots, \Delta_n$, where $\Delta_k = [\lambda_k, \mu_k)$. According to (1),

$$\lambda_k E(\Delta_k) \leqq AE(\Delta_k) \leqq \mu_k E(\Delta_k)$$

for every $\Delta_k = [\lambda_k, \mu_k)$. Summing over all $k = 1, 2, \ldots, n$ and taking into account

$$\sum_{k=1}^n E(\Delta_k) = E,$$

it follows that $$\sum_{k=1}^n \lambda_k E(\Delta_k) \leqq A \leqq \sum_{k=1}^n \mu_k E(\Delta_k).$$

Now select $\nu_k \in [\lambda_k, \mu_k)$, then

---

[28] According to the definition of E_λ, elements with $(A - \lambda E)x = 0$ belong to the orthogonal complement of the subspace L_{E_λ}. If, however, E_λ is defined so that these elements are mapped on L_{E_λ} (which can be done without violating 1), 2) and 4)), then E_λ is continuous on the right.

$$\sum_{k=1}^{n}(\lambda_k - \nu_k)E(\Delta_k) \leqq A - \sum_{k=1}^{n}\nu_k E(\Delta_k) \leqq \sum_{k=1}^{n}(\mu_k - \nu_k)E(\Delta_k).$$

We put $\max_k (\mu_k - \lambda_k) = \delta$ and find from the preceding inequalities

$$-\delta E \leqq A - \sum_{k=1}^{n}\nu_k E(\Delta_k) \leqq \delta E$$

and consequently $-\delta(x, x) \leqq (\|A - \sum_{k=1}^{n}\nu_k E(\Delta_k)\| x, x) \leqq \delta(x, x)$.

This implies that
$$\left\| A - \sum_{k=1}^{n}\nu_k E(\Delta_k) \right\| \leqq \delta,$$

i.e.,
$$A = \lim_n \sum_{k=1}^{n}\nu_k E(\Delta_k) = \int_m^{M+\varepsilon}\lambda d E_\lambda,$$

q.e.d. For a completely continuous operator, this equality becomes (6), §32.

Remark. Because the convergence of a sequence $\{A_n\}$ of operators to A in the sense of uniform convergence implies strong operator convergence and the convergence of the quadratic forms $(A_n x, x)$ to (Ax, x), it follows from Theorem 3 that

1) $$Ax = \lim_n \sum_{k=1}^{n}\nu_k E(\Delta_k)x = \int_m^{M+\varepsilon}\lambda dE_\lambda x,$$

and

2) $$(Ax, x) = \lim_n \sum_{k=1}^{n}\nu_k(E(\Delta_k)x, x) = \int_m^{M+\varepsilon}\lambda d(E_\lambda x, x)$$

for all $x \in H$.

§37. FUNCTIONS OF OPERATORS. RESOLVENTS

DEFINITION OF $F(A)$. We now define integrals of the form

$$\int_m^{M+\varepsilon} F(\lambda)dE_\lambda,$$

where $F(\lambda)$ is an arbitrary complex step function on the closed interval $[m, M]$ and $\{E_\lambda\}$ the decomposition of unity generated by the self adjoint operator A. If λ_0 is a discontinuity of this function, then we agree that $F(\lambda_0) = F(\lambda_0 - 0)$. We extend $F(\lambda)$ to the half-open interval $[m, M + \varepsilon)$ by defining there $F(\lambda) = F(M)$. Hence $F(\lambda) = \nu_k$ on $\Delta_k = [\lambda_k, \mu_k)\,(k = 1, 2, \ldots, n)$ where

$$\sum_{k=1}^{n}\Delta_k = [m, M + \varepsilon).$$

By definition

$$\int_m^{M+\varepsilon} F(\lambda)dE = \sum_{k=1}^{n}\nu_k E(\Delta_k).$$

It is easy to see that the equality

$$\int_m^{M+\varepsilon} F(\lambda)dE_\lambda = \sum_{k=1}^m c_k E(\Delta_k')$$

also holds, where Δ_k' are arbitrary subintervals of the Δ_k whose sum is the interval $[m, M+\varepsilon)$. The operator $\int_m^{M+\varepsilon} F(\lambda)dE_\lambda$ is designated by $F(A)$. It is a *function of the operator A* depending on the function $F(\lambda)$ of the real variable λ. In this manner, we obtain a correspondence of the step functions of a real variable with the operator functions of A.

This correspondence has the following properties.

1. If
$$F(\lambda) = aF_1(\lambda) + bF_2(\lambda)$$
then
$$F(A) = aF_1(A) + bF_2(A) \quad \text{(Additivity).}$$

2. If
$$F(\lambda) = F_1(\lambda)\,F_2(\lambda)$$
then
$$F(A) = F_1(A)\,F_2(A) \quad \text{(Multiplicativity).}$$

3. $\bar{F}(A) = [F(A)]^*$, where $\bar{F}$ designates the complex conjugate function.
4. $\|F(A)\| \leqq \max |F(\lambda)|$.
5. $AB = BA$ implies $F(A)B = BF(A)$ for every bounded linear operator B.

For the proof of 1. and 2., we decompose $[m, M+\varepsilon)$ into subintervals Δ_k on which the two functions $F_1(\lambda)$ and $F_2(\lambda)$ are constant. Then, we have for $F(\lambda) = aF_1(\lambda) + bF_2(\lambda)$,

$$F(A) = \sum_{k=1}^n (ac_k^{(1)} + bc_k^{(2)})\,E(\Delta_k)$$

$$= a\sum_{k=1}^n c_k^{(1)}\,E(\Delta_k) + b\sum_{k=1}^n c_k^{(2)}\,E(\Delta_k) = aF_1(A) + bF_2(A)$$

and for $F(\lambda) = F_1(\lambda)F_2(\lambda)$, by the orthogonality of $E(\Delta_k)$ and $E(\Delta_l)$ for $k \neq l$,

$$F(A) = \sum_{k=1}^n c_k^{(1)}\,c_k^{(2)}\,E(\Delta_k) = \left(\sum_{k=1}^n c_k^{(1)}E(\Delta_k)\right)\left(\sum_{l=1}^n c_l^{(2)}\,E(\Delta_l)\right) = F_1(A)\,F_2(A).$$

Furthermore,

$$(F(A)\,x, x) = \left(\sum_{k=1}^n c_k E(\Delta_k)\,x, x\right) = \left(x, \sum_{k=1}^n \bar{c}_k E(\Delta_k)x\right) = (x, \bar{F}(A)\,x),$$

which implies
$$[F(A)]^* = \bar{F}(A).$$
Finally,

$$|(F(A)\,x, x)| = \left|\sum_{k=1}^n c_k(E(\Delta_k)\,x, x)\right| \leqq \sum_{k=1}^n |c_k|\,(E(\Delta_k)\,x, x) \leqq \max_\lambda |F(\lambda)|\,(x, x).$$

This implies

$$\|F(A)\| = \sup_{\|x\|=1} |(F(A)\,x, x)| \leqq \max_\lambda |F(\lambda)|.$$

Property 5. is obvious.

It follows from the definition of $F(A)$ that, in particular, $E(\Delta) = \chi_\Delta(A)$ where $\chi_\Delta(\lambda)$ is the characteristic function of the half open interval Δ.

Now let $F(\lambda)$ be an arbitrary function continuous on $[m, M]$. We extend it to

$[m, M+\varepsilon)$ by putting $F(\lambda) = F(M)$ for $\lambda \in (M, M+\varepsilon)$. There exists a sequence of step functions $F_n(\lambda)$ converging uniformly to $F(\lambda)$ on $[m, M+\varepsilon)$. We consider the corresponding functions $F_n(A)$ of the operator.

$$\| F_m(A) - F_n(A) \| \leqq \max_{\lambda} | F_m(\lambda) - F_n(\lambda) | \to 0 \text{ as } m, n \to \infty.$$

By virtue of the completeness of the space of operators, there exists an operator

$$B = \lim_n F_n(A).$$

The integral

$$\int_m^{M+\varepsilon} F(\lambda) dE_\lambda$$

is, by definition of B, the operator function generated by the limiting function $F(\lambda)$ and is designated by $F(A)$. We easily convince ourselves that this process is independent of the special choice of the sequence $F_n(\lambda)$ converging to $F(\lambda)$. Now, it is easy to verify that the properties 1. through 5. are preserved in the case of continuous functions. In particular,

$$A^n = \int_m^{M+\varepsilon} \lambda^n dE_\lambda, \quad n = 0, 1, 2, \ldots .$$

Resolvents. The correspondence between the functions of a real variable and the operator functions can be used for the investigation of properties of self adjoint operators, especially of spectral properties.

THEOREM 1. *For a given λ_0 there exists the resolvent $R_{\lambda_0} = (A - \lambda_0 E)^{-1}$, if one of the following conditions is fulfilled*:

1) λ_0 *is not real,*

2) λ_0 *is real and lies in the exterior of the closed interval* $[m, M]$,

3) *If $\lambda_0 \in [m, M]$, there exists a half-open interval* $[\alpha, \beta)$ with $\alpha < \lambda_0 < \beta$ *and inside which E_λ is constant.*

Then

$$R_{\lambda_0} = \int_m^{M+\varepsilon} \frac{dE_\lambda}{\lambda - \lambda_0}.$$

Proof. In the first two cases, the function

$$f(\lambda) = \frac{1}{\lambda - \lambda_0}$$

is continuous in $[m, M+\varepsilon)$ for ε sufficiently small. We have

$$\int_m^{M+\varepsilon} \frac{dE_\lambda}{\lambda - \lambda_0} \int_m^{M+\varepsilon} (\lambda - \lambda_0) dE_\lambda = \int_m^{M+\varepsilon} dE_\lambda = E.$$

And by

$$\int_m^{M+\varepsilon} (\lambda - \lambda_0) dE_\lambda = A - \lambda_0 E,$$

$$\int_m^{M+\varepsilon} \frac{dE_\lambda}{\lambda - \lambda_0} = R_{\lambda_0}.$$

In the third case, we decompose $[m, M+\varepsilon)$ into three half open intervals $[m, \alpha), [\alpha, \beta), [\beta, M+\varepsilon)$. Let

$$\varphi(\lambda) = \frac{1}{\lambda - \lambda_0} \text{ on } [m, \alpha) \text{ and } [\beta, M+\varepsilon)$$

and let $\varphi(\lambda)$ be linear on $[\alpha, \beta)$ where $\varphi(\alpha) = \dfrac{1}{\alpha - \lambda_0}$, $\varphi(\beta) = \dfrac{1}{\beta - \lambda_0}$. For every function $\psi(\lambda)$, we have because of the constancy of E_λ

$$\int_\alpha^\beta \psi(\lambda) dE_\lambda = O,$$

and therefore we may also write

$$\int_m^{M+\varepsilon} \varphi(\lambda)\, dE_\lambda = \int_m^{M+\varepsilon} \frac{dE_\lambda}{\lambda - \lambda_0}.$$

Thus, as in the first two cases,

$$\int_m^{M+\varepsilon} \frac{dE_\lambda}{\lambda - \lambda_0} \int_m^{M+\varepsilon} (\lambda - \lambda_0)\, dE_\lambda = E.$$

This implies the existence of

$$R_{\lambda_0} = \int_m^{M+\varepsilon} \frac{dE_\lambda}{\lambda - \lambda_0}.$$

THEOREM 2. *If R_{λ_0} exists for a real λ_0, then λ_0 lies in the interior of an interval $[\alpha, \beta)$ for which E_λ is constant.*

For an arbitrary $x \in H$, we form

$$(A - \lambda_0 E)\, x = \int_m^{M+\varepsilon} (\lambda - \lambda_0)\, dE_\lambda x$$

and apply the operator $R_{\lambda_0} E(\varDelta)$ on both sides, where $\varDelta = [\alpha, \beta)$ is a certain half open interval containing λ_0 in its interior. We obtain

$$E(\varDelta)\, x = R_{\lambda_0} \left(\int_\alpha^\beta (\lambda - \lambda_0)\, dE_\lambda x \right).$$

This implies
$$\| E(\varDelta)\, x \| \leqq \| R_{\lambda_0} \| \left\| \int_\alpha^\beta (\lambda - \lambda_0)\, dE_\lambda x \right\|.$$

As we easily verify, this becomes

$$\left\| \int_\alpha^\beta (\lambda - \lambda_0)\, dE_\lambda x \right\| \leqq K \| E(\Delta)\, x \|,$$

where $K = \max\{\beta - \lambda_0, \lambda_0 - \alpha\}$. Consequently

$$\| E(\Delta)x \| \leqq K \| R_{\lambda_0} \| \, \| E(\Delta)x \|.$$

Now we choose an interval $[\alpha, \beta)$ so small that $K \| R_{\lambda_0} \| < \frac{1}{2}$. Then we have

$$\| E(\Delta)x \| \leqq \tfrac{1}{2} \| E(\Delta)x \|.$$

This, however, is only possible if $E(\Delta)x = 0$, and since $x \in H$ was arbitrary, $E(\Delta) = 0$. Therefore, for every half open interval $\Delta' \subset \Delta$, a fortiori, $E(\Delta') = 0$. This means that E_λ is constant in $[\alpha, \beta)$.

From Theorems 1 and 2, we obtain the

COROLLARY. *The set of regular real points of a self adjoint operator A is open. Consequently, the spectrum of A is a closed set on the real number line.*

The closedness of the spectrum is a general property of the bounded linear operators on Banach spaces.

EIGENVALUES OF A SELF ADJOINT OPERATOR.

THEOREM 3. *λ_0 is an eigenvalue of a self adjoint operator A if and only if λ_0 is a point of discontinuity of E_λ.*

Necessity. Let, for a certain $x_0 \neq 0$,

$$Ax_0 - \lambda_0 x_0 = 0.$$

Then

$$((A - \lambda_0 E)^2 x_0, x_0) = 0$$

and consequently

$$\int_m^{M+\varepsilon} (\lambda - \lambda_0)^2 \, d(E_\lambda x_0, x_0) = 0.$$

Since the integrand is non-negative and the weight increases monotonically,

$$\int_\alpha^\beta (\lambda - \lambda_0)^2 d(E_\lambda x_0, x_0) = 0$$

for arbitrary intervals $[\alpha, \beta)$. In particular, for every $\varepsilon > 0$,

$$\int_{\lambda_0+\varepsilon}^{M+\varepsilon} (\lambda - \lambda_0)^2 \, d(E_\lambda x_0, x_0) = 0.$$

Since $(\lambda - \lambda_0)^2 \geqq \varepsilon^2$ on the interval of integration, a fortiori,

$$\varepsilon^2 \int_{\lambda_0+\varepsilon}^{M+\varepsilon} d(E_\lambda x_0, x_0) = \varepsilon^2 [(x_0, x_0) - (E_{\lambda_0+\varepsilon} x_0, x_0)] = 0.$$

Consequently

$$(x_0, x_0) - (E_{\lambda_0+\varepsilon} x_0, x_0) = 0,$$

i.e.,

$$E_{\lambda_0+\varepsilon} x_0 = x_0. \tag{1}$$

Analogously,

$$\int_m^{\lambda_0-\varepsilon} (\lambda - \lambda_0)^2 \, d(E_\lambda x_0, x_0) = 0,$$

from which we obtain, on invoking $E_m = 0$,

$$E_{\lambda_0 - \varepsilon} x_0 = 0. \tag{2}$$

It follows from (1) and (2) that

$$(E_{\lambda_0+\varepsilon} - E_{\lambda_0 - \varepsilon}) x_0 = x_0,$$

and since ε was arbitrary,

$$(E_{\lambda_0+0} - E_{\lambda_0}) x_0 = x_0.$$

Thus λ_0 is a point of discontinuity of E_λ, where the eigenelement x_0 lies in the subspace corresponding to the projection operator $E_{\lambda_0+0} - E_{\lambda_0}$.

Sufficiency. Let $E_{\lambda_0+0} \neq E_{\lambda_0}$ and let x_0 be an arbitrary element of the subspace corresponding to $E_{\lambda_0+0} - E_{\lambda_0}$. Then

$$(E_{\lambda_0+0} - E_{\lambda_0}) x_0 = 0,$$

i.e., x_0 lies in the orthogonal complement of $L_{E_{\lambda_0}}$ in $L_{E_{\lambda_0+0}}$. Therefore

$$E_{\lambda_0+0} x_0 = x_0, \quad E_{\lambda_0} x_0 = 0.$$

Hence, certainly,

$$E_\lambda x_0 = x_0 \text{ for } \lambda > \lambda_0$$

and consequently

$$E(\Delta) x_0 = x_0 \text{ for } \Delta = [\lambda_0, \lambda_0 + \varepsilon).$$

Then

$$A x_0 = A E(\Delta) x_0 = \int_{\lambda_0}^{\lambda_0+\varepsilon} \lambda \, dE_\lambda x_0,$$

$$\lambda_0 x_0 = \lambda_0 E(\Delta) x_0 = \int_{\lambda_0}^{\lambda_0+\varepsilon} \lambda_0 \, dE_\lambda x_0$$

and therefore

$$A x_0 - \lambda_0 x_0 = \int_{\lambda_0}^{\lambda_0+\varepsilon} (\lambda - \lambda_0) \, dE_\lambda x_0.$$

This implies

$$\| A x_0 - \lambda_0 x_0 \| \leqq \varepsilon \| E(\Delta) x_0 \|,$$

and since ε was arbitrary,

$$\| A x_0 - \lambda_0 x_0 \| = 0.$$

At the same time we have found that the whole subspace corresponding to the operator $E_{\lambda_0+0} - E_{\lambda_0}$ consists of the eigenelements of A which belong to λ_0.

With this, we close the discussion of the spectral theory of linear operators in Hilbert spaces.

CHAPTER VI

SEVERAL TOPICS IN NON-LINEAR FUNCTIONAL ANALYSIS

In this chapter we will study differentiation and integration in abstract spaces and discuss several applications.

§38. DIFFERENTIATION OF ABSTRACT FUNCTIONS OF REAL VARIABLES

Let E be a normed linear space and K the closed interval [0, 1] of the real number line. We consider an operator $x = x(t)$, which need not be linear and maps K into E. In the following, we will call such an operator an abstract function on the interval [0, 1].

For these functions, we shall define and deduce properties of the fundamental operations of analysis: differentiation and integration.

DIFFERENTIATION. We consider the function $x = x(t)$ with $x \in E$ and $t \in [0, 1]$. We define the derivative $x'(t)$ by

$$\frac{d}{dt}x(t) = x'(t) = \lim_{\Delta t \to 0} \frac{1}{\Delta t}[x(t+\Delta t) - x(t)], \tag{1}$$

in case the limit in the right-hand side exists. It follows that

$$x'(t) = \frac{1}{\Delta t}[x(t+\Delta t) - x(t)] + \alpha(t, \Delta t),$$

where $$\alpha(t, \Delta t) \to 0 \text{ as } \Delta t \to 0.$$

Therefore $$x(t+\Delta t) - x(t) = x'(t)\Delta t - \alpha(t, \Delta t)\Delta t. \tag{2}$$

As $\Delta t \to 0$, the right side of the equation (2) tends to 0. Hence if $x(t)$ has a derivative with repect to t, then $x(t)$ is continuous at the point t.

As an example we mention differentiation of a vector function $x_n(t)$. If we interpret t as time, then $x_n'(t)$ is the velocity vector.

To the transition from the mechanics of a system of points to the mechanics of a continuum there corresponds the transition from an n-dimensional vector $x_n(t)$ to a time (t) dependent element $x(t)$ of a certain function space. Then $x'(t)$ is the velocity.

In one-dimensional problems (string, rod, etc.) we consider $x(t)$ and $x'(t)$ as elements of the space $C[0, 1]$.

We easily recognize the following properties of differentiation:

1. $[x(t) + y(t)]' = x'(t) + y'(t)$.
2. $[\lambda x(t)]' = \lambda x'(t)$ for all numbers λ.
3. Let there be defined for the elements $x \in E$ a left-sided (right-sided) multi-

plication with elements $y \in E$. Suppose also that it is continuous and distributive with respect to addition and commutative with scalar multiplication. Then

$$[yx(t)]' = yx'(t), \tag{3}$$

i.e., constant factors can be removed from under the differentiation symbol.

Properties 1. and 2. are clear, 3. follows from the continuity and distributivity of left-sided multiplication, namely

$$\begin{aligned} yx'(t) &= y \lim_{\Delta t \to 0} \frac{1}{\Delta t} [x(t+\Delta t) - x(t)] \\ &= \lim_{\Delta t \to 0} y \left\{ \frac{1}{\Delta t} [x(t+\Delta t) - x(t)] \right\} \\ &= \lim_{\Delta t \to 0} \frac{1}{\Delta t} [yx(t+\Delta t) - yx(t)] = \frac{d}{dt} [yx(t)]. \end{aligned}$$

Analogously, we obtain
$$[x(t)y]' = x'(t)y \tag{3a}$$
for right-sided multiplication.

Example. Let $x = x(t)$ with $x \in E$ and let A be an operator from $(E \to E_1)$.

$$[Ax(t)]' = Ax'(t). \tag{4}$$

If $A = A(t) \in (E \to E_1)$ and $x \in E$, then

$$[A(t)x]' = A'(t)x. \tag{4a}$$

In particular, for a linear functional $f \in \bar{E}$:

$$\{f[x(t)]\}' = f[x'(t)], \tag{5}$$
$$\{f(t)(x)\}' = f'(t)(x). \tag{5a}$$

DERIVATIVES OF HIGHER ORDER. We will now define derivatives of higher order. As in the case of ordinary functions,[29] we can give two definitions of the nth derivative of an abstract function $x = x(t)$.

1.
$$\bar{\Delta}^n_{\Delta t} x(t) = \sum_{k=0}^{n} (-1)^{n-k} \binom{n}{k} x(t + k\Delta t),$$

is the nth difference of $x(t)$ at the point t. Further, we call

$$\Delta^n_{\Delta t} x(t) = \bar{\Delta}^n_{\Delta t} x \left(t - \frac{n}{2} \Delta t\right)$$

the nth *central difference*.

The expression
$$x^{[n]}(t) = \lim_{\Delta t \to 0} \frac{1}{(\Delta t)^n} \Delta^n_{\Delta t} x(t) \tag{6}$$

shall be called — under the hypothesis that this limit exists — the *nth difference derivative* of the function $x(t)$ at the point t.

If the limiting process in (6) is uniform in the neighborhood of every point t, then $x^{[n]}(t)$ is called the *uniform* nth difference-derivative.

2. The nth derivative $x^{(n)}(t)$ is defined by differentiating successively n times:

[29] See Appendix II.

$$x'(t)_0 = \frac{d}{dt}\, x(t),\; x''(t)_0 = \frac{d}{dt}\,[x'(t)_0], \ldots, x^{(n)}(t)_0 = \frac{d}{dt}\,[x^{(n-1)}(t)_0].$$

THEOREM. *If the continuous nth derivative $x^{(n)}(t)_0$ exists in a neighborhood of a point t, then in this neighborhood the uniform nth difference-derivative $x^{[n]}(t)$ exists too, and*

$$x^{[n]}(t) = x^{(n)}(t)_0.$$

Conversely, if in a neighborhood of the point t the uniformly continuous, uniform difference-derivative $x^{[n]}(t)$ exists, then there exists in this neighborhood the nth derivative $x^{(n)}(t)_0$.

These statements hold for functions whose ranges consist of numbers. The transition to abstract functions is accomplished by a method which is often applied in functional analysis. We shall carry this out for the first statement.

For an arbitrary linear functional $f \in \bar{E}$,

$$\varphi(t) = f[x(t)]$$

is a function whose domain of definition and range consist of numbers. Because of (5), we obtain

$$f[x'(t)_0] = \{f[x(t)]\}' = \varphi'(t)$$
$$f[x''(t)_0] = \{f[x'(t)_0]\}' = \{\varphi'(t)\}' = \varphi''(t),$$
$$\cdots\cdots\cdots\cdots\cdots\cdots$$
$$f[x^{(n)}(t)_0] = \{f[x^{(n-1)}(t)_0]\}' = \{\varphi^{n-1}(t)\}' = \varphi^{(n)}(t).$$

Furthermore

$$f\left[\frac{1}{(\Delta t)_n}\, \Delta^n_{\Delta t}\, x(t)\right] = \frac{1}{(\Delta t)^n} \sum_{k=0}^{n} (-1)^{n-k} \binom{n}{k} \varphi\left(t + \left(k - \frac{n}{2}\right) \Delta t\right)$$
$$= \frac{1}{(\Delta t)^n}\, \Delta^n_{\Delta t} \varphi(t) = \varphi^{(n)}(t + \theta \Delta t) = f[x^{(n)}(t + \theta \Delta t)]^{30}$$

where $-\frac{1}{2} \leqq \theta \leqq \frac{1}{2}$. Since $x^{(n)}(t)_0$, according to the hypothesis, is continuous in a neighborhood of the point t,

$$\| x^{(n)}(t + \theta \Delta t)_0 - x^{(n)}(t)_0 \| \leqq \varepsilon_{\Delta t},$$

where $\varepsilon_{\Delta t} \to 0$ as $\Delta t \to 0$ uniformly in a neighborhood of the point t.

From this we have

$$\left| f\left[\frac{1}{(\Delta t)^n}\, \Delta^n_{\Delta t} x(t)\right] - f[x^{(n)}(t)_0] \right| \leqq \varepsilon_{\Delta t} \|f\|. \tag{7}$$

The inequality (7) holds for arbitrary $f \in \bar{E}$, therefore

$$\left\| \frac{1}{(\Delta t)^n}\, \Delta^n_{\Delta t}\, x(t) - x^{(n)}(t)_0 \right\| \leqq \varepsilon_{\Delta t}$$

and consequently $\qquad x^{(n)}(t)_0 = \lim_{\Delta t \to 0} \dfrac{1}{(\Delta t)^n} \Delta^n_{\Delta t}\, x(t).$

[30] See Appendix II.

In this case, the convergence is uniform in a neighborhood of every point t, q.e.d.

PARTIAL DERIVATIVES. We introduce the concept of the *partial derivative* of an abstract function. To do this, we consider a function of n real variables $t_1, t_2, \dots, t_n$ with a range lying in a normed linear space E: $y = f(t_1, t_2, \dots, t_n) \in E$. We can interpret $t_1, t_2, \dots, t_n$ as components of the n-dimensional vector $T = \sum_{i=1}^{n} t_i e_i$, where the e_i are orthonormal basis vectors, i.e., n-dimensional mutually orthogonal unit vectors.

We now define the *nth partial difference-derivative*

$$\frac{\partial^n}{\partial t_1 \partial t_2 \dots \partial t_n} f(t_1, t_2, \dots, t_n)$$

at the point $T_0 = \sum_{i=1}^{n} t_i^{(0)} e_i$.

To do this, we form the nth partial difference:

$$\Delta^n_{t_1, t_2, \dots, t_n; \Delta t} f(T_0) = \sum_{i_1, i_2, \dots, i_k} (-1)^{n-k} f[T_0 + \Delta t(e_{i_1} + e_{i_2} + \dots + e_{i_k})].$$

Here $0 \leqq i_1 < i_2 < \dots < i_k \leqq n\, (i_1, i_2, \dots, i_k)$ is a subset of $(1, 2, \dots, n)$. The summation extends over all such subsets. For the empty set we put $k = 0$ and T_0 as the argument. The nth central partial difference at the point T_0

$$\bar{\Delta}^n_{t_1, t_2, \dots, t_n; \Delta t} f(T_0)$$

is the nth partial difference at the point

$$T_0' = T_0 - \frac{1}{2} \Delta t \sum_{i=1}^{n} e_i.$$

Therefore

$$\Delta^n_{t_1, t_2, \dots, t_n; \Delta t} f(T_0) = \bar{\Delta}^n_{t_1, t_2, \dots, t_n; \Delta t} f\left(T_0 - \frac{1}{2} \Delta t \sum_{i=1}^{n} e_i \right).$$

Then the limit as $\Delta t \to 0$ of

$$\frac{1}{(\Delta t)^n} \Delta^n_{t_1, t_2, \dots, t_n; \Delta t} f(T_0),$$

in case it exists, is called the *nth partial difference derivative*

$$\frac{\partial^n}{\partial t_1 \partial t_2 \dots \partial t_n} f(t_1^{(0)}, t_2^{(0)}, \dots, t_n^{(0)})$$

of the function f at the point $T_0 = (t_1^{(0)}, t_2^{(0)}, \dots, t_n^{(0)})$.

Parallel to this, we can define the *nth partial derivative*. It is the result of carrying out successive differentiations of the function $f(t_1, t_2, \dots, t_n)$ with respect to $t_{k_n}, t_{k_{n-1}}$, and finally t_{k_1}, where $k_1, k_2, \dots, k_n$ is an arbitrary permutation of the indices $1, 2, \dots, n$. Of course, we have to assume that the derivatives

$$\frac{\partial}{\partial t_{k_n}} f(t_1, t_2, \dots, t_n), \frac{\partial}{\partial t_{k_{n-1}}} \left(\frac{\partial}{\partial t_{k_n}} f(t_1, t_2, \dots, t_n) \right),$$

$$\dots, \frac{\partial}{\partial t_{k_i}} \left\{ \frac{\partial}{\partial t_{k_{i+1}}} \left(\dots \frac{\partial}{\partial t_{k_n}} f(t_1, t_2, \dots, t_n) \right) \right\}, \dots$$

obtained successively exist in a neighborhood of T_0.

THEOREM. *If in a neighborhood of the point* $T_0 = (t_1^{(0)}, t_2^{(0)}, \ldots, t_n^{(0)})$ *there exists an nth partial derivative of the function f and if this derivative is continuous at* T_0, *then there also exists at* T_0 *the nth partial difference-derivative and both derivatives coincide.*

Proof. Let l be an arbitrary linear functional of $\bar{E}$. Then

$$l[f(t_1, t_2, \ldots, t_n)] = lf(t_i)$$

is a function of $t_1, t_2, \ldots, t_n$, whose range consists of numbers. We have [31]

$$\Delta^n_{t_1,t_2,\ldots,t_n;\Delta t}\, lf(t_i^{(0)}) = \frac{\partial}{\partial t_{k_1}}\left\{\cdots \frac{\partial}{\partial t_{k_n}} lf(t_i^{(0)} + \theta_i \Delta t)\right\}$$

for every permutation $k_1, k_2, \ldots, k_n$ of the indices $1, 2, \ldots, n$ and

$$l\left[\frac{1}{(\Delta t)^n}\,\Delta^n_{t_1,t_2,\ldots,t_n;\Delta t}\, f(t_i^{(0)})\right] = \frac{1}{(\Delta t)^n}\,\Delta^n_{t_1,t_2,\ldots,t_n;\Delta t}\, lf(t_i^{(0)})$$

$$= \frac{\partial}{\partial t_{k_1}}\left\{\cdots \frac{\partial}{\partial t_{k_n}} lf(t_i^{(0)} + \theta_i \Delta t)\right\}$$

$$= l\left[\frac{\partial}{\partial t_{k_1}}\left\{\cdots \frac{\partial}{\partial t_{k_n}} f(t_i^{(0)} + \theta_i \Delta t)\right\}\right],$$

because the linear functional l can be removed from under the derivative sign.

According to the hypothesis, the nth partial derivative of $f(t_i)$ is continuous at the point $(t_i^{(0)})$, so that for an arbitrary $\varepsilon_{\Delta t} > 0$,

$$\left\|\frac{\partial}{\partial t_{k_1}}\left\{\cdots \frac{\partial}{\partial t_{k_n}} f(t_i^{(0)} + \theta_i \Delta t)\right\} - \frac{\partial}{\partial t_{k_1}}\left\{\cdots \frac{\partial}{\partial t_{k_n}} f(t_i^{(0)})\right\}\right\| \leqq \varepsilon_{\Delta t}$$

for sufficiently small $|\Delta t|$. Hence

$$\left|l\left[\frac{1}{(\Delta t)^n}\,\Delta^n_{t_1,t_2,\ldots,t_n;\Delta t} f(t_i^{(0)})\right\} - l\left[\frac{\partial}{\partial t_{k_1}}\left\{\cdots \frac{\partial}{\partial t_{k_n}} f(t_i^{(0)})\right\}\right]\right|$$

$$= \left|l\left[\frac{\partial}{\partial t_{k_1}}\left\{\cdots \frac{\partial}{\partial t_{k_n}} f(t_i^{(0)} + \theta_i \Delta t)\right\}\right] - l\left[\frac{\partial}{\partial t_{k_1}}\left\{\cdots \frac{\partial}{\partial t_{k_n}} f(t_i^{(0)})\right\}\right]\right|$$

$$\leqq \|l\|\left\|\frac{\partial}{\partial t_{k_1}}\left\{\cdots \frac{\partial}{\partial t_{k_n}} f(t_i^{(0)} + \theta_i \Delta t)\right\} - \frac{\partial}{\partial t_{k_1}}\left\{\cdots \frac{\partial}{\partial t_{k_n}} f(t_i^{(0)})\right\}\right\| \leqq \|l\|\,\varepsilon_{\Delta t}.$$

This inequality is valid for every $l \in \bar{E}$, therefore

$$\left\|\frac{1}{(\Delta t)^n}\,\Delta^n_{t_1,t_2,\ldots,t_n;\Delta t} f(t_i^{(0)}) - \frac{\partial}{\partial t_{k_1}}\left\{\cdots \frac{\partial}{\partial t_{k_n}} f(t_i^{(0)})\right\}\right\| \leqq \varepsilon_{\Delta t}.$$

It follows that

---

[31] See Appendix II.

$$\frac{1}{(\Delta t)^n} \Delta^n_{t_1,t_2,\dots,t_n;\Delta t} f(t_i^{(0)}) \to \frac{\partial}{\partial t_{k_1}} \left\{ \dots \frac{\partial}{\partial t_{k_n}} f(t_i^{(0)}) \right\}$$

as $\Delta t \to 0$, and we have proved the existence of the limit of the expression

$$\frac{1}{(\Delta t)^n} \Delta^n_{t_1,t_2,\dots,t_n;\Delta t} f(t_i^{(0)})$$

and its equality with the nth partial derivative.

COROLLARY. *Two nth partial derivatives which correspond to different permutations of the indices* $1, 2, \dots, n$, *coincide at the points where they both are continuous. This means that the nth partial derivative does not depend on the order of the differentiations.*

The equality

$$\frac{\partial}{\partial t_{k_1}} \left\{ \dots \frac{\partial}{\partial t_{kn}} f(t_i) \right\} = \lim_{\Delta t \to 0} \frac{1}{(\Delta t)^n} \Delta^n_{t_1, t_2,\dots,t_n;\Delta t} f(t_i).$$

holds in this case. The right hand side, however, does not depend on the permutation $k_1, \dots, k_n$.

In the sequel, if we speak about the nth partial derivative in a certain region, we shall assume also that it is continuous in this region. Therefore, both definitions give the same result and we shall designate this derivative by the symbol

$$\frac{\partial^n}{\partial t_1 \partial t_2 \dots \partial t_n} f(t_i).$$

If h_i are arbitrary elements of the space E, and if

$$y = f\left(\sum_{i=1}^{n} t_i h_i\right) \in E,$$

holds, then y is a function of the n parameters $t_1, t_2, \dots, t_n$. The function

$$f\left(\sum_{i=1}^{n} t_i h_i\right)$$

has, for arbitrary $h_1, \dots, h_n$, an nth partial derivative

$$\left[\frac{\partial^n}{\partial t_1 \partial t_2 \dots \partial t_n} f\left(\sum_{i=1}^{n} t_i h_i\right)\right]_{t_1 = \dots = t_n = 0}.$$

If $h_1 = \dots = h_n$, then it is not difficult to show that

$$\left[\frac{\partial^n}{\partial t_1 \partial t_2 \dots \partial t_n} f\left(x + \sum_{i=1}^{n} t_i h_i\right)\right]_{t_1 = \dots = t_n = 0} = \frac{d^n}{dt^n} f(x + th)_{t=0}$$

holds.

§39. INTEGRATION OF ABSTRACT FUNCTIONS AND SOLUTIONS OF DIFFERENTIAL EQUATIONS

INTEGRATION. We consider a partition $A = A[t_0, t_1, \dots, t_n]$ of the closed interval $[a, b]$ into closed subintervals $[t_i, t_{i+1}]$ with $t_0 = a < t_1 < \dots < t_n = b$.

The partition $B = B[t_0', t_1', \dots, t_m']$ is called a *refinement* of $A = A[t_0, t_1, \dots, t_n]$, if every interval $[t_i', t_{i+1}']$ is contained in an interval $[t_l, t_{l+1}]$. In the partition B, therefore, every interval of A is again partitioned. If every interval $[t_i, t_{i+1}]$ of the partition A has a length which does not exceed a positive number $\tau, t_{i+1} - t_i \leqq \tau$, then A is called a *τ-partition* of the interval $[a, b]$ and is designated by A_τ.

We call
$$S(A, x(t)) = \sum_{i=0}^{n-1} x(t_i)(t_{i+1} - t_i) \tag{1}$$
an *integral sum* of the abstract function $x(t)$ with $x \in E$ and $t \in [a, b]$ for the partition $A = A[t_0, t_1, \ldots, t_n]$.

THEOREM. *Let $x(t)$ be a continuous function in a complete space E with $t \in [a, b]$ and A_{τ_n} a sequence of partitions of the interval $[0, 1]$ such that $\tau_n \to 0$ as $n \to \infty$. If we now form the integral sums $S(A_{\tau_n}, x(t))$, then they tend to a limit $S = \lim_{n\to\infty} S(A_{\tau_n}, x(t))$ as $n \to \infty$. The limit S does not depend on the choice of the A_{τ_n}.*

The proof of the theorem is based on the following three lemmas.

LEMMA 1. *Every continuous function $x(t) \in E$ with $t \in [a, b]$ is uniformly continuous, i.e., for every positive number ε we can determine a number τ such that for $t', t'' \in [a, b]$,*
$$\|x(t') - x(t'')\| < \varepsilon, \tag{2}$$
whenever $|t' - t''| < \tau$.

Proof. We use $\varphi(t, t')$ to designate the function which is defined on the square $a \leqq t \leqq b, a \leqq t' \leqq b$, by
$$\varphi(t, t') = \|x(t) - x(t')\|.$$
The function φ is continuous on this square, therefore uniformly continuous. Consequently, we can determine for every $\varepsilon > 0$, a number τ such that
$$|\varphi(t, t'') - \varphi(t', t'')| < \varepsilon \text{ for } |t - t'| < \tau \tag{3}$$
is fulfilled.

Furthermore, for $t' = t, \varphi(t, t) = \|x(t) - x(t)\| = 0$. Hence according to (3)
$$\varphi(t, t') = |\varphi(t, t') - \varphi(t, t)| < \varepsilon \text{ for } |t - t'| < \tau$$
or
$$\|x(t') - x(t)\| < \varepsilon \text{ for } |t - t'| < \tau,$$
q.e.d.

LEMMA 2. *If the partition W of the closed interval $[a, b]$ is a refinement of the τ-partition $V = V_\tau$ then*
$$\|S(V, x(t) - S(W, x(t))\| \leqq \varepsilon(b - a). \tag{4}$$

Proof. If the point t' belongs to the interval $[t_i, t_{i+1}]$ of the τ-partition V_τ, then $|t' - t_i| \leqq t_{i+1} - t_i \leqq \tau$. Hence, on account of (2),
$$\|x(t') - x(t_i)\| \leqq \varepsilon. \tag{5}$$

Let n, m be the numbers of the intervals $[t_l, t_{l+1}]$, $[t'_j, t'_{j+1}]$ of V, W, respectively. Since W is a refinement of V, each of the points $t_l, l = 1, 2, \ldots, n$ coincides with one of the points $t'_j : t'_{k_l} = t_l$ (we have, therefore, the ordering $0 = k_0 < k_1 < \ldots < k_n = m, m \geqq n$). Therefore, the interval $[t_l, t_{l+1}]$ of V is partitioned into $k_{l+1} - k_l$ intervals $[t'_j, t'_{j+1}]$ of partition W, where $j = k_l, k_l + 1, \ldots, k_{l+1} - 1$. According to (5), we have for every j
$$\|x(t'_j) - x(t_l)\| \leqq \varepsilon.$$

Therefore

$$S(V,x(t)) = \sum_{l=0}^{n-1} x(t_l)(t_{l+1}-t_l) = \sum_{l=0}^{n-1} x(t_l) \sum_{j=k_l}^{k_{l+1}-1} (t'_{j+1}-t'_j),$$

$$S(W,x(t)) = \sum_{j=0}^{m-1} x(t'_j)(t'_{j+1}-t'_j) = \sum_{l=0}^{n-1} \sum_{j=k_l}^{k_{l+1}-1} x(t'_j)(t'_{j+1}-t'_j),$$

$$\|S(V,x(t)) - S(W,x(t))\| = \left\| \sum_{l=0}^{n-1} \sum_{j=k_l}^{k_{l+1}-1} [x(t_l)-x(t'_j)](t'_{j+1}-t'_j) \right\|$$

$$\leqq \sum_{l=0}^{n-1} \sum_{j=k_l}^{k_{l+1}-1} \|x(t_l)-x(t'_j)\|(t'_{j+1}-t'_j)$$

$$\leqq \sum_{l=0}^{n-1} \sum_{j=k_l}^{k_{l+1}-1} \varepsilon(t'_{j+1}-t'_j) = \varepsilon(b-a),$$

q.e.d.

LEMMA 3. *Let V_τ and $V_{\tau'}$ be arbitrary τ, τ' partitions, respectively, of the closed interval $[a,b]$. Then*

$$\|S(V_\tau,x(t)) - S(V_{\tau'},x(t))\| \leqq (\varepsilon_\tau + \varepsilon_{\tau'})(b-a). \tag{6}$$

Proof. We can always select a partition W of the interval $[a,b]$ which is a refinement of both V_τ and $V_{\tau'}$. Then, according to (4),

$$\|S(V_\tau,x(t)) - S(W,x(t))\| \leqq \varepsilon_\tau(b-a)$$

and

$$\|S(V_{\tau'},x(t)) - S(W,x(t))\| \leqq \varepsilon_\tau(b-a),$$

from which

$$\|S(V_\tau,x(t)) - S(V_{\tau'},x(t))\| \leqq (\varepsilon_\tau + \varepsilon_{\tau'})(b-a)$$

follows, q.e.d.

Now we prove the theorem formulated at the beginning. For a given ε, we consider a sequence of τ_n-partitions V_{τ_n} of the interval $[a,b]$, where $\tau_n \to 0$ as $n \to \infty$. Because of Lemma 3, the inequality

$$\|S(V_{\tau_n},x(t)) - S(V_{\tau_{n+p}},x(t))\| \leqq (\varepsilon_{\tau_n} + \varepsilon_{\tau_{n+p}})(b-a)$$

holds for the corresponding integral sums $S(V_{\tau_n},x(t))$.

Since it holds for every ε, the sequence $S(V_{\tau_n},x(t))$ is a fundamental sequence and because $S(V_{\tau_n},x(t)) \in E$ and E is complete, this sequence has a limit S in E.

Now let $V_{\tau_n'}$ be another sequence of partitions of the interval $[a,b]$ where $\tau' \to 0$. Using the preceding argument as a basis, $S(V_{\tau_n'},x(t))$ converges to a limit S_1 as $n \to \infty$. We shall prove that $S_1 = S$.

To do this, we consider the sequence $V_{\tau_1}, V_{\tau'_1}, V_{\tau_2}, V_{\tau'_2}, \ldots$ The integral sums $S(V_{\tau_n},x(t)), S(V_{\tau_n'},x(t))$, respectively, corresponding to this sequence, form a convergent sequence. If its limit is equal to S_2, then it must be equal to the limits S and S_1. Consequently, $S = S_1 = S_2$, and the theorem is proved.

We call S the *integral* of $x(t)$ over $[a,b]$ and designate it by $\int_a^b x(t)dt$.

We can also deal with a complicated definition of integral (e.g., Stieltjes-integral).

A concrete example for this was dealt with on p. 161, where the spectral representation of a self adjoint operator was given.

Properties of the Integral.

1) $$\int_a^b [x(t)+y(t)]dt = \int_a^b x(t)dt + \int_a^b y(t)dt.$$

2) For a fixed scalar λ: $$\int_a^b \lambda x(t)dt = \lambda \int_a^b x(t)dt.$$

These properties are evident.

3) $$\left\| \int_a^b x(t)dt \right\| \leqq \int_a^b \|x(t)\|\, dt. \tag{7}$$

For, $$\|S(V_\tau, x(t))\| = \left\| \sum_{i=0}^{n-1} x(t_i)(t_{i+1}-t_i) \right\|$$
$$\leqq \sum_{i=0}^{n-1} \|x(t)\|(t_{i+1}-t_i) = S(V_\tau, \|x(t)\|). \tag{8}$$

As $\tau \to 0$, the integral sums tend to the integrals

$$\int_a^b x(t)dt \text{ and } \int_a^b \|x(t)\|\, dt,$$

and the inequality (8) goes over to (7), q.e.d.

4) If there is defined a right-sided (or left-sided) multiplication for the elements $x \in E$ with elements $y \in E_1$, (see p. 169), then

$$\int_a^b x(t)\, y\, dt = \int_a^b x(t)\, dt\, y. \tag{9}$$

For, $$S(V_\tau, x(t)y) = \sum_{i=0}^{n-1} x(t_i)y(t_{i+1}-t_i)$$
$$= \left(\sum_{i=0}^{n-1} x(t_i)(t_{i+1}-t_i) \right) y = S(V_\tau, x(t))y \tag{10}$$

holds for a τ-partition $V_\tau(t_0, t_1, \ldots, t_n)$. As $\tau \to 0$, equation (10) goes over to (9).

Analogously, for left-sided multiplication,

$$\int_a^b yx(t)dt = y \int_a^b x(t)\, dt \tag{9'}$$

holds.

Examples. 1. If A is a linear operator mapping E into E_1, and $x(t) \in E$, then

$$\int_a^b Ax(t)\, dt = A \int_a^b x(t)\, dt.$$

2. If $A = A(t)$ is an operator from $(E \to E_1)$, which depends continuously on t, and if $x \in E$, then

$$\int_a^b A(t)\,x\,dt = \left(\int_a^b A(t)\,dt\right) x.$$

In the first example, the operator A is a constant left-sided factor and in the second example, x is a constant right-sided factor.

If, in particular, f is a linear functional from $\bar{E}$, then

$$f\left(\int_a^b x(t)\,dt\right) = \int_a^b f(x(t))\,dt, \int_a^b f(t)(x)\,dt = \left(\int_a^b f(t)\,dt\right)(x). \tag{10'}$$

5) If the function $x = x(t)$, $x \in E$, $t \in [a, b]$, has a continuous derivative with respect to t, $x'(t) = \frac{d}{dt}\,x(t)$, then

$$\int_a^b x'(t)\,dt = x(b) - x(a). \tag{11}$$

For, according to (10'), we obtain, for an arbitrary linear functional l,

$$l\left(\int_a^b x'(t)\,dt\right) = \int_a^b l\,[x'(t)\,]\,dt = \int_a^b \frac{d}{dt}\{l[x(t)\,]\}\,dt.$$

$[x\ (t)\ ]$ is, however, a function of t whose range consists of numbers. Therefore:

$$\int_a^b \{l\,[x(t)]\}'\,dt = l\,[x(b)] - l\,[x(a)].$$

Hence,

$$l\left(\int_a^b x'(t)\,dt\right) = l\,[x(b) - x(a)]. \tag{12}$$

The equation (12) holds for an arbitrary linear functional $l \in \bar{E}$. With this, (11) is proved.

SOLVING DIFFERENTIAL EQUATIONS. We consider the differential equation

$$\frac{dx}{dt} = f(t, x). \tag{13}$$

$x = x(t)$ and $f(t, x)$ are elements of a normed space E; let $t \in [a, b]$. We assume that $f(t, x)$ is continuous in t and that, with respect to x, the Lipschitz condition

$$\|f(t, x_1) - f(t, x_2)\| \leqq M\,\|x_1 - x_2\| \tag{14}$$

is satisfied.

We designate by $C^E[a, b]$ the space of continuous functions $x(t)$ with $t \in [a, b]$ and $x(t) \in E$. We introduce a norm in $C^E[a, b]$ by

$$\|x\|_C = \max_{a \leqq t \leqq b} \|x(t)\|.$$

Besides E, $C^E[a, b]$ is a complete space. We prove this assertion in the same way as that on p. 118 for the special case $E = R$ and $C^E[a, b] = C[0, 1]$.

Besides the equality (13), we consider

$$x(t) = x_0 + \int_{t_0}^{t} f(t, x(t))\,dt, \quad a \leqq t_0 \leqq t \leqq t_0 + \delta \leqq b. \tag{15}$$

We designate the right side of this equality by $A(x)$. A is an operator which transforms $x = x(t) \in C^E[t_0, t_0 + \delta]$ into an element of the same space. We have

$$\|A(x) - A(y)\|_C = \left\| \int_{t_0}^{t} [f(t, x(t)) - f(t, y(t))]\, dt \right\|_C$$

$$\leq \int_{t_0}^{t_0+\delta} \|f(t, x(t)) - f(t, y(t))\|\, dt.$$

It follows, according to (14), that

$$\|A(x) - A(y)\|_C \leq M \int_{t_0}^{t_0+\delta} \|x(t) - y(t)\|\, dt$$

$$\leq M\delta \max_{t_0 \leq t \leq t_0+\delta} \|x(t) - y(t)\| = M\delta \|x(t) - y(t)\|_C. \quad (16)$$

If $M\delta < 1$, then the operator A determines, according to (16), a contracting mapping of the space $C^E[a, b]$ into itself and consequently there exists exactly one solution of (15) (see §7).

The equation (15) is equivalent to (13) for the initial value $x(t_0) = x_0$. Therefore (13) has exactly one solution in the interval $[t_0, t_0 + \delta]$.

In particular, this equation has for an arbitrary initial value $x(a) = x_0$ exactly one solution $x(t)$ on the interval $[a, a + \delta]$. $x(t)$ can be continued to the whole interval $[a, b]$. For, if $a + \delta < b$ and $x(a + \delta) = x_1$, then we construct by repetition of this process a solution on the interval $[a + \delta, a + 2\delta]$ with the initial value x_1, etc.

Examples. 1. If E is the n–dimensional space, then we obtain the well-known existence theorem for a system of n differential equations.

2. If E is one of the spaces l_p (l_2), c, m, etc., then we obtain the existence theorem for a solution of the corresponding classes of infinite systems of differential equations.

3. We consider the integro-differential equation

$$\frac{dy}{dt} = f(t, y) + \int_a^b K(t, s)\, y(s)\, ds. \quad (17)$$

Let $f(t, y)$ satisfy the Lipshitz condition (14) and the kernel $K(t, s)$ be bounded, i.e., $|K(t, s)| < M_1$.

If the right side (17) is designated by $F(t, y)$, we obtain the inequality

$$|F(t, y) - F(t, z)| \leq M|y - z| + M_1(b - a)|y - z| = (M + M_1(b - a))|y - z|.$$

$F(t, y)$ satisfies the Lipschitz condition. Consequently, (17) has a unique solution for an arbitrary initial value $y = y_0$.

§40. HOMOGENEOUS FORMS AND POLYNOMIALS

MULTIPLICATION OF ELEMENTS. Often we have to deal with a multiplication where the factors and product are elements of different spaces.

We consider three linear spaces E_1, E_2, and E_3 and introduce a *multiplication* which associates with any two elements $x \in E_1$ and $y \in E_2$ an element $z = xy \in E_3$. Let this multiplication have the following properties:

1) $(x_1+x_2)y=x_1y+x_2y$,
2) $x(y_1+y_2)=xy_1+xy_2$ and
3) $x_n\to x, y_n\to y, x_ny_n\to xy$.

Multiplication xy is, for every fixed x an additive and continuous operation with respect to y. It is therefore a linear operation, defined on E_2, whose range lies in E_3. Similarly, multiplication xy is, for every fixed y, a linear operation mapping E_1 into E_3. It follows from this that

$$(\lambda x)y=\lambda(xy) \text{ and } x(\lambda y)=\lambda(xy).$$

Examples. 1. Let A, B be linear operators mapping E_1 into E_2, E_2 into E_3, respectively. Then BA is a linear operator transforming E_1 into E_3.

2. If $x(t)$ is a function from $L_p(0,1)$, and $y(t)\in L_q(0,1)$, then their product

$$z(t)=x(t)y(t)$$

is an element of $L_r(0,1)$, where $\dfrac{1}{r}=\dfrac{1}{p}+\dfrac{1}{q}$.

3. We can interpret the inner product of two elements x and y of a real Hilbert space as product of x and y: $x\in H$, $y\in H$, $xy=(x,y)\in(-\infty,\infty)=R$.

4. If $x\in E$ and A is a linear operator $\in(E\to E')$, then $Ax\in E'$ can be interpreted as the product of A and x:

$$A\in(E\to E'),\ x\in E,\ Ax\in E'.$$

We readily prove that for the enumerated examples all properties of the multiplication explained above are fulfilled.

THEOREM. *If E_1, E_2 and E_3 are linear normed spaces, then there exists a positive constant N such that*

$$\|xy\|_{E_3}\leqq N\|x\|_{E_1}\|y\|_{E_2}.$$

Let us assume the opposite. Then, for every natural number n, we can find elements $x_n\in E_1$ and $y_n\in E_2$ such that

$$\|x_ny_n\|\geqq n^2\|x_n\|\,\|y_n\|.$$

We put
$$x_n'=\frac{1}{n\|x_n\|}x_n,\quad y_n'=\frac{1}{n\|y_n\|}y_n$$

and obtain
$$\|x_n'y_n'\|=\frac{1}{n^2\|x_n\|\,\|y_n\|}\|x_ny_n\|\geqq 1.$$

On the other hand,
$$\|x_n'\|=\frac{1}{n},\quad \|y_n'\|\frac{1}{n}.$$

Therefore, $\|x_n'\|$ and $\|y_n'\|$ tend to 0, whereas $\|x_n'y_n'\|$ remains greater than 1. This, however, contradicts the continuity of multiplication.

n-TERMED LINEAR FORMS. Let $E_1, E_2, \ldots, E_n, E_0$ be linear normed spaces. A function $a_n(h_1,h_2,\ldots,h_n)$ with $h_i\in E_i$ and the range in E_0 is called an *n-termed linear form*, if it is linear and homogeneous in every one of the variables $h_1, h_2, \ldots, h_n$.

Bilinear forms will serve as an example. They were dealt with in §30. For them $E_1=E, E_2=E, E_0=R$.

We agree to write $a_n(h_1,h_2,\ldots,h_n)=a_nh_1h_2\ldots h_n$. We define the norm of a_n by

$$\| a_n \| = \sup \frac{\| a_n h_1 h_2 \dots h_n \|}{\| h_1 \| \| h_2 \| \dots \| h_n \|}. \tag{1}$$

Obviously

$$\| a_n h_1 \dots h_n \| \leqq \| a_n \| \| h_1 \| \| h_2 \| \dots \| h_n \|.$$

If

$$a_n h_1 \dots h_n = b_n h_1 \dots h_n$$

for arbitrary $h_1, h_2, \dots$, we shall write $a_n = b_n$. The totality of the forms $a_n = a_n h_1 \dots h_n$ with $h_i \in E_i$ and $a_n h_1 \dots h_n \in E_0$ forms a linear normed space, if we interpret the sum and the multiplication by a scalar in the customary sense and if we define the norm of a_n by the equality (1). We can therefore interpret the form $a_n h_1 \dots h_n$ as follows:

$$a_n \in (E_1 \to \{E_2 \to [E_3 \to \dots \to (E_n \to E_0) \dots]\})$$
$$a_n h_1 \in (E_2 \to [E_3 \to \dots \to (E_n \to E_0) \dots]),$$
$$\dots \dots \dots \dots \dots \dots$$
$$a_n h_1 \dots h_{n-1} \in (E_n \to E_0).$$

$a_n h_1 h_2 \dots h_n$ is, therefore, to be interpreted as a product formed from left to right with the elements $a_n, h_1, h_2, \dots, h_n$:

$$a_n h_1 h_2 \dots h_n = \{ \dots [(a_n h_1) h_2] h_3 \dots \} h_n.$$

A form is called *symmetric* if $E_1 = E_2 = \dots = E_n = E$ and $a_n h_1 h_2 \dots h_n = a_n h_{i_1} h_{i_2} \dots h_{i_n}$ hold, where $i_1, i_2, \dots, i_n$ is an arbitrary permutation of the indices $1, 2, \dots, n$.

As an example of a symmetric form we mention the bilinear form (Ax, y), where A is a self-adjoint operator on the real Hilbert space.

With every form $b_n h_1 h_2 \dots h_n$, a *symmetric form*

$$S b_n h_1 h_2 \dots h_n = \frac{1}{n!} \sum_{(i_1, i_2, \dots, i_n)} b_n h_{i_1} h_{i_2} \dots h_{i_n} \tag{2}$$

can be associated. The sum is to be taken over all permutations $i_1, i_2, \dots, i_n$ of the indices $1, 2, \dots, n$. If $a_n h_1 h_2 \dots h_n$ is already symmetric, then $S a_n = a_n$.

The form $a_n h \dots h$, which we obtain from the symmetric form $a_n h_1 \dots h_n$ by putting $h_1 = \dots = h_n = h$, is called a *homogeneous form of the nth degree.* We call a homogeneous form of the second degree quadratic (see §30).

We introduce for the homogeneous form of the nth degree, the abbreviation

$$a_n \underbrace{h \dots h}_{n\text{-times}} = a_n h^n.$$

PROPERTIES OF n-TERMED HOMOGENEOUS LINEAR FORMS.

1.
$$a_n (th)^n = t^n a_n h^n.$$

2. If a_n is a symmetric form, then $a_n h_1 h_2 \dots h_n$ is distributive and commutative with respect to every $h_1, h_2, \dots, h_n$. Therefore, for example, we obtain, according to the multinomial theorem

$$a_n (t_1 h_1 + t_2 h_2 + \dots + t_k h_k)^n$$
$$= \sum_{n_1 + \dots + n_k = n, n_i \geqq 0} \frac{n!}{n_1! \dots n_k!} t_1^{n_1} \dots t_k^{n_k} a_n h_1^{n_1} \dots h_k^{n_k} \tag{3}$$

3. $$\frac{\partial^n}{\partial t_1 \partial t_2 \ldots \partial t_n} \{a_n(t_1 h_1 + \ldots + t_n h_n)^n\} = n!\, a_n h_1 \ldots h_n. \tag{4}$$

For, the operator $\frac{\partial^n}{\partial t_1 \partial t_2 \ldots \partial t_n}$ makes all terms on the right side of the equation (3) with $k < n$ zero, except the summand which contains all factors $t_1, t_2, \ldots, t_n$, i.e., the term with $n_1 = n_2 = \ldots = n_n = 1$. It is equal to $n! t_1 \ldots t_n a_n h_1 \ldots h_n$ and consequently (4) holds.

4. If the equality $$a_n h^n = b_n h^n \tag{5}$$
holds for arbitrary $h \in E$, then the corresponding linear forms
$$a_n h_1 h_2 \ldots h_n \text{ and } b_n h_1 h_2 \ldots h_n \tag{6}$$
coincide for all $h_1, h_2, \ldots, h_n \in E$, i.e., $a_n = b_n$. For, according to (5),
$$a_n(t_1 h_1 + t_2 h_2 + \ldots + t_n h_n)^n = b_n(t_1 h_1 + t_2 h_2 + \ldots + t_n h_n)^n,$$
for arbitrary $h_1, h_2, \ldots, h_n$, from which
$$\frac{\partial^n}{\partial t_1 \ldots \partial t_n} a_n (t_1 h_1 + \ldots + t_n h_n)^n = \frac{\partial^n}{\partial t_1 \ldots \partial t_n} b_n (t_1 h_1 + \ldots + t_n h_n)^n$$
By virtue of (4), we obtain (6).

5. If a linear operator $A \in (E_1 \to E_2)$ and a homogeneous form $a_n h^n$ of the nth degree with $h \in E$ and $a_n h^n \in E_1$, are given, then also $A(a_n h^n)$ is a homogeneous form of the nth degree.

Let $a_n h_1 h_2 \ldots h_n$ be the corresponding symmetric form. Since $A(a_n h_1 h_2 \ldots h_n)$ depends linearly on every h_i, then $A(a_n h_2 \ldots h_n)$ is an n-termed linear form with respect to $h_1, h_2, \ldots, h_n$ with a range in E_2. This form is symmetric. Therefore $A(a_n h^n)$ is a homogeneous form of degree n.

6. The product $(a_n h^n)(b_m h^m)$ of homogeneous forms $a_n h^n$ and $b_m h^m$ of degrees n and m is a homogeneous form of degree $n + m$.

The product $(a_n h_1 \ldots h_n)(b_m h_{n+1} \ldots h_{n+m})$ of the corresponding n- and m-termed linear forms, respectively, is a linear form of $n + m$ variables. We consider now the associated symmetric form $c_{n+m} h_1 \ldots h_{n+m} = S[(a_n h_1 \ldots h_n)(b_m h_{n+1} \ldots h_{n+m})]$ and put $h_1 = \ldots = h_{n+m} = h$. We obtain the homogeneous form $c_{n+m} h^{n+m}$ of the $(n+m)$th degree. By virtue of (2), it is equal to $(a_n h^n)(b_m h^m)$.

POLYNOMIALS. A sum $P_n(h) = \sum_{k=1}^{n} a_k h^k$ of homogeneous forms with $h \in E$, where all $a_k h^k$ are elements of the space E_1, is called a *polynomial of nth degree in h*.

We mention several simple properties of polynomials.

1. If
$$y = P_n(h) = \sum_{k=1}^{n} a_k h^k \in E_1, \quad z = Q_m(h) = \sum_{l=1}^{m} b_l h^l \in E_2$$
are polynomials in h of degree n, m, respectively, and if a product of two arbitrary elements in E_1 and E_2 is defined, then
$$yz = P_n(h)\, Q_m(h)$$
is a polynomial in h of degree $n + m$.

As a matter of fact, because of distributivity of multiplication

$$yz = \left(\sum_{k=1}^{n} a_k h^k\right)\left(\sum_{l=1}^{m} b_l h^l\right) = \sum_{k=1}^{n} \sum_{l=1}^{m} (a_k h^k)(b_l h^l).$$

Each term $(a_k h^k)(b_l h^l)$ because of property 6. is a form of degree $k+l \leqq n+m$. The term $(a_n h^n)(b_m h^m)$ is a form of degree $n+m$. Therefore the whole product is a polynomial of degree $n+m$, q.e.d.

2. Let $P_n(h)$ and $Q_m(g)$ be polynomials of nth degree in h, mth degree in g, respectively. For $h = Q_m(g), P_n(h) = P_n[Q_m(g)]$ is a polynomial in g of degree nm.

We prove this by induction on n. For $n = 1$,

$$P_1(h) = Ah$$

where A is a linear operator in h. If

$$Q_m(g) = \sum_{k=1}^{m} b_k g^k$$

holds, then by virtue of property 5. of homogeneous forms

$$Ah = A\,Q_m(g) = \sum_{k=1}^{m} Ab_k g^k$$

is a polynomial in g of degree m. The assertion is therefore true for $n = 1$.

Suppose it holds for all polynomials $P_k(h)$ with $k \leqq n-1$.

We consider a polynomial $P_n(h) = \sum_{k=1}^{n} a_n h^n$ and obtain

$$P_n(h) = P_{n-1}(h) + a_n h^n = P_{n-1}(h) + (a_n h^{n-1})h.$$

$P_{n-1}(h)$ is a polynomial in h of degree $n-1$. If

$$h = Q_m(g),$$

then by the induction hypothesis,

$$P_{n-1}(h) = R_{(n-1)m}(g) \text{ and } a_n h^{n-1} = R^*_{(n-1)m}(g), \text{ where}$$

$R_{(n-1)m}(g)$ and $R^*_{(n-1)m}(g)$ are polynomials in g of degree $(n-1)m$; it follows that

$$P_n(h) = R_{(n-1)m}(g) + R^*_{(n-1)m}(g)\,Q_m(g).$$

Because of property 1. for polynomials, a product $R^*_{(n-1)m}(g)\,Q_m(g)$ of two polynomials is a polynomial $R_{nm}(g)$. Therefore

$$P_n(h) = P_n[Q_m(g)] = R_{(n-1)m}(g) + R_{nm}(g) = R^*_{nm}(g),$$

and $R^*_{nm}(g)$ is a polynomial of g of degree nm, q.e.d.

§41. THE DIFFERENTIAL OF AN ABSTRACT FUNCTION

NOTATION. Let E and E_1 be linear normed spaces, and let $y = f(x)$ be an abstract function defined on E with range in E_1. We introduce the following notations:

If $\beta = \varphi(\alpha)\,(\alpha, \beta \in R)$ and if $\beta \to 0$ as $\alpha \to 0$, then we write

$$\beta = o(\alpha).$$

If we have the inequality $|\beta| \leqq N|\alpha|$, where N is constant, then we write

$$\beta = O(\alpha).$$

Let $x \in E, y_1 = y_1(x) \in E_1, y_2 = y_2(x) \in E_1$. For short, we write

$$y_1 \overset{x}{\underset{n}{=}} y_2$$

for the relation

$$\| y_1(x) - y_2(x) \| = o(\| x \|^n).$$

I.e., $y_1 = y_2$ with accuracy as far as quantities of order above n compared with $\| x \|$.

We have

1. If $y_1 \overset{x}{\underset{n}{=}} y_2$ and $y_2 \overset{x}{\underset{n}{=}} y_3$, then $y_1 \overset{x}{\underset{n}{=}} y_3$.

2. $y_1 \overset{x}{\underset{n}{=}} y_2$, $x = f(x_1)$ and $\| x_1 \| = O(\| x \|)$ imply $y_1 \overset{x_1}{\underset{n}{=}} y_2$.

3. If $P(h)$ and $Q(h)$ are polynomials in h with equal coefficients for the powers $h, h^2, \ldots, h^n$, i.e., if

$$P(h) - Q(h) = \sum_{k=n+1}^{p} a_k h^k,$$

then

$$P(h) \overset{h}{\underset{n}{=}} Q(h).$$

We assume that the multiplications which occur in the following formulas have meaning.

4. If $\| A(x) \|$ is bounded in a neighborhood of $x = \theta$ and $y_1(x) \overset{x}{\underset{n}{=}} y_2(x)$, then

$$A(x)\, y_1(x) \overset{x}{\underset{n}{=}} A(x)\, y_2(x).$$

Analogously,

$$y_1(x)\, A(x) \overset{x}{\underset{n}{=}} y_2(x)\, A(x).$$

5. $A(x) \overset{x}{\underset{n}{=}} A_1(x)$, and $B(x) \overset{x}{\underset{n}{=}} B_1(x)$

imply

$$A(x)\, B(x) \overset{x}{\underset{n}{=}} A_1(x)\, B_1(x).$$

For, by virtue of property 4.,

$$A(x)\, B(x) \overset{x}{\underset{n}{=}} A(x)\, B_1(x) \overset{x}{\underset{n}{=}} A_1(x)\, B_1(x).$$

From this, however, we have property 5, because of 1.

We now give two definitions of the *differential of an abstract function.*

STRONG DIFFERENTIAL (Fréchet differential). Let h and x be arbitrary elements of the space E. Suppose further that there exists a linear operator $l \in (E \to E_1)$ (it depends, in general, on x) such that

$$f(x + h) - f(x) = lh + \alpha(x, h) \tag{1}$$

and moreover let $\dfrac{\| \alpha(x, h) \|}{\| h \|} \to 0$ for $\| h \| \to 0$. (2)

We therefore write $f(x + h) - f(x) \overset{h}{\underset{l}{=}} lh$. (3)

Then lh is called *the differential of the function $f(x)$ at the point x for the increment h.* It is designated by $df(x, h)$.

Example. Let E be a normed ring with unit element (see pp. 90-91). We consider the abstract function $f(x) = x^{-1}$. If for a given $x \in E$ there also exists x^{-1}, then for sufficiently small $\| \Delta x \|$, according to Theorem 3 of §20, there also exists the element $f(x + \Delta x) = (x + \Delta x)^{-1}$. Because $y^{-1} - x^{-1} = x^{-1}(x - y)y^{-1}$, if we put $y = x + \Delta x$, we obtain

$$f(x + \Delta x) - f(x + \Delta x)^{-1} - x^{-1}$$
$$= -x^{-1}\Delta x(x + \Delta x)^{-1} = -x^{-1}\Delta x x^{-1} + \alpha(x, \Delta x)$$

with

$$\alpha(x, \Delta x) = -x^{-1}\Delta x[(x + \Delta x)^{-1} - x^{-1}] = x^{-1}\Delta x x^{-1}\Delta x(x + \Delta x)^{-1}.$$

We have

$$\| \alpha(x, \Delta x) \| \leqq \| x^{-1}\|^2 \| (x + \Delta x)^{-1}\| \| \Delta x\|^2 = \gamma \| \Delta x \|.$$

$\gamma \to 0$ as $\| \Delta x \| \to 0$, therefore

$$df(x, \Delta x) = -x^{-1}\Delta x x^{-1}.$$

If Δx commutes with x^{-1}, then $df(x, \Delta x) = -x^{-2}\Delta x$.

We designate the linear operator l, which in general depends on x, by $f'(x)$. Therefore

$$df(x, h) = f'(x)h \text{ with } f'(x) \in (E \to E_1).$$

The operator $f'(x)$ can be considered as a function of x. It is defined on a certain subset of E.

We call $f'(x)$ the *first strong derivative of the function $f(x)$ at the point x.* We can write the equality (3) in the form

$$f(x + h) \stackrel{h}{\underset{1}{=}} f(x) + f'(x)h. \tag{4}$$

The right side is a polynomial in h of degree 1, which approximates $f(x + h)$ in $\| h \|$ to a higher order than the first.

WEAK DIFFERENTIAL. We call the expression

$$df(x, h)_1 = \frac{d}{dt} f(x + th) \Big|_{t=0} = \lim_{t \to 0} \frac{f(x + th) - f(x)}{t}$$

the *weak differential* of the function $f(x)$ at the point x. Of course, we assume that the limit on the right exists in the sense of norm-convergence.[32]

Example. We consider the space $C'[a, b]$ of all functions $y(t)$ continuously differentiable on the interval $a \leqq t \leqq b$, with the norm

$$\| y \| = \max_{a \leqq t \leqq b} (|y(t)|, |y'(t)|)$$

and the functional of a simple problem of variational calculus

$$I(y) = \int_a^b f(t, y, y')\, dt.$$

The definitions of the above differential correspond here to the usual definitions of variation.

---

[32] Sometimes we call $\lim_{t \to 0} \dfrac{f(x + th) - f(x)}{t}$ a weak differential if the limit is understood in the sense of *weak* convergence of the elements.

We recognize similar situations also for other functionals of the calculus of variations. Indeed, the concept of the differential of an abstract function has its origin in the calculus of variations.

THEOREM 1. *If the strong differential exists, then the weak differential exists and* $df(x,h) = df(x,h)_1$.

Proof. We have

$$f(x+th) - f(x) = df(x, th) + \alpha(x, th) = t\,df(x,h) + \alpha(x, th).$$

According to (2), $\quad \| \alpha(x, th) \| = o(\| th \|) = o(|t| \, \| h \|) = \gamma(t)$

as $t \to 0$ to a higher than first order. Therefore

$$\frac{f(x+th) - f(x)}{t} = df(x,h) + \frac{\alpha(x, th)}{t} \to df(x,h)$$

as $t \to 0$. Hence

$$df(x,h) = \lim_{t \to 0} \frac{f(x+th) - f(x)}{t} = df(x,h)_1.$$

The directional derivative need not be linear in h. If, however, this is the case, then $df(x,h)_1 = l_1 h = f'(x)_1 h$, where $f'(x)_1$ is a linear operator in h and $f'(x)_1 \in (E \to E_1)$. We then call $f'(x)$ the *weak derivative of the function* $f(x)$ *at the point* x.

THEOREM 2. *If the weak differential* $df(x,h)_1$ *exists in* $\| x' - x \| \leqq r$ *and is uniformly continuous in x and continuous in h, then there exists in this neighborhood the strong differential* $df(x,h)$ *and we have* $df(x,h) = df(x,h)_1$.

Proof. For $\| h \| \leqq r$, the differential $df(x_t, h)_1$ exists at all points $x_t = x + th$, $0 \leqq t \leqq 1$. Since

$$df(x_t, h)_1 = \lim_{\Delta t \to 0} \frac{f(x_t + \Delta t h) - f(x_t)}{\Delta t}$$

and

$$x_t + \Delta t h = x + (t + \Delta t) h = x_{t + \Delta t}$$

it follows that

$$df(x_t, h)_1 = \lim_{\Delta t \to 0} \frac{f(x_{t+\Delta t}) - f(x_t)}{\Delta t} = \frac{d}{dt} f(x_t) = \frac{d}{dt} f(x+th).$$

We now show that $df(x,h)_1$ is additive with respect to h:

$$df(x, h + h_1)_1 = df(x,h)_1 + df(x, h_1)_1. \tag{5}$$

We first remark that from

$$df(x_t, h)_1 = \frac{d}{dt} f(x+th)$$

follows

$$f(x+th) - f(x) = \int_0^t \frac{d}{d\tau} f(x + \tau h)\, d\tau = \int_0^t df(x + \tau h, h)_1 \, d\tau = t\, df(x,h)_1 + \varepsilon_1, \tag{6}$$

whereby

$$\varepsilon_1 = \int_0^t [df(x + \tau h, h)_1 - df(x,h)_1]\, d\tau.$$

Correspondingly we obtain

$$f(x+t(h+h_1))-f(x)=t\,df(x,h+h_1)_1+\varepsilon_2 \tag{7}$$

with
$$\varepsilon_2=\int_0^t \{df(x+\tau(h+h_1),h+h_1)_1-df(x,h+h_1)_1\}\,d\tau,$$

also
$$f(x+t(h+h_1))-f(x+th)=t\,df(x,h_1)_1+\varepsilon_3 \tag{8}$$

with
$$\varepsilon_3=\int_0^t \{df(x+th+\tau h_1,h_1)_1-df(x,h_1)_1\}\,d\tau.$$

Since $df(x,h)$ is continuous in x, we have for arbitrary $\eta>0$ with sufficiently small $t>0$ and $0\leqq\tau\leqq t$

$$\|df(x+\tau h,h)_1-df(x,h)_1\|\leqq\frac{\eta}{3},$$

$$\|df[x+\tau(h+h_1),h+h_1]_1-df(x,h+h_1)_1\|\leqq\frac{\eta}{3}$$

$$\|df(x+th+\tau h_1,h_1)_1-df(x,h_1)_1\|\leqq\frac{\eta}{3}.$$

Therefore

$$\|\varepsilon_1\|=\left\|\int_0^t\{df(x+\tau h,h)_1-df(x,h)_1\}\,d\tau\right\|\leqq\frac{1}{3}\,\eta t$$

and correspondingly

$$\|\varepsilon_2\|\leqq\frac{1}{3}\,\eta t,\quad \|\varepsilon_3\|\leqq\frac{1}{3}\,\eta t.$$

From (6), (7), and (8), it follows,

$$0=[f(x+th)-f(x)]+[f(x+t(h+h_1))-f(x+th)]-[f(x+t(h+h_1))-f(x)]$$
$$=t[df(x,h)_1+df(x,h_1)_1-df(x,h+h_1)_1]+(\varepsilon_1+\varepsilon_3-\varepsilon_2),$$

i.e.,
$$df(x,h)_1+df(x,h_1)_1=df(x,h+h_1)_1=\frac{1}{t}(\varepsilon_2-\varepsilon_3-\varepsilon_1)$$

and thus

$$\|df(x,h)_1+df(x,h_1)_1-df(x,h+h_1)_1\|\leqq\frac{1}{t}(\|\varepsilon_1\|+\|\varepsilon_2\|+\|\varepsilon_3\|)\leqq\eta.$$

Since η was arbitrary,

$$\|df(x,h)_1+df(x,h_1)_1-df(x,h+h_1)_1\|=0,$$

which proves (5).

In addition, $df(x,h)_1$ was assumed to be continuous in h. Consequently $df(x,h)$ is a linear operator with respect to h: $df(x,h)_1=f'(x)_1h$. Since $df(x+h)_1=f'(x)_1h$ is continuous with respect to x, then $f'(x)_1$ is continuous with respect to x.

We now prove that

$$f(x+h)-f(x)\overset{h}{\underset{1}{=}}f'(x)_1h. \tag{4'}$$

Then $df(x, h)_1$, being (in h) the linear part of the increment $f(x+h)-f(x)$, coincides with $df(x, h)$. We obtain

$$f(x+h)-f(x)=\int_0^1 \frac{d}{dt}f(x+th)dt=\int_0^1 df(x+th, h)_1 dt$$

$$=\left\{\int_0^1 f'(x+th)_1 dt\right\} h=f'(x)h+\varepsilon \tag{9}$$

with $$\varepsilon=\left\{\int_0^1 [f'(x+th)_1-f'(x)_1]\,dt\right\} h.$$

By virtue of the uniform continuity of $f'(x)_1$, for $0 \leqq t \leqq 1$

$$\|f'(x+th)_1-f'(x)_1\|=\alpha(\|h\|)\to 0 \text{ for } \|h\|\to 0.$$

We conclude from this

$$\|\varepsilon\| \leqq \left\|\int_0^1 [f'(x+th)_1-f'(x)_1]dt\right\| \|h\|$$

$$\leqq \int_0^1 \|f'(x+th)_1-f'(x)_1\|\,dt\,\|h\|=\alpha(\|h\|)\,\|h\|, \tag{9'}$$

and this proves (4). Hence

$$df(x, h)_1=df(x, h) \text{ and } f'(x)_1=f'(x),$$

q.e.d.

In the sequel, we shall always assume that for all differentiable functions considered, $df(x, h)_1=df(x, h)$.

Remark. If in the neighborhood $\|x-x_0\| \leqq r$ of x_0 the inequality

$$\|f'(x)\| \leqq N$$

holds and if x_1, x_2, belong to this neighborhood, then

$$\|f(x_2)-f(x_1)\| \leqq N\|x_2-x_1\|.$$

Since the region $\|x-x_0\| \leqq r$ is convex, with x_1 and x_2 the segment

$$x_t=(1-t)x_1+tx_2, 0 \leqq t \leqq 1$$

also belongs to this neighborhood. Therefore

$$\|f'(x_t)\| \leqq N.$$

Since, however, $$f(x_2)-f(x_1)=\int_0^1 f'(x_t)(x_2-x_1)dt$$

we have $$\|f(x_2)-f(x_1)\| \leqq \int_0^1 \|f'(x_t)\|\,dt\,\|x_2-x_1\| \leqq N\|x_2-x_1\|.$$

§42. DIFFERENTIALS AND DERIVATIVES OF HIGHER ORDER

TAYLOR'S FORMULA. Considering an approximation of the function $f(x+h)$ by a polynomial in h of the first degree, we have defined the strong differential of the first order.

Suppose there exists a polynomial in h

$$P_n(h) = a_1h + a_2h^2 + \ldots + a_nh^n$$

of degree n such that

$$f(x+h) - f(x) \overset{h}{\underset{n}{=}} P_n(h). \tag{1}$$

I.e., let

$$f(x+h) - f(x) = P_n(h) + \alpha_n(h)$$

where

$$\left.\begin{aligned} &\| \alpha_n(h) \| \leqq \varepsilon(\| h \|) \, \| h^n \|, \\ &\varepsilon(\| h \|) \to 0 \text{ for } \| h \| \to 0. \end{aligned}\right\} \tag{2}$$

The polynomial $P_n(h)$ is called Taylor's sum of degree n for the function $f(x+h)$. The nth term multiplied by $n!$ is called the *nth strong differential of the function $f(x)$ at the point x*, and the function $f(x)$ itself is called *n times differentiable*. (M. A. Gavurin studied nth order of differentiability.)

We designate the nth strong differential by $d^nf(x, h)$, and then

$$d^nf(x, h) = n!\, a_nh^n.$$

The symmetric n-termed linear form belonging to $d^nf(x, h)$ is $d^nf(x, h_1, \ldots, h_n) = n!\, a_nh_1h_2 \ldots h_n$. This linear form $n!\, a_n$ is called the *nth strong derivative of the function $f(x)$ at the point x* and is designated by $f^{(n)}(x)$. Thus

$$d^nf(x, h) = f^{(n)}(x)\, h^n.$$

We prove

$$d^nf(x, h) = \frac{d^n}{dt^n} f(x+th) \Big|_{t=0}. \tag{3}$$

From (1), we have

$$f(x+th) = f(x) + \sum_{k=1}^{n-1} t^k a_k h^k + t^n a_n h^n + \alpha(th)$$

where

$$\lim_{t \to 0} \frac{\alpha(th)}{t^n} = \theta.$$

We get

$$\frac{d^n}{dt^n} \sum_{k=1}^{n-1} (t^k a_k h^k) = \theta,$$

$$\frac{d^n}{dt^n} (t^n a_n h^n) = n!\, a_n h^n$$

and finally

$$\frac{d^n}{dt^n} (\alpha(th)) \Big|_{t=0} = \lim_{\Delta t \to 0} \frac{\Delta^n(\alpha(th)\,|_{t=0}}{(\Delta t)^n}$$

$$= \lim_{\Delta t \to 0} \frac{1}{(\Delta t)^n} \sum_{k=0}^{n} (-1)^k \binom{n}{k} \alpha \left[\left(k - \frac{n}{2} \right) \Delta th \right].$$

According to (2)

$$\frac{\alpha\left[\left(k-\frac{n}{2}\right)\Delta th\right]}{(\Delta t)^n}\to\theta$$

as $\Delta t\to 0$ and therefore

$$\frac{d^n}{dt^n}\,[\alpha(th)]_{t=0}=\theta.$$

Hence we have obtained

$$d^n f(x,h)=n!a_n h^n=\frac{d^n}{dt^n}\,f(x+th)\bigg|_{t=0}.$$

Formula (1) takes the form

$$f(x+h)-f(x)\underset{n}{\overset{h}{=}}f'(x)h+\frac{1}{2}f''(x)h^2+\ldots+\frac{1}{n!}f^{(n)}(x)h^n.$$

If $d^nf(x,h)$ exists in a certain region and if the relation (2) (in which $a_n(h)$ in addition depends on x) is fulfilled in this region uniformly for all x, then we call $d^nf(x,h)$ a *uniform strong differential.* In this case, we have on the right side of (3) the uniform difference derivative.

We now give another definition of the nth differential. Let the first differential $df(x,h)=f'(x)h$ exist in a neighborhood of the point x. Let the function $f'(x)$ and consequently $df(x,h)$ be differentiable with respect to x. We arrive at the *second differential* $d[df(x)\,h,h_1]=d[f'(x)\,h,h_1]=[df'(x,h_1)]h$. If we write

$$df'(x,h_1)=f''(x)_1h$$

then we call $f''(x)_1$ *second derivative.* We have

$$d[df(x,h)h_1]=f''(x)_1h_1h=\frac{d}{dt}\,df(x+th_1,h)\bigg|_{t=0}$$

$$=\frac{\partial}{\partial t}\left[\frac{\partial}{\partial t_1}f(x+th+t_1h_1)\bigg|_{t_1=0}\right]\bigg|_{t=0}$$

$$=\frac{\partial^2}{\partial t\,\partial t_1}\,f(x+th+t_1h_1)\bigg|_{t=t_1=0}.$$

Analogously, we define the nth differential, having assumed that the $(n-1)th$ differential

$$d\{d[\ldots df(x,h_1)h_2\ldots]h_{n-1}\}=f^{n-1}(x)_1h_{n-1}h_{n-2}\ldots h_1$$

exists and that it is differentiable as a function of x, i.e., that $f^{n-1}(x)_1$ is differentiable. Finally we obtain

$$\begin{aligned}d(d\{d[\ldots df(x,h_1)h_2\ldots]h_{n-1}\}\,h_n)&=d[f^{n-1}\,(x)_1h_{n-1}h_{n-2}\ldots h_1,h_n]\\&=df^{(n-1)}\,(x,h_n)_1h_{n-1}h_{n-2}\ldots h_1.\end{aligned}\qquad(4)$$

If we write $df^{n-1}(x,h_n)_1=f^n(x)_1h_n$, then we obtain the *nth derivative* $f^n(x)_1$ and

$$d(d\{d[\ldots df(x,h_1)h_2\ldots]h_{n-1}\}\,h_n)=f^n(x)_1h_1h_2\ldots h_n.$$

From equation (4), we have

$$d^nf(x,h_1,h_2\ldots h_n)_1=f^n(x)_1h_1h_2\ldots h_n=\frac{\partial^n}{\partial t_1\partial t_2\ldots\partial t_n}f\left(x+\sum_{i=1}^n t_ih_i\right)\bigg|_{t_1=t_2=\ldots=t_n=0}.$$

Putting $h_1=h_2=\ldots h_n=h$,

$$d^n f(x,h)_1 = f^n(x)_1 h^n = \left.\frac{d^n}{dt^n} f(x+th)\right|_{t=0}. \tag{5}$$

From (3) and (5), there follows the conjecture:

If, in a region Q, an nth uniform strong differential $d^n f(x,h)$ exists and is continuous in x, then there also exists in Q the nth differential $d^n f(x,h)_1$,

$$d^n f(x,h)_1 = d^n f(x,h)$$

or

$$f^{(n)}(x) = f^{(n)}(x)_1.$$

If conversely, there exists the nth differential and $f^n(x)$ is uniformly continuous in x, then there also exists in Q the nth uniform strong differential, and it coincides in Q with the nth differential.

We prove this conjecture $d^n f(x,h)_1 = f^n(x)_1 h^n$ by induction. For $n=1$ this assertion coincides with Theorem 2 and the formulas (9′), (9) in §41 and is proved there. Let the statement hold for $n-1$. Since

$$f^n(x)_1 = [f'(x)_1]^{(n-1)}$$

holds, we obtain

$$f'(x+h) = f'(x)_1 + f''(x)_1 h + \frac{1}{2} f'''(x)_1 h^2 + \ldots + \frac{1}{(n-1)!} f^n(x)_1 h^{n-1} + \alpha(x,h),$$

where $\|\alpha(x,h)\| \leqq \varepsilon(\|h\|)\,\|h\|^{n-1}$, $\varepsilon(\|h\|) \to 0$ as $\|h\| \to 0$.

From this, for $0 \leqq t \leqq 1$, it follows that

$$f'(x+th) = f'(x)_1 + t f''(x)_1 h + \frac{1}{2} t^2 f'''(x)_1 h^2 + \ldots$$
$$+ \frac{1}{(n-1)!} t^{n-1} f_n(x)_1 h^{n-1} + \alpha(th),$$

where $\|\alpha(th)\| \leqq \varepsilon_n t^n \|h\|^n \leqq \varepsilon_n \|h\|^n$, $\varepsilon_n \to 0$ as $\|h\| \to 0$.

Therefore

$$f(x+h) - f(x) = \int_0^1 f'(x+th)_1 h\, dt$$

$$= \int_0^1 \Big[f'(x)_1 + t f''(x)_1 h + \frac{1}{2} t^2 f'''(x)_1 h^2 + \ldots$$
$$+ \frac{1}{(n-1)!} t^{n-1} f^n(x)_1 h^{n-1} \Big] h dt + R_n$$

where $R_n = \int_0^1 \alpha(th)\, h\, dt$.

Hence

$$f(x+h) = f(x) + f'(x)_1 h + \frac{1}{2!} f''(x)_1 h^2 + \frac{1}{3!} f'''(x)_1 h^3 + \ldots$$
$$+ \frac{1}{n!} f^{(n)}(x)_1 h^n + R_n$$

where $\| R_n\| \leqq \int_0^1 \| \alpha(th) \| \| h \| dt \leqq \varepsilon_n \| h \|^n$, $\varepsilon_n \to 0$ as $\| h \| \to 0$.

Thus the sum on the right side of the last expression for $f(x+h)$ is a Taylor sum for the function $f(x)$, and we have $f^n(x)_1 h^n = d^n f(x,h)$, in other words

$$d^n f(x,h)_1 = d^n f(x,h).$$

In the following, we shall interpret the nth derivative and the nth differential in the strong sense. We now investigate the nth derivative of a composite function and a product.

1) Let $z = \varphi(y)$, $y = \psi(x)$, $x \in E$, $y \in E_1$, $z \in E_2$, and $z = \varphi(\psi(x)) = f(x)$. In addition, let $y_0 = \psi(x_0)$, $z_0 = \varphi(y_0) = f(x_0)$. If $\varphi(y)$ and $\psi(x)$ are n times differentiable at x_0, and y_0, respectively, then $f(x)$ is also n times differentiable at x_0.

Proof. According to the hypothesis, there exists in E a polynomial of nth degree $P_n(h)$ such that

$$\psi(x_0+h) - \psi(x_0) \overset{h}{\underset{n}{=}} P_n(h).$$

On the other hand, there exists in E_1 a polynomial of the nth degree $Q_n(g)$ such that

$$\varphi(y_0+g) - \varphi(y_0) \overset{g}{\underset{n}{=}} Q_n(g).$$

In particular, for $\quad g = \psi(x_0+h) - \psi(x_0),$

and consequently for $\quad \psi(x_0+h) = \psi(x_0) + g = y_0 + g$

we have

$$f(x_0+h) - f(x_0) = \varphi[\psi(x_0+h)] - \varphi[\psi(x_0)] = \varphi(y_0+g) - \varphi(y_0) \overset{g}{\underset{n}{=}} Q_n(g). \quad (6)$$

But $\quad g = \psi(x_0+h) - \psi(x_0) \overset{h}{\underset{n}{=}} P_n(h)$

therefore $\quad Q_n(g) \overset{h}{\underset{n}{=}} Q_n(P_n(h)).$

Because of property 2 for polynomials, $Q_n(P_n(h))$ is a polynomial in h. This polynomial can be approximated with necessary precision by $R_n(h)$ of nth degree in h:

$$Q_n(g) \overset{h}{\underset{n}{=}} Q_n[P_n(h)] \overset{h}{\underset{n}{=}} R_n(h). \quad (7)$$

Furthermore since the function ψ is differentiable at the point x_0,

$$\| g \| = \| \psi(x_0+h) - \psi(x_0) \| = O(\| h \|).$$

Therefore the symbol $\overset{g}{\underset{n}{=}}$ can be replaced by $\overset{h}{\underset{n}{=}}$, and the equality (6) takes the form

$$f(x_0+h) - f(x_0) \overset{h}{\underset{n}{=}} Q_n(g).$$

From this and (7),

$$f(x_0+h)-f(x_0)\underset{n}{\overset{h}{=}}R_n(h). \tag{8}$$

Because, therefore, a polynomial $R_n(h)$ fulfills the relation (8), our theorem holds.

If $\varphi(x)$ and $\psi(x)$ are n-times continuously differentiable with respect to x, then $f(x)=\varphi[\psi(x)]$ has this property too. In this case the coefficients of the polynomials $P_n(h)$ and $Q_n(h)$ are continuous functions of x. Therefore, the coefficients of $Q_n(P_n(h))$ are also continuous in x. Consequently, the polynomial $R_n(h)$ can be selected such that the coefficients are continuous.

2) Let $x\in E, y=f(x)\in E_1, z=\varphi(x)\in E_2$, and let there be defined for the elements $y\in E_1$ and $z\in E_2$ a product in E_3. If $f(x)$ and $\varphi(x)$ are n-times continuously differentiable, then $F(x)=f(x)\varphi(x)$ is also n-times continuously differentiable.

Proof. By hypothesis,

$$f(x+h)\underset{n}{\overset{h}{=}}f(x)+P_n(h),\quad \varphi(x+h)\underset{n}{\overset{h}{=}}\varphi(x)+Q_n(h)$$

where $P_n(h)$ and $Q_n(h)$ are polynomials of the nth degree in h:

$$P_n(h)=\sum_{k=1}^{n}a_k(x)h^k,\ Q_n(h)=\sum_{k=1}^{n}b_k(x)h^k$$

and $a_k(x)$ and $b_k(x)$ are continuous functions. Therefore

$$f(x+h)\,\varphi(x+h)\underset{n}{\overset{h}{=}}f(x)\varphi(x)+[f(x)\,Q_n(h)+\varphi(x)P_n(h)+P_n(h)\,Q_n(h)].$$

The expression in brackets is a polynomial with continuous coefficients. We neglect the terms which contain h to a higher power than the nth and obtain $R_n(h)=\sum_{k=1}^{n}c_k(x)h^n$ where $c_k(x)$ are continuous. Therefore

$$f(x)\,Q_n(h)+\varphi(x)P_n(h)+P_n(h)\,Q_n(h)\underset{n}{\overset{h}{=}}R_n(h).$$

Therefore
$$f(x+h)\,\varphi(x+h)\underset{n}{\overset{h}{=}}f(x)\,\varphi(x)+R_n(h). \tag{9}$$

This relation shows that $F(x)=f(x)\ \varphi\ (x)$ is an n-times continuously differentiable function of x.

In conclusion, we remark that for functionals defined in complex linear spaces, existence of the first differential in the neighborhood of a point implies the existence of all differentials of higher order in this neighborhood. We can also represent a functional by a "Taylor series" [50].

§43. DIFFERENTIATION OF FUNCTIONS OF TWO VARIABLES

We investigate a function $\varphi(x,y)$ of two variables $x\in E, y\in E_1$, with $\varphi(x,y)\in E_2$. We can consider (x,y) as an element of the direct sum $E\dot{+}E_1$. The function $\varphi(x,y)$ is called *n-times differentiable at point* (x_0,y_0) if the relation

$$\varphi(x_0+h,y_0+g)-\varphi(x_0,y_0)\underset{n}{\overset{(h,g)}{=\!=}}a_1(h,g)+a_2(h,g)^2+\ldots+a_n(h,g)^n$$

holds. Here $a_k(h,g)^k$ are homogeneous forms of degree k in the element $(h,g) \in E \dot{+} E_1$. Obviously $a_n(h,g)^n$ is a sum of linear forms of the type $a_n h_1 h_2 \ldots h_n$, where every h_i is equal to h or g.

THEOREM. *If $\varphi(x,y)$ and $y=f(x)$ are n-times differentiable at $(x_0, y_0), x_0$, respectively, where $y_0=f(x_0)$, then $\varphi(x,f(x))$ is a function of x which is n-times differentiable at $x=x_0$.*

Proof. If
$$g=f(x_0+h)-f(x_0),$$
then
$$f(x_0+h)=y_0+g,$$
and
$$g \overset{h}{\underset{n}{=}} P_n(h)=a_1h+a_2h^2+\ldots+a_nh^n. \tag{1}$$
Therefore
$$\|g\|=0(\|h\|)$$
and also
$$\|(h,g)\|=\|h\|+\|g\|=0(\|h\|). \tag{2}$$
Since $\varphi(x,y)$ is an n-times differentiable function, then
$$\varphi(x_0+h, y_0+g)-\varphi(x_0,y_0) \overset{h}{\underset{n}{=}} A_1(h,g)+A_2(h,g)^2+\ldots+A_n(h,g)^n, \tag{3}$$
for, by virtue of (2), we can replace $\overset{(h,g)}{\underset{n}{=}}$ in (3) by $\overset{h}{\underset{n}{=}}$. Because
$$A_k(h,g) \overset{h}{\underset{n}{=}} A_k(h,P_n(h))^k$$
follows from (1), we conclude that
$$\varphi(x_0+h, y_0+g)-\varphi(x_0,y_0) \overset{h}{\underset{n}{=}} \sum_{k=1}^{n} A_k(h,P_k(h))^k, \tag{4}$$
where $A_k(h,P_k(h))^k$ is a sum of terms of the form $a_k h_1 h_2 \ldots h_k$, where every h_i is equal to h or to $P_k(h)$. Therefore $A_k(h,P_k(h))^k$ and consequently the entire sum in (4) is a polynomial in h. If we omit the terms of higher than nth degree in h, we obtain a polynomial $R_n(h)$, for which
$$R_n(h) \overset{h}{\underset{n}{=}} \sum_{k=1}^{n} A_k(h,P_k(h))^k$$
holds. Thus
$$\varphi(x_0+h, y_0+g)-\varphi(x_0,y_0) \overset{h}{\underset{n}{=}} R_n(h),$$
q.e.d.

We now introduce the concept of the *partial derivative* of a function of two variables:
$$d\varphi[(x_0,y_0);(h,g)]=A_1(h,g)=a_1h+a_2g,$$
$$d^2\varphi[(x_0,y_0);(h,g)]=A_2(h,g)^2=a_{11}h^2+a_{12}hg+a_{21}gh+a_{22}g^2$$
and so on. Using the notation
$$a_1=\varphi_x'(x_0,y_0), \qquad a_2=\varphi_y'(x_0,y_0),$$
$$a_{11}=\varphi_{xx}''(x_0,y_0), \qquad a_{12}=\varphi_{xy}''(x_0,y_0),$$
$$a_{21}=\varphi_{yx}''(x_0,y_0), \qquad a_{22}=\varphi_{yy}''(x_0,y_0)$$

we obtain

$$a_1 h = \varphi'_x h = \frac{d}{dt}\varphi(x_0 + th, y_0)_{t=0},$$

$$a_2 g = \varphi'_y g = \frac{d}{dt}\varphi(x_0, y_0 + tg)_{t=0},$$

$$a_{11}h^2 = \varphi''_{xx}h^2 = \frac{d^2}{dt^2}\varphi(x_0 + th, y_0)_{t=0},$$

$$a_{12}hg = \varphi''_{xy}hg = \frac{\partial^2}{\partial t_1 \partial t_2}\varphi(x_0 + t_1 h, y_0 + t_2 g)_{t_1=t_2=0},$$

$$a_{21}gh = \varphi''_{yx}gh = \frac{\partial^2}{\partial t_2 \partial t_1}\varphi(x_0 + t_1 h, y_0 + t_2 g)_{t_1=t_2=0},$$

$$a_{22}g^2 = \varphi''_{yy}g^2 = \frac{d^2}{dt^2}\varphi(x_0. y_0 + tg)_{t=0}$$

and so on.

If $\varphi(x, y)$ has a continuous second differential, then $a_{12}hg = a_{21}gh$, for the operators $\frac{\partial}{\partial t_1}$ and $\frac{\partial}{\partial t_2}$ are commutative if $\frac{\partial^2}{\partial t_1 \partial_2}\varphi(x_0 + t_1 h, y_0 + t_2 g)$ is a continuous function of t_1 and t_2.

§44. THEOREMS ON IMPLICIT FUNCTIONS

We shall prove a generalization of the theorem on implicit functions.

To do this, we consider the direct sum $E \dotplus E_1$ of the spaces E and E_1 as well as an operator $\varphi(x, y)$ transforming $E \dotplus E_1$ into E:

Hence, let $x \in E, y \in E_1, \varphi(x, y) \in E$ and

1) $\varphi(x_0, y_0) = \theta,$ (1)

2) let $\varphi(x, y)$ have a continuous derivative $\varphi'_y(x, y)$ in a neighborhood of the point (x_0, y_0), and

3) let $[\varphi'_y(x_0, y_0)]^{-1}$ exist.

In consequence of the continuity of the operator $\varphi'_y(x, y)$, the operator $[\varphi'_y(x, y)]^{-1}$ is defined for all (x, y) in a certain η-neighborhood of the point (x_0, y_0) and its norm is bounded.

$$\| [\varphi'_y(x_0 + h, y_0 + g)]^{-1} \| \leqq K \text{ for } \|h\| \leqq \eta, \|g\| \leqq \eta. \tag{2}$$

THEOREM 1. *If the hypotheses* 1) — 3) *are fulfilled, then there exists for every* $\tau_1, 0 \leqq \tau_1 \leqq \eta$, *a positive number* $\tau_0 \leqq \eta$ *and a continuous operator* $y = f(x)$, $x \in E, y \in E_1$, *which is defined in the neighborhood* $\|x - x_0\| \leqq \tau_0$ *of* x_0, *such that*

$$y = f(x) \tag{3}$$

is equivalent to

$$\varphi(x, y) = \theta \tag{4}$$

in a certain neighborhood of (x_0, y_0). *I.e., every pair* (x, y) *with* $\|x - x_0\| \leqq \tau_0$, *satisfying* (3), *fulfills* (4) *also, and conversely every pair* (x, y) *satisfying* (4) *for*

$\|x-x_0\| \leqq \tau_0, \|y-y_0\| \leqq \tau_1$ *fulfills* (3). *The operator* $f(x)$ *is continuous in* x *and* $f(x_0)=y_0$.

Proof. We put $x=x_0+h, y=y_0+g$, choosing h such that $\|h\| \leqq \eta$ is fixed and want to find a solution $g=g(h)$ of

$$\varphi(x_0+h, y_0+g)=\theta \tag{5}$$

or the equivalent equality

$$g=\psi(h,g) \tag{6}$$

where

$$\begin{aligned}\psi(h,g) &= g-[\varphi'(x_0+h,y_0)]^{-1}\varphi(x_0+h,y_0+g)\\ &= -[\varphi_y'(x_0+h,y_0)]^{-1}[\varphi(x_0+h,y_0+g)-\varphi_y'(x_0+h,y_0)g].\end{aligned} \tag{7}$$

After having defined $g(h)$, we define the function $y=f(x)$ for $x=x_0+h$, $y=y_0+g$, by $y_0=f(x_0)$ and

$$y=f(x)=f(x_0+h)=f(x_0)+g(h).$$

We want to solve the equation (6) by iteration, putting

$$g^{(0)}(h)=g^{(0)}=\theta, \quad g^{(n)}(h)=g^{(n)}=\psi(h,g^{(n-1)}).$$

By virtue of (7), $\psi(h,\theta)=-[\varphi_y'(x_0+h,y_0)]^{-1}\varphi(x_0+h,y_0)$.
On account of the continuity of the functions $[\varphi_y']^{-1}$ and φ,

$$g^{(1)}(h)=\psi(h,\theta)\to-[\varphi_y'(x_0,y_0)]^{-1}\varphi(x_0,y_0)=\theta \tag{8}$$

as $h\to\theta$. Because of (7), we obtain further

$$\psi_g'(h,g)=-[\varphi_y'(x_0+h,y_0)]^{-1}[\varphi_y'(x_0+h,y_0+g)-\varphi_y'(x_0+h,y_0)]. \tag{9}$$

Since the function φ_y' is continuous at (x_0,y_0), then for every positive ε it is possible to define positive numbers $\tau_0, \tau_1<\eta$ such that for $\|h\|\leqq\tau_0$ and $\|g\|\leqq\tau_1$

$$\|\varphi_y'(x_0+h,y_0+g)-\varphi_y'(x_0+h,y_0)\|\leqq\varepsilon.$$

From this, (9), and (2), we have

$$\|\psi_g'(h,g)\|\leqq K\varepsilon \text{ for } \|h\|\leqq\tau_0 \text{ and } \|g\|\leqq\tau_1.$$

If therefore $\|h\|\leqq\tau_0, \|g'\|\leqq\tau_1, \|g''\|\leqq\tau_1$, then

$$\|\psi(h,g')-\psi(h,g'')\|\leqq K\varepsilon\|g'-g''\|. \tag{10}$$

It is possible to fix positive numbers $\tau_0\leqq\eta$ and $\tau_1\leqq\eta$ such that $K\varepsilon\leqq\dfrac{1}{2}$ and because of (8) $\|\psi(h,\theta)\|\leqq\dfrac{\tau_1}{2}$ for $\|h\|\leqq\tau_0$.

Then

$$\|\psi(h,g')-\psi(h,g'')\|\leqq\frac{1}{2}\|g'-g''\|,$$

so that the operator $\psi(h,g)$ yields a contracting mapping with the coefficient $\dfrac{1}{2}$ in the sphere $\|g\|\leqq\tau_1$.

If $\|g\|\leqq\tau_1$ and $g'=\psi(h,g)$ with $\|h\|\leqq\tau_0$, then furthermore

$$\begin{aligned}\|g'\| &= \|\psi(h,g)\|\leqq\|\psi(h,g)-\psi(h,\theta)\|+\|\psi(h,\theta)\|\\ &\leqq K\varepsilon\|g\|+\|\psi(h,\theta)\|\leqq\frac{1}{2}\tau_1+\frac{1}{2}\tau_1=\tau_1,\end{aligned}$$

so that by the operator $\psi(h,g)$ the sphere $\|g\|\leqq\tau_1$ is mapped into itself.

If we consider this sphere as a metric space, then the hypotheses for a contracting mapping are fulfilled (see §7; Remark 2). Therefore the sequence $g^{(n)} = g^{(n)}(h)$ converges to the solution $g = g(h)$ of (6) or (5). This solution is unique in the sphere $\|g\| \leqq \tau_1$ and depends continuously on h. We note also the inequality

$$\|g(h)\| \leqq 2\|g^{(1)}(h)\|. \tag{11}$$

Remark 1. For $\|h\| \leqq \tau_0$ it is possible to put $\tau_1 = \max(2\|g^{(1)}(h)\|, \|h\|)$; as $\|h\| \to 0$ we have $\tau_1 \to 0$, and $\varepsilon \to 0$.

Remark 2. If there exists in the neighborhood of the point (x_0, y_0) a continuous second derivative $\varphi''_{yy}(x, y) = \varphi''_{yy}(x, y)_2$ and if

$$\|\varphi''_{yy}(x, y)\| \leqq K_2, \tag{12}$$

then we obtain as the right side of the inequality (10), $\varepsilon = K_2\tau$.

If now there exists φ'_x in the η-neighborhood of the point (x_0, y_0) and if

$$\|\varphi'_x(x_0 + h, y_0)\| \leqq l \text{ for } \|h\| \leqq \eta,$$

then
$$\|\varphi(x_0 + h, y_0)\| = \|\varphi(x_0 + h, y_0) - \varphi(x_0, y_0)\| \leqq l\|h\|.$$
(2), (11) and
$$g^{(1)}(h) = -[\varphi'_y(x_0 + h, y_0)]^{-1}\varphi(x_0 + h, y_0)$$
imply
$$\|g^{(1)}(h)\| \leqq Kl\|h\|, \|g(h)\| \leqq 2Kl\|h\|.$$
Therefore, by Remark 1, $\tau_1 = m\|h\|$ with $m = \max(1; 2Kl)$, and by the fulfillment of (12) $\varepsilon = K_2 m\|h\|$.

THEOREM 2. *We assume that in the η-neighborhood of (x_0, y_0), φ'_x, φ'_y and for $k > 1$ also φ''_{yy} are bounded and continuous. If, furthermore $\varphi(x, y)$ is a k-times differentiable function in an η-neighborhood of the point (x_0, y_0) from $E \dotplus E_1$, then $y = f(x)$ is also k-times differentiable in a τ_0-neighborhood of the point $x_0 \in E$.*

Proof. We first show that $f(x)$ is a k-times differentiable function at the point x_0, i.e. that $g(h)$ is k-times differentiable for $h = \theta$. For this, it is necessary to prove the existence of a polynomial

$$g_k(h) = \sum_{l=1}^{k} a_l h^l$$

with
$$g(h) \overset{h}{\underset{k}{=}} g_k(h). \tag{13}$$

We construct successively polynomials $g_n(h)$ of degree n in h, $n = 0, 1, 2, \ldots, k$, such that

$$g(h) \overset{h}{\underset{n}{=}} g_n(h). \tag{14}$$

From the construction of the $g_n(h)$ the proof of the theorem will follow. We put $g_0(h) = \theta_1$. Let $n \leqq k$ and let the polynomial g_{n-1} be constructed: $g(h) \overset{h}{\underset{n-1}{=}} g_{n-1}(h)$. We introduce the notation

$$\begin{aligned} t_n(h) &= \psi(h, g_{n-1}(h)) \\ &= g_{n-1}(h) - [\varphi'_y(x_0 + h, y_0)]^{-1}\varphi(x_0 + h, y_0 + g_{n-1}(h)) \end{aligned} \tag{15}$$

($t^{(1)}(h)$ coincides with $g^{(1)}(h)$ from Theorem 1.

Because φ is k-times differentiable with respect to both arguments and $g_{n-1}(h)$

is a polynomial in h, then $\varphi(x_0+h, y_0+g_{n-1}(h))$, as a function of h, is also k-times differentiable with respect to h for $h \to \theta, g_{n-1}(h) \to \theta$ and $\varphi(x_0+h, y_0+g_{n-1}(h))$ becomes $\varphi(x_0, y_0) = \theta$. $\varphi_y'(x_0+h, y_0)$ is $k-1$-times differentiable. Therefore $[\varphi_y'(x_0+h, y_0)]^{-1}$ has also this property. But then $t_n(h)$ is differentiable $k \geqq n$ times. Consequently, there exists a polynomial $g_n(h)$ such that

$$t_n(h) \overset{h}{\underset{n}{=}} g_n(h) \tag{16}$$

holds.

We now show that $g(h) \overset{h}{\underset{n}{=}} t_n(h)$. From this and (16), the relation (14) follows. Since $g = g(h)$ is a solution of the equation $g = \psi(h, g)$, $g(h) = \psi(h, g(h))$, we obtain according to (10)

$$\| g(h) - t_n(h) \| = \| \psi(h, g(h)) - \psi(h, g_{n-1}(h)) \| \leqq K\varepsilon \| g(h) - g_{n-1}(h) \|. \tag{17}$$

Let $n = 1$. Then

$$\| g(h) - t_1(h) \| \leqq K\varepsilon \| g(h) - g_0(h) \| = K\varepsilon \| g(h) \| \leqq 2K^2 l\varepsilon \| h \|,$$

consequently, $$g(h) \overset{h}{\underset{1}{=}} t_1(h).$$

Let $n > 1$, then $k \geqq n \geqq 2$. By hypothesis, there exists in an η-neighborhood of the point (x_0, y_0) a bounded second derivative φ_{yy}'', and because of Remark 2 to Theorem 1 and (17),

$$\| g(h) - t_n(h) \| \leqq K_3 \| h \| \| g(h) - g_{n-1}(h) \|.$$

By hypothesis, $g(h) \overset{h}{\underset{n-1}{=}} g_{n-1}(h)$, i.e.,

$$\| g(h) - g_{n-1}(h) \| = \alpha \| h \|^{n-1}, \alpha \to 0 \text{ as } \| h \| \to 0.$$

Therefore, $$\| g(h) - t_n(h) \| \leqq \alpha K_3 \| h \|^n,$$

i.e., $$g(h) \overset{h}{\underset{n}{=}} t_n(h) \overset{h}{\underset{n}{=}} g_n(h),$$

q.e.d.

We have proved that the function $y = f(x)$ is k-times differentiable at the point $x = x_0$. If we vary x in the sphere $\| x - x_0 \| \leqq \eta$, all properties of $f(x)$ on which this result was based remain unchanged. Hence $f(x)$ is k-times differentiable on the whole sphere.

Remark. We have used boundedness of the derivatives φ_x', φ_y' and φ_y'' in the sphere $\| x - x_0 \| \leqq \eta$. For the other derivatives boundedness is not required.

Finally, we show that

$$df(x_0, h) = g_1(h) = [\varphi_y'(x_0, y_0)]^{-1} \varphi_x'(x_0, y_0) \tag{18}$$

or $$f'(x_0) = [\varphi_y'(x_0, y_0)]^{-1} \varphi_x'(x_0, y_0)$$

holds. Indeed $g_1(h)$ is a linear operator (a polynomial of first degree in h), by definition, with

$$\psi(h, g_0) = \psi(h, \theta) \overset{h}{\underset{1}{=}} g_1(h).$$

Because
$$\varphi(x_0 + h, y_0) = \varphi(x_0 + h, y_0) - \varphi(x_0, y_0) \overset{h}{\underset{1}{=}} g_1(h)$$

holds,
$$[\varphi_y'(x_0, y_0)]^{-1} \varphi(x_0 + h, y_0)$$

$$\overset{h}{\underset{1}{=}} [\varphi_y'(x_0 + h, y_0)]^{-1} \varphi_x'(x_0, y_0) \overset{h}{\underset{1}{=}} [\varphi_y'(x_0, y_0)]^{-1} \varphi_x'(x_0, y_0).$$

From this, we have (18).

§45. APPLICATIONS OF THE THEOREM ON IMPLICIT FUNCTIONS

DEPENDENCE OF THE SOLUTION OF THE EQUATION. We investigate the space $D_{U,E}$ of functions $f(x)$, defined on the subset U of E and whose range lies also in E, i.e., if $x \in U$, then $f(x) \in E$. Let the functions $f(x)$ be differentiable, and let $f(x)$, $f'(x)$ be continuous with bounded norm.

Suppose
$$\|f\| = \sup (\|f(x)\| + \|f'(x)\|).$$

Then $D_{U,E}$ is a linear normed space.

We consider the equation

$$f(x) = \theta, x \in U \text{ and } f \in D_{U,E}$$

and assume that $f(x_0) = \theta$ for an $x_0 \in U$ and that $f' \in (E \to E)$ has an inverse. Then there holds the following

THEOREM 1. *There exist constants $\tau > 0, \tau_1 > 0$ such that for an arbitrary $f_1 \in D_{U,E}$ with $\|f_1 - f\| \leqq \tau$ the equation $f_1(x) = \theta$ has a solution $x = x_0 + \Delta x$ with $\|\Delta x\| \leqq \tau_1$ and $\Delta x \to \theta$ as $f_1 \to f$.*

Proof. We look upon $f(x)$ as a function of $x \in E$ and $f \in D_{U,E}$: $f(x) = \Phi(f, x)$. Then $\Phi_x'(f, x) = f'(x)$. By hypothesis $f(x_0) = \theta$, and $[f'(x_0)]^{-1}$ exists. This means that $\Phi(f, x_0) = \theta$ and that $[\Phi_x'(f, x_0)]^{-1}$ exists. By virtue of the theorem on implicit functions we have: For every $\tau > 0$ there exists a $\tau_1 > 0$ such that the equation $\Phi(f + \Delta f, x + \Delta x) = 0$ has a solution $x_0 + \Delta x$, whenever $\|\Delta f\| \leqq \tau_1$. This means that $f_1(x) = \theta, f_1 = f + \Delta f$ and $x = x_0 + \Delta x$, where $\Delta x \to \theta$ as $\Delta f \to \theta$. This completes the proof.

We remark that $\Phi(f,x) = f(x)$ has the differential $\Phi_f'(f, x)\Delta f = \Delta f(x)$. From the existence of $\Phi_f'(f, x)$ it follows that Δx is a differentiable function of Δf and

$$\Delta x \overset{\Delta f}{\underset{1}{=}} [\Phi_x'(f, x_0)]^{-1} [\Phi_f'(f, x_0) \Delta f] = [f'(x_0)]^{-1} \Delta f(x_0).$$

The right side of this equation is a uniform differential of the function $\Delta x = \varphi(\Delta f)$ for $\Delta x = \theta$, i.e., equal to $d\varphi(\theta, \Delta f)$.

APPLICATION TO EIGENELEMENTS. We consider the direct sum $H_1 = H \dot{+} R$, where H is a real Hilbert space and R the real number line, and we designate by (x, t) the elements of H_1 with $x \in H$ and $t \in R$. Let f be a non-linear operator defined on H_1 with range in the same space.

$$\{y, t_1\} = f(x, t) \text{ with } x, y \in H \text{ and } t, t_1 \in R.$$

Let it have the special form

$$y = Ax - tx, t_1 = (x, x) - 1.$$

Here we let A be a completely continuous linear self adjoint operator from $(H \to H)$.

The equation $f(x, t) = \theta_{H_1}$ becomes

$$Ax - tx = \theta, (x, x) = 1,$$

i.e., t is an eigenvalue and x the corresponding normalized eigenelement of the operator A. If $\{x, t\}$ takes on the increment $\{\Delta x, \Delta t\}$ then $f(x, t)$ takes on the increment

$$\{A\,\Delta x - t\,\Delta x - \Delta t\, x - \Delta t\,\Delta x, 2(x, \Delta x) + (\Delta x, \Delta x)\}.$$

The linear part of this increment is

$$df(x, t; \Delta x, \Delta t) = \{A\,\Delta x - t\,\Delta x - \Delta t\, x, 2(x, \Delta x)\}.$$

$f'(x, t)$ is consequently a linear operator from $(H_1 \to H_1)$. It transforms $\{\Delta x, \Delta t\}$ into $\{(A\Delta x - t\,\Delta x - \Delta t x, 2(x, \Delta x)\}$.

If t is a simple eigenvalue, then there exists an inverse operator

$$[f'(x, t)]^{-1},$$

i.e., for every $\{z, r\} \in H_1$, the equations

$$A\,\Delta x - t\,\Delta x - \Delta t\, x = z, \quad 2(x, \Delta x) = r \tag{1}$$

have a solution $\{\Delta x, \Delta t\}$. If we put $\Delta x = ax + (\Delta x)_1$ with $a = (\Delta x, x)$ and $(\Delta x_1, x) = 0$ analogously $z = bx + z_1$ where $b = (z,x)$ and $(z_1, x) = 0$. Furthermore, since $(A - t\,E)\,x = \theta$, we have

$$(A\,\Delta x) - t\,\Delta x) - \Delta t\, x = (A - t\,E)\,(\Delta x)_1 - \Delta t\, x$$

and $$((A - t\,E)\,(\Delta x)_1, x) = ((A - t\,E)x, (\Delta x)_1) = 0.$$

From this and (1), we obtain

$$(A - t\,E)\,(\Delta x)_1 = z_1, \tag{2}$$

$$\Delta t = -b = -(z, x), 2a = r. \tag{3}$$

Since the right side of (2) is orthogonal to x (i.e., orthogonal to all eigenelements belonging to the eigenvalue t), this equation has a unique solution

$$(\Delta x)_1 = (A - t\,E)_1^{-1} z_1 = (A - t\,E)_1^{-1}\,[z - (z, x)\,x]$$

orthogonal to x. Here we have designated by $(A - t\,E)_1$, the operator $A - t\,E$ restricted to the subspace of all elements orthogonal to x.

Therefore, it is necessary for every $\{z, r\} \in H_1$ that

$$\Delta x = \frac{r}{2}\,x + (A - t\,E)_1^{-1}\,[z - (z, x)\,x], \ \Delta t = -(z, x),$$

and we verify that this satisfies (1). From this, the existence of the operator $[f'(x, t)]^{-1}$ follows.

Because of Theorem 1, there exist constants τ and τ_1 with $\|A\| \leqq \tau$, an eigenvalue $t + \Delta t$, and a normalized eigenelement $x + \Delta x$ for the operator $A + \Delta A$:

$$[(A + \Delta A) - (t + \Delta t)\,E]\,(x + \Delta x) = \theta, \ \ (x + \Delta x, x + \Delta x) = 1.$$

Here the eigenvalue and eigenelement are determined uniquely as long as $\|\Delta x\| + |\Delta t| \leqq \tau_1$.

We find the first approximation $\{\Delta x_1, \Delta t_1\}$ of $\{\Delta x, \Delta t\}$ from the equations

$$A\,\Delta x_1 - t\,\Delta x_1 - \Delta t_1 x + \Delta A\, x = \theta \text{ and } (x, \Delta x_1) = 0.$$

If we multiply the first of these equations by the scalar x and realize that

$$((A - t\,E)\,\Delta x_1, x) = ((A - t\,E)\,x, \Delta x_1) = 0$$

then we obtain
$$\Delta t_1 = (\Delta A\, x, x).$$

Furthermore
$$(A - t\,E)\,\Delta x_1 = \Delta t_1\, x - \Delta A\, x. \tag{4}$$

The equation (4) always has a solution, because its right side is orthogonal to x:

$$(\Delta t_1\, x, x) - (\Delta A\, x, x) = \Delta t_1 - (\Delta A\, x, x) = 0.$$

Consequently
$$\Delta x_1 = (A - t\,E)_1^{-1}\,[\Delta t_1 x - \Delta A\, x].$$

EQUATIONS DEPENDING ON PARAMETERS.

THEOREM 2. *Let $y = f(x, t)$ be a function of the element $x \in E$ and the numerical parameter t. Let its range be also in E. Suppose also that $f(x, t)$ is n-times differentiable with respect to x and t. Assume finally that the equation $f(x, 0) = \theta$ has a solution $x = x_0$, and that the operator $[f'_x(x_0, 0)]^{-1}$ exists. Then there exist constants $\tau > 0$ and $\tau_1 > 0$ such that for $|t| < \tau$ the equation*

$$f(x, t) = \theta \tag{5}$$

has one and only one solution x_t with $\|x_t - x_0\| \leqq \tau_1$. x_t is, as a function of t, n-times differentiable.

This is an immediate consequence of the theorem on implicit functions.

Example. Let A_t be a completely continuous linear operator from $(H \to H)$ which is n-times differentiable with respect to t. Let the operator A_0 have a simple eigenvalue λ_0 with the normalized eigenelement x_0:

$$A_0 x_0 - \lambda_0 x_0 = \theta, \quad (x_0, x_0) = 1.$$

We consider the direct sum $H \dotplus R$ and define a function $\Phi_t(x, \lambda)$ with

$$(x, \lambda) \in H \dotplus R, \quad \Phi_t(x, \lambda) \in H + R$$
$$\Phi_t(x, \lambda) = (A_t x - \lambda x,\ (x, x) - 1).$$

This equation has for a normalized eigenelement x_t and an eigenvalue λ_t of the operator A_t the form

$$\Phi_t(x, \lambda) = \theta_{H \dotplus R}\,.$$

Because A_t is n-times differentiable with respect to x, $\Phi_t(x, \lambda)$ has the same property. As in the previous section, we convince ourselves that

$$\left[\frac{\partial}{\partial x}\,\Phi_t(x_0, \lambda_0, 0)\right]^{-1}$$

exists. Thus we can apply Theorem 2. Consequently

$$\Phi_t(x, \lambda) = \theta_{H \dotplus R}.$$

i.e.,
$$A_t\, x - \lambda x = \theta,\ (x, x) = 1 \tag{6}$$

has the solution x_t. It is n-times differentiable with respect to t.

VARIATIONAL EQUATIONS. Under the hypotheses of Theorem 2, the solution $x(t)$ of (5) is a differentiable function of t. The derivative $x'(t)$ of the function $x(t)$ with respect to t for $t = 0$ is called the variation

$$\delta x = \left.\frac{dx}{dt}\right|_{t=0}$$

Correspondingly for $y = y(x, t)$

$$\delta y = \left.\frac{\partial y}{\partial t}\right|_{t=0}.$$

The equation $y(x(t), t) = \theta$ is an identity. Differentiation with respect to t yields

$$y'_x(x(t), t)\, x'(t) + \frac{\partial y(x(t), t)}{\partial t} = \theta.$$

For $t = 0$ we obtain
$$y'_x(x_0, 0)\, \delta x + \delta y = \theta. \tag{7}$$

This equation is called the equation of variation for the equation $y(x, t) = \theta$.

Because $[y'_x(x_0, 0)]^{-1}$ exists,

$$\delta x = -[y'_x(x_0, 0)]^{-1}\, \delta y. \tag{8}$$

For example, $(A - \lambda E)\Delta x + \Delta A x - \Delta\lambda x = \theta$ can be considered to be a variational equation for (6), if $\Delta A, \Delta x, \Delta\lambda$ are replaced by $\delta A, \delta x, \delta\lambda$. We then have

$$\delta x = -(A - \lambda E)_1^{-1}\, (\delta A x - \delta\lambda x), \quad \delta\lambda = (\delta A x, x).$$

APPLICATIONS TO DIFFERENTIAL EQUATIONS. We return to the differential equation
$$\frac{dx}{dt} = f(x, t) \text{ with } x(0) = x_0. \tag{9}$$

Here $x = x(t), x_0$ and $f(x, t)$ are elements of the space E. (9) is equivalent to the integral equation

$$x(t) - x_0 - \int_0^t f(x, \tau)\, d\tau = \theta. \tag{10}$$

We designate the left side of the equation (10) by $F(x(t), x_0)$. The functions $x(t)$ are from $C_1^E[0, 1]$[33], $F(x(t), x_0)$ is an operator mapping the direct sum

$$E \dot{+} C_1^E[0, 1]$$

into $C_1^E[0, 1]$. If $f(x, \tau)$ is n-times differentiable with respect to x, and if $\dfrac{\partial^n f}{\partial x^n}$ is continuous in τ, then $f(x(\tau), \tau)$ is an n-times differentiable operator $\in (C_1^E[0, 1] \to C_1^E[0, 1])$. It follows from this, that $F(x(\tau), x_0)$ is an operator n-times differentiable with respect to $x(\tau)$. Because x_0 is the only summand occurring in F, F is an n-times differentiable function in $E \dot{+} C_1^E[0, 1]$.

If $X = x(t)$ takes the increment $\Delta X = \Delta x(t)$, then

$$F'_X \Delta X = \Delta x(t) - \int_0^t f'_x(x(\tau), \tau)\, \Delta x(\tau) d\tau, \tag{11}$$

is a linear part of the increment of $F(X, x_0)$ with respect to ΔX. The right side of (11) is an operator $(C_1^E[0, 1] \to C_1^E[0, 1])$. It has an inverse. For, for every $z(t)$ in $C_1^E[0, 1]$,

$$F'_X \Delta X = z(t)$$

[33] C_1^E is the set of continuously differentiable functions $x(t)$ with $t \in [0, 1]$ and $x(t) \in E$.

or
$$\Delta x(t) - \int_0^t f'_x(x(\tau), \tau) \Delta x(\tau) d\tau = z(t)$$

is equivalent to the differential equation for $\Delta x(t)$

$$\frac{d\Delta x(t)}{dt} = f'_x(x(t), t) \Delta x(t) + z'(t).$$

The last equation has, because of the existence theorem, a uniquely determined solution (the right side is linear with respect to $\Delta x(t)$ — the Lipschitz condition with respect to Δx is automatically fulfilled). This solution is the inverse operator

$$\Delta x(t) = \Delta X = (F'_X)^{-1} z(t).$$

The hypotheses of the theorem on the implicit functions are fulfilled.

The solution $X = x(t)$ of the equation (11) can be considered as a function of the initial value x_0: $X = X(x_0)$ and $X(x_0)$ is n-times differentiable with respect to x_0.

In particular, if E is a k-dimensional space, then we obtain the theorem on the continuously differentiable dependence on the initial values of the solution.

Now let all quantities appearing in (9) be functions of a parameter ξ. Then (9) reads

$$\frac{dx_\xi}{dt} = f(x_\xi, \xi, t) \tag{9'}$$

with the initial condition $x_\xi(0) = x_{0\xi}$.

This equation can be replaced by the equivalent one:

$$F_\xi(x_\xi(t), t) = x_\xi(t) - x_{0\xi} - \int_0^t f_\xi(x_\xi(\tau), \tau) d\tau = \theta. \tag{10'}$$

Let us now assume that for $\xi = 0$ all quantities occurring in (10′) go over to the corresponding ones without the subscript ξ, i.e., (10′) becomes (10). The left side of (10′) can be considered to be an operator applied on the element $x_\xi(t)$ of $C_1^E [0, 1]$ and the numerical parameter ξ. From the existence of the inverse operator $(F'_x)^{-1}$ there follows application of variational equations. The variational equation for (10′) reads

$$\delta x(t) - \delta x_0 - \int_0^t f'_x(x(\tau), \tau) \delta x(\tau) d\tau = \theta$$

or
$$\delta x'(t) = f'_x(x(t), t) \delta x(t)$$

with the initial condition $\delta x(0) = \delta x_0$.

This is the variational equation for (9′).

Applications of the theorem on implicit function stem from G. P. Akilov.

§46. TANGENT MANIFOLDS

CASE OF DIRECT SUMS. Let the function $\varphi(x)$ map the Banach space E into the Banach space E_1: $x \in E, \varphi(x) \in E_1$. We consider the totality N of the elements x satisfying the equation

$$\varphi(x) = \theta.$$

Let $\varphi(x_0) = \theta$, i.e., $x_0 \in N$, and let the function $\varphi(x)$ be differentiable at the point x_0:

$$\varphi(x_0 + h) \underset{1}{\overset{h}{=}} \varphi'(x_0)h.$$

If the operator $\varphi'(x_0) \in (E \to E_1)$ maps the space E onto the whole space E_1, then we shall call the element x_0 *regular*.

For the following, let x_0 always be regular. The totality of the elements $h \in E$ for which

$$\varphi'(x_0)h = 0$$

holds, we shall designate by T_0. T_0 is a subspace of E.

The set of the elements $x_0 + h$ with $h \in T_0$ is called the *linear tangent manifold* T_{x_0} of N at the point x_0.

First, let E be the direct sum of T_0 and a subspace T_1. Every element $y \in E$ is therefore of the form

$$y = h + g \text{ with } h \in T_0 \text{ and } g \in T_1.$$

The linear operator $\varphi'(x_0)$ maps T_1 onto the whole of E_1. Indeed, $\varphi'(x_0)$ maps E onto the whole space E_1, therefore we can find for every $z \in E_1$ an element $y \in E$ such that

$$\varphi'(x_0)y = z.$$

But $y = h + g$ with $h \in T_0$, $g \in T_1$ and $\varphi'(x_0)h = \theta$. Hence

$$\varphi'(x_0)y = \varphi'(x_0)g = z.$$

We define a linear operator $A \in (T_1 \to E_1)$ by $Ag = \varphi'(x_0)g$. A maps, according to what we just proved, T_1 onto the whole of E_1. If $g, g_1 \in T_1$ and $Ag = Ag_1$, then $g = g_1$. For, because of

$$A(g - g_1) = \theta, \text{ i.e.}, \varphi'(x_0)(g - g_1) = \theta,$$

therefore, $g - g_1 \in T_0$. But since E is the direct sum of T_0 and T_1, $g - g_1 = \theta$, $g = g_1$.

According to the theorem of Banach, the operator A has an inverse linear operator A^{-1}.

THEOREM 1. *If E is the direct sum of the subspaces T_0 and T_1, then there exists a topological (i.e., biunique and bicontinuous) mapping, which transforms a neighborhood of X_0 in the set N and a neighborhood of the same element x_0 in the linear tangent manifold T_{x_0} into each other. Corresponding points have a distance which is small of higher order than their distance from x_0.*

Proof. In the neighborhood of x_0 the elements have the form

$$x = x_0 + h + g \text{ with } h \in T_0, g \in T_1.$$

We write the defining equation of N in the form

$$\Phi(h, g) = \varphi(x_0 + h + g) = 0. \tag{1}$$

For $h = \theta$ and $g = \theta$, (1) is fulfilled. Further, the partial derivative of the function $\Phi(h, g)$ which corresponds to the increment Δg has for $h = \theta$, $g = \theta$, the form

$$\Phi_g'(\theta, \theta)\Delta g = \varphi'(x_0)\Delta g = A\Delta g. \tag{2}$$

Operator $A = \Phi_g'(\theta, \theta)$ has an inverse. Then by the theorem on implicit functions in the neighborhood of point $h = \theta$, $g = \theta$, equation (1) is equivalent to

$$g = \psi(h),$$

where $\psi(h)$ is a differentiable function with $\psi(\theta) = \theta$. Hence we can represent every point $x \in N$ in a neighborhood of x_0 by

$$x = x_0 + h + \psi(h), h \in T_0, \psi(h) \in T_1.$$

Therefore we have constructed a mapping which associates with every point $x' = x_0 + h$ in a neighborhood of x_0 in T_{x_0} biuniquely and bicontinuously an $x'' = x_0 + h + \psi(h)$ from a neighborhood of x_0 in N. The mapping is consequently topological. From

$$\Phi'_h(\theta, \theta)h + \Phi'_g(\theta, \theta)\psi'(\theta)h = \Phi'_h(\theta, \theta)h + A\psi'(\theta)h = \theta_{E_1}$$

we obtain

$$d\psi(\theta, h) = \psi'(\theta)h = -A^{-1}\Phi'_h(\theta, \theta)h = -A^{-1}\varphi'(x_0)h = -A^{-1}\theta_{E_1} = \theta_{E_1}.$$

Hence
$$\psi(h) \underset{1}{\overset{h}{=}} \psi(\theta) + \psi'(\theta)h = \theta,$$

i.e., $\|\psi(h)\| = o(\|h\|)$. $o(\|h\|)$ becomes smaller as $\|h\| \to 0$ of higher than first order in $\|h\|$. $\|\psi(h)\|$, however, is the distance of the point $x_0 + h$ of the linear tangent manifold T_{x_0} from the corresponding point $x = x_0 + h + \psi(h)$ of N. This distance tends to 0 of higher order than $\|h\|$ or $\|h + \psi(h)\|$, i.e., than the distance of the points $x_0 + h$ and $x_0 + h + \psi(h)$ from the point of tangency x_0, q.e.d.

GENERAL CASE. In general we cannot assume the existence of a subspace T_1, such that $E = T_0 \dot{+} T_1$; as we shall see, however, Theorem 1 is valid in a bit weakened form.

We form the factor space E/T_0 of cosets with respect to T_0 (see pp. 57, 67). Every element T of the factor space E/T_0 is a set of elements of E. If h_1 and h_2 are in T, then $h_1 - h_2 \in T_0$, i.e., $\varphi'(x_0)(h_1 - h_2) = \theta$. Therefore

$$\varphi'(x_0)h_1 = \varphi'(x_0)h_2.$$

The operator $\varphi'(x_0) \in (E \to E_1)$ transforms the arbitrary elements h_1 and h_2 of T into one of E_1. If conversely

$$\varphi'(x_0)h_1 = \varphi'(x_0)h_2,$$

then
$$\varphi'(x_0)(h_1 - h_2) = \theta, \text{ i.e., } h_1 - h_2 \in T_0,$$

hence h_1 and h_2 belong to the same class T of E/T_0. Consequently, the operator $\varphi'(x_0)$ gives rise to a linear operator A which maps E/T_0 into E_1, and

$$AT = \varphi'(x_0)h,$$

if $T \in E/T_0$ and h is an arbitrary point of T. According to these considerations, AT does not depend on the choice of $h \in T$.

Let y be an arbitrarily chosen element of E_1. By hypothesis, the operator $\varphi'(x_0)$ maps the space E onto the whole of E_1. Hence we can find an element $h \in E$ such that $\varphi'(x_0)h = y$. h belongs, however, to a certain T of E/T_0. By definition, $AT = \varphi'(x_0)h = y$. The operator A has therefore an inverse A^{-1}. According to the theorem of Banach, A^{-1} is also linear.

THEOREM 2. *We can associate with every point of the set N a point of the tangent manifold T_{x_0} and, conversely, there belongs to every element of T_{x_0} one of N, such* *that the distance between corresponding points becomes small of higher order than their distance from the point of tangency (this correspondence is not unique).*

We prove this theorem in a way similar to that for implicit functions.

Proof. Let $h \in T_0$. We construct a sequence $\{T_n\}$ of elements of E/T_0 and a sequence $\{g_n\}$ of points $g_n \in T_n$: let $g_0 = \theta \in T_0$ and let all g_i, T_i with $i = 1, 2, \ldots, n-1$ be constructed. Then we define T_n and g_n as follows:

$$T_n = T_{n-1} - A^{-1}\varphi(x_0 + h + g_{n-1}) \tag{3}$$

and we select from the coset T_n a g_n where

$$\|g_n - g_{n-1}\| \leqq 2\|T_n - T_{n-1}\|.$$

Such a choice is possible because

$$\|T_n - T_{n-1}\| = \inf_{g \in T_n} \|g - g_{n-1}\|.$$

Since g_{n-1} lies in T_{n-1}, we obtain by definition of the operator A

$$AT_{n-1} = \varphi'(x_0)g_{n-1}.$$

Therefore we can rewrite (3) in the form

$$T_n = -A^{-1}[\varphi(x_0 + h + g_{n-1}) - \varphi'(x_0)g_{n-1}].$$

Since

$$T_{n-1} = -A^{-1}[\varphi(x_0 + h + g_{n-2}) - \varphi'(x_0)g_{n-2}],$$

we have

$$T_n - T_{n-1} = -A^{-1}[\varphi(x_0 + h + g_{n-1} - \varphi(x_0 + h + g_{n-2}) - \varphi'(x_0)(g_{n-1} - g_{n-2})].$$

We put

$$g_t = g_{n-2} + t(g_{n-1} - g_{n-2})$$

and get

$$\varphi(x_0 + h + g_{n-1}) - \varphi(x_0 + h + g_{n-2}) = \int_0^1 \varphi'(x_0 + h + g_t)\, dt (g_{n-1} - g_{n-2}).$$

Therefore,

$$T_n - T_{n-1} = -A^{-1} \int_0^1 [\varphi'(x_0 + h + g_t) - \varphi'(x_0)]\, dt (g_{n-1} - g_{n-2}). \tag{4}$$

Let

$$\|h\| = r, \|g_{n-2}\| \leqq r, \|g_{n-1}\| \leqq r.$$

Then

$$\|g_t\| \leqq r;$$

hence also

$$\|h + g_t\| \leqq 2r.$$

Because of the continuity of $\varphi'(x)$ at the point x_0, there exists for every $\varepsilon > 0$ a number $r > 0$ such that

$$\|\varphi'(x) - \varphi'(x_0)\| \leqq \varepsilon \text{ for } \|x - x_0\| \leqq 2r.$$

Therefore

$$\|\varphi'(x_0 + h + g_t) - \varphi'(x_0)\| \leqq \varepsilon,$$

and it follows from (4) and the definition of g_n that

$$\|T_n - T_{n-1}\| \leqq \|A^{-1}\| \int_0^1 \|\varphi'(x_0 + h + g_t) - \varphi'(x_0)\| \, dt \, \|g_{n-1} - g_{n-2}\|$$
$$\leqq \|A^{-1}\| \varepsilon \|g_{n-1} - g_{n-2}\|,$$
$$\|g_n - g_{n-1}\| \leqq 2\|T_n - T_{n-1}\| \leqq 2\|A^{-1}\| \varepsilon \|g_{n-1} - g_{n-2}\|.$$

For sufficiently small ε,

$$2\|A^{-1}\| \varepsilon \leqq \frac{1}{2},$$

therefore
$$\|g_n - g_{n-1}\| \leqq \frac{1}{2} \|g_{n-1} - g_{n-2}\|.$$

Now let $\|h\| = r$ and $\|g_i\| \leqq r$ for $i = 1, 2, \ldots, n-1$. Then

$$\|g_n - g_{n-1}\| \leqq \frac{1}{2} \|g_{n-1} - g_{n-2}\| \leqq \ldots \leqq \frac{1}{2^{n-1}} \|g_1 - g_0\| = \frac{1}{2^{n-1}} \|g_1\|$$

and
$$\|g_n\| = \|g_1 + (g_2 - g_1) + \ldots + (g_n - g_{n-1})\|$$
$$\leqq \|g_1\| + \|g_2 - g_1\| + \ldots + \|g_n - g_{n-1}\|$$
$$\leqq \|g_1\| \left(1 + \frac{1}{2} + \ldots + \frac{1}{2^{n-1}}\right) \leqq 2\|g_1\|.$$

Since $g_0 = \theta, T_1 = -A^{-1} \varphi(x_0 + h)$ and

$$\|g_1\| \leqq 2\|T_1\| \leqq 2\|A^{-1}\| \|\varphi(x_0 + h)\|.$$

Furthermore

$$\varphi(x_0 + h) = \varphi(x_0) + \varphi'(x_0)h + \varepsilon(h) \text{ with } \|\varepsilon(h)\| = o(\|h\|).$$

Now $\varphi(x_0) = \theta$ and h lies in T_0; consequently, also

$$\varphi'(x_0)h = \theta$$

and
$$\varphi(x_0 + h) = \varepsilon(h).$$

From this we have
$$\|g_1\| \leqq 2\|A^{-1}\| \|\varepsilon(h)\|. \tag{5}$$

For sufficiently small $r > 0$ and $\|h\| \leqq r$,

$$\|\varepsilon(h)\| \leqq \frac{1}{4\|A^{-1}\|} \|h\|.$$

From this we get
$$\|g_1\| \leqq \frac{1}{2} \|h\| \leqq \frac{1}{2} r$$

and finally
$$\|g_n\| \leqq r.$$

Therefore,

$$\|g_n - g_{n-1}\| \leqq \frac{1}{2} \|g_{n-1} - g_{n-2}\|$$

holds for all n.

Hence the sequence $\{g_n\}$ converges to an element $g \in E$ with $\|g\| \leqq \|h\|$ and furthermore because of (5)

$$\|g\| \leqq 2\|g_1\| \leqq 4\|A^{-1}\| \ \|\varepsilon(h)\|. \tag{6}$$

Correspondingly, the T_n of E/T_0 converge to a T of E/T_0 and $g \in T$.

Since as $n \to \infty$, $g_n \to g$ and $T_n \to T$, the equation (3) goes over to

$$T = T - A^{-1}\varphi(x_0 + h + g),$$

as

$$A^{-1}\varphi(x_0 + h + g) = \theta$$

or

$$\varphi(x_0 + h + g) = \theta.$$

Consequently

$$x_0 + h + g \in N.$$

We associate $x_0 + h \in T_{x_0}$ with this point.

The inequality (6) shows that

$$\|g\| = o(\|h\|)$$

holds, i.e., the distance $\|g\|$ between $x_0 + h \in T_{x_0}$ and the corresponding point $x_0 + h + g \in N$ becomes small of higher order than the distance $\|h\|$ from the point x_0 of tangency.

Let the point $x = x_0 + d$ lie in N, i.e.,

$$\varphi(x_0 + d) = \varphi(x_0) = \theta.$$

We obtain

$$\varphi(x_0 + d) - \varphi(x_0) = \varphi'(x_0)d + \varepsilon(d) = \theta$$

where $\|\varepsilon(d)\| = o(\|d\|)$. Therefore

$$\varphi'(x_0)d = -\varepsilon(d).$$

We designate by T the class from E/T_0 which contains d. Then $\varphi'(x_0)d = AT$ or $AT = -\varepsilon(d)$, which implies

$$T = -A^{-1}\varepsilon(d), \quad \|T\| \leqq \|A^{-1}\| \, \|\varepsilon(d)\|.$$

Among the elements of the space E which belong to T, let g be an element for which

$$\|g\| \leqq 2\|T\| \leqq 2\|A^{-1}\| \, \|\varepsilon(d)\|$$

holds. Since $g \in T$ and $d \in T$, it follows that

$$d - g \in T_0, \quad x_0 + d - g \in T_{x_0}.$$

We associate $x = x_0 + d$ with this point. For the distance $\|g\|$ of these points we obtain the approximation

$$\|g\| = o(\|d\|).$$

This completes the proof.

LOCALLY LINEAR SPACES. Connected with the concept of the tangent manifold is the concept of a *locally linear space*. It is important for certain research [40].

We consider two metric spaces X and Y, and let φ be a topological mapping of X into Y. Let $\varphi(x) \in Y$ correspond to the point $x \in X$. The mapping φ is called *almost isometric at the point* $x_0 \in X$, if the distance of two arbitrary elements x_1 and x_2 in X is connected with the distance of their images in Y by the inequality

$$\rho(x_1, x_2)(1 - \varepsilon) \leqq \rho(\varphi(x_1), \varphi(x_2)) \leqq \rho(x_1, x_2)(1 + \varepsilon)$$

where ε tends to 0 as $\rho(x_0, x_1) + \rho(x_0, x_2)$ tends to 0.

Example. Again let N be the set in E given by $\varphi(x) = \theta$, T_{x_0} the tangent manifold belonging to N at the regular point $x_0 \in N$ and $E = T_0 \dot{+} T$, where T_0 is the totality of the elements h with $\varphi'(x_0)h = \theta$. For $r > 0$ we associate with every element

$$x_0 + h \text{ of } T_{x_0} \text{ with } \|h\| \leqq r$$

the element

$$x_0 + h + \psi(h) \text{ of } N$$

(see Theorem 1). We obtain a topological mapping ψ of the neighborhood of the point θ in T_0 onto a neighborhood of x_0 in N, in which θ and x_0 are in one-to-one correspondence.

The mapping ψ is almost isometric at $\theta \in T_0$ (or $x \in N$). For, if h and h_1 are in T_0, $\|h\| \leqq r$ and $\|h_1\| \leqq r$; then the point $x = x_0 + h + \psi(h) \in N$ belongs to h, $x_1 = x_0 + h_1 + \psi(h_1) \in N$ to h_1. We have

$$\|x - x_1\| = \|h - h_1 + [\psi(h) - \psi(h_1)]\|,$$

which implies

$$\|h - h_1\| - \|\psi(h) - \psi(h_1)\| \leqq \|x - x_1\| \leqq \|h - h_1\| + \|\psi(h) - \psi(h_1)\|. \tag{7}$$

The function $\psi(h)$ has a continuous derivative $\psi(h)$ with $\psi'(\theta) = \theta$. Therefore, for $\|h\| \leqq \tau, \|\psi'(h)\| \leqq \varepsilon$. Furthermore

$$\|\psi(h) - \psi(h_1)\| = \left\| \left\{ \int_0^1 \psi'(h_1 + t(h - h_1))\, dt \right\} (h - h_1) \right\|$$

$$\leqq \int_0^1 \|\psi'(h_1 + t(h - h_1))\|\, dt \, \|h - h_1\|.$$

If

$$\|h\| + \|h_1\| = \tau,$$

then

$$\|h\| \leqq \tau \text{ and } \|h_1\| \leqq \tau$$

for $0 \leqq t \leqq 1$, and consequently

$$\|h_1 + t(h - h_1)\| \leqq \tau \text{ and } \|\psi(h) - \psi(h_1)\| \leqq \varepsilon \|h - h_1\|.$$

Therefore (7) becomes

$$\|h - h_1\|(1 - \varepsilon) \leqq \|x - x_1\| \leqq \|h - h_1\|(1 + \varepsilon).$$

This proves that our mapping is almost isometric.

We now define locally linear space.

Let a metric space X be given. If every sufficiently small neighborhood of a point $x \in X$ can be mapped almost isometrically onto a neighborhood of θ of a Banach space, then X is called *locally linear*.

The preceding example shows that *in a Banach space every set, all of whose points are regular, forms a locally linear space.*

We can extend the concept of a differential to functions defined on such spaces. The spaces of admissible functions of the various classical variational problems are locally linear. The variations of the functionals considered there are differentials of functions in locally linear spaces [11], [40].

§47. EXTREMA

We now offer applications to the calculus of variations. Let $f(x)$ be a functional defined on the space E. $f(x)$ has a *minimum* at $x_0 \in E$ (*maximum*), if $f(x) \geqq f(x_0)$, ($f(x) \leqq f(x_0)$) for all points in a neighborhood of x_0. Minima and maxima are called *extrema*.

THEOREM 1. *If the differentiable functional $f(x)$ has an extremum at x_0, then $f'(x_0) = 0$, i.e., $df(x_0, h) = 0$ for all $h \in E$.*

Proof. First of all,

$$f'(x_0)h = \frac{d}{dt} f(x_0 + th)_{t=0}.$$

But $f(x_0 + th)$ is, however, a function of t, its values are real and for $t = 0$ it assumes an extremum; therefore

$$df(x_0, h) = f'(x_0)h = \frac{d}{dt} f(x_0 + th)_{t=0} = 0.$$

Because h was an arbitrary element of E, we are through.

We now seek extrema with side conditions. Let $\varphi(x)$ be a function defined on E and with range in $E_1, x \in E$, $\varphi(x) \in E_1$. Moreover, let $f(x)$ be a functional defined on E.

$f(x)$ has a *minimum* at x_0 (*maximum*) under the side condition $\varphi(x_0) = \theta$, if $f(x) \geqq f(x_0)$, $(f(x) \leqq f(x_0))$ holds in a neighborhood of x_0 for which $\varphi(x) = \theta$.

THEOREM 2. *If $f(x)$ has a minimum at x_0 under the condition $\varphi(x) = \theta$, where x_0 is an element of N, then there exists a linear functional l defined on E_1, $l \in \bar{E}_1$ and, for the functional*

$$F(x) = f(x) - l\varphi(x),$$
$$F'(x_0) = 0, \text{ i.e., } dF(x_0, h) = 0$$

for all $h \in E$.

Proof. We show first that $df(x_0, h) = 0$ for all $h \in T_0$. Let us assume that

$$h \in T_0 \text{ and } df(x_0, h) = c \neq 0.$$

For every t there corresponds to the point $x_0 + th$, according to Theorem 2 of §46, a point

$$x_0 + th + g$$

of the set N with $\|g\| = o(\|th\|)$. According to the definition of the differential, we get

$$f(x_0 + th + g) = f(x_0) + df(x_0, th + g) + \varepsilon$$
$$= f(x_0) + f'(x_0)th + f'(x_0)g + \varepsilon = f(x_0) + ct + f'(x_0)g + \varepsilon.$$

As $t \to 0, f'(x_0)g + \varepsilon$ tends to 0 of higher order than ct. Therefore the sign of the difference

$$f(x_0 + th + g) - f(x_0)$$

coincides with that of ct, thus t and this difference change their signs simultaneously. Then, however, x cannot be an extremum of $f(x)$. Consequently, the assumption $c \neq 0$ is false.

Hence under the hypotheses of the theorem,

$$df(x_0, h) = 0$$

holds for all h with $\varphi'(x_0)h = \theta$, i.e.,

$$d\varphi(x_0, h) = \theta.$$

Therefore
$$df(x_0, h_1) = df(x_0, h_2),$$

if h_1 and h_2 belong to the same coset $T \in E/T_0$. We introduce the functional

$$\chi(T) = df(x_0, h),$$

where $h \in T$ is arbitrary. We have

$$|\chi(T)| = |df(x_0, h)| = |f'(x_0)h| \leqq \|f'(x_0)\| \|h\|,$$

and we obtain

$$|\chi(T)| \leqq \|f'(x_0)\| \|T\|,$$

by taking the greatest lower bound. Consequently, $\chi(T)$ is a linear functional on E/T_0. On the other hand,

$$T = A^{-1}y,$$

where $y \in E$ with $\varphi'(x_0)h = y$ for all $h \in T$ (see p. 204). Consequently

$$df(x_0, h) = \chi(T) = \chi(A^{-1}y) = l(y).$$

Because of $y = \varphi'(x_0)h = d\varphi(x_0, h)$

$$df(x_0, h) = l[d\varphi(x_0, h)].$$

If we put

$$F(x) = f(x) - l[\varphi(x)],$$

$$dF(x_0, h) = 0$$

follows for all $h \in E$, q.e.d.

Example. An isoperimetric problem [41]. We want to find an extremum of the functional $f_0(x)$ subject to the conditions $f_i(x_0) = 0$, $i = 1, 2, \ldots, n$. Let f_0, f_i be functionals defined on E. We consider the f_i's to be components of an n–dimensional vector $\vec{f}$ and we designate by $\vec{f}(x)$ the vector with the components $f_i(x)$, $i = 1, 2, \ldots, n$. Let there be an extremum at the point $x_0 \in E$. According to Theorem 2, there exists a linear functional $l(\vec{f})$ such that $dF(x_0, h) = 0$ for $F(x) = f_0(x) - l[\vec{f}(x)]$. In n–dimensional space of vectors $\vec{f}$, however,

$$l(\vec{f}) = \sum_{i=1}^{n} \lambda_i f_i,$$

holds where the λ_i are constants. Thus

$$F(x) = f_0(x) - \sum_{i=1}^{n} \lambda_i f_i(x).$$

Therefore, at point x_0

$$f_0'(x_0) - \sum_{i=1}^{n} \lambda_i f_i'(x_0) = 0.$$

We thus have the theorem regarding the Euler multipliers.

APPENDIX

I. AUXILIARY INEQUALITIES

We say that a measurable function $x(t)$ defined on the closed interval $[0, 1]$ belongs to the class $L_p(0, 1)$, or otherwise is a function with integrable th$p$ power, if

$$\int_0^1 |x(t)|^p \, dt < \infty.$$

The integral is to be interpreted in the sense of Lebesgue; p means a positive number.

In the following, we assume $p \geqq 1$. If $p = 1$, then we have the class of integrable functions, which is designated by $L(0, 1)$.

We prove that, if $x(t) \in L_p(0, 1)$ and $y(t) \in L_p(0, 1)$, then

$$x(t) + y(t) \in L_p(0, 1).$$

For two numbers a and b,

$$|a + b| \leqq |a| + |b|.$$

We investigate two cases:

1) If $|a| > |b|$, then

$$|a + b| \leqq 2|a|$$

and

$$|a + b|^p \leqq 2^p |a|^p \leqq 2^p (|a|^p + |b|^p).$$

2) If $|b| \geqq |a|$, then

$$|a + b|^p \leqq 2^p |b|^p \leqq 2^p (|a|^p + |b|^p).$$

In either case we obtain

$$|a + b|^p \leqq 2^p (|a|^p + |b|^p).$$

We put $a = x(t), b = y(t)$ and have

$$|x(t) + y(t)|^p \leqq 2^p (|x(t)|^p + |y(t)|^p).$$

Because

$$\int_0^1 |x(t)|^p \, dt < \infty \text{ and } \int_0^1 |y(t)|^p \, dt < \infty,$$

also

$$\int_0^1 |x(t) + y(t)|^p \, dt < \infty,$$

q.e.d.

Let there be given the set of number sequences $x = \{\xi_i\}$ with

$$\sum_{i=1}^{\infty} |\xi_i|^p < \infty.$$

We denote this set by l_p. As in the preceding case, we convince ourselves that if $x \in l_p$ and $y \in l_p$, then

$$x+y\in l_p,$$

i.e., if

$$x=\{\xi_i\}, y=\{\eta_i\}, \sum_{i=1}^{\infty}|\xi_i|^p<\infty \text{ and } \sum_{i=1}^{\infty}|\eta_i|^p<\infty$$

then

$$\sum_{i=1}^{\infty}|\xi_i+\eta_i|^p<\infty.$$

THE HÖLDER INEQUALITY. In many investigations, the Buniakovskii inequality is used. In what follows we prove a generalization of this inequality due to Hölder.

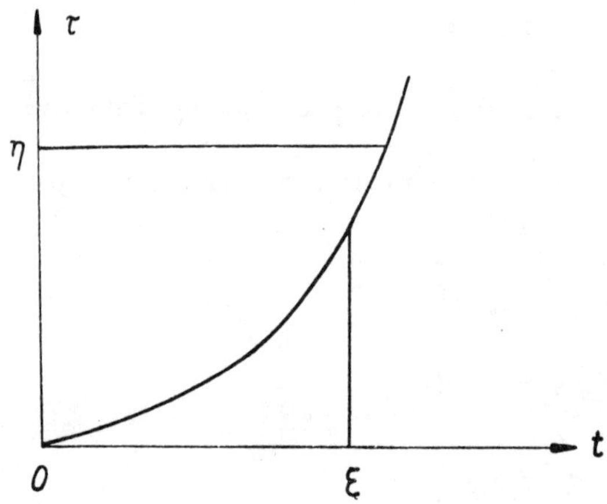

Fig. 4

We consider the function $\tau=t^a$ with $a>0$. Since $\tau'=$ $=at^{a-1}>0$ for $t>0$, $\tau=t^a$ is an increasing function for positive t. For these t, the function $t=\tau^{1/a}$ is defined.

We represent the function $\tau=t^a$ graphically, choose two real positive numbers ξ and η, mark the corresponding points on the t- and τ-axes, respectively, and draw straight lines parallel to the axes.

We obtain two "triangles" bounded by the curved lines and their areas are given by

$$S_1=\frac{\xi^{a+1}}{a+1} \text{ and } S_2=\frac{\eta^{\frac{1}{a}+1}}{\frac{1}{a}+1}.$$

On the other hand, it is clear that

$$S_1+S_2\geqq\xi\eta$$

and the equals sign holds only for $\eta=\xi^a$. Thus we have

$$\xi\eta\leqq\frac{\xi^{a+1}}{a+1}+\frac{\eta^{\frac{1}{a}+1}}{\frac{1}{a}+1}.$$

We put $a+1=p$ and $\frac{1}{a}+1=q$. Then p and q are connected by

$$\frac{1}{p}+\frac{1}{q}=1.$$

Such numbers p and q are called conjugates of each other. Obviously, if $p>1$, then $q>1$. Thus we obtain for arbitrary ξ and η and for all pairs of conjugates p and q

$$\xi\eta \leqq \frac{\xi^p}{p} + \frac{\eta^q}{q}. \tag{1}$$

We take functions $x(t) \in L_p(0, 1)$ and $y(t) \in L_q(0, 1)$ and put

$$\xi = \frac{|x(t)|}{\left(\int_0^1 |x(t)|^p \, dt\right)^{\frac{1}{p}}} \quad \eta = \frac{|y(t)|}{\left(\int_0^1 |y(t)|^q \, dt\right)^{\frac{1}{q}}}.$$

With these definitions inserted in (1), we find

$$\frac{|x(t)|\,|y(t)|}{\left(\int_0^1 |x(t)|^p \, dt\right)^{\frac{1}{p}} \left(\int_0^1 |y(t)|^q \, dt\right)^{\frac{1}{q}}} \leqq \frac{|x(t)|^p}{p\int_0^1 |x(t)|^p \, dt} + \frac{|y(t)|^q}{q\int_0^1 |y(t)|^q \, dt}.$$

On the right side we have integrable functions. Hence the left side of the inequality is integrable. We integrate and obtain

$$\frac{\int_0^1 |x(t)|\,|y(t)| \, dt}{\left(\int_0^1 |x(t)|^p dt\right)^{\frac{1}{p}} \left(\int_0^1 |y(t)|^q dt\right)^{\frac{1}{q}}} \leqq \frac{1}{p} + \frac{1}{q} = 1$$

or

$$\int_0^1 |x(t)\, y(t)| \, dt \leqq \left(\int_0^1 |x(t)|^p \, dt\right)^{\frac{1}{p}} \left(\int_0^1 |y(t)|^q \, dt\right)^{\frac{1}{q}}. \tag{2}$$

This is the Hölder inequality for integrals. For the special case $p = q = 2$, it represents the Buniakovskii inequality.

Now let $x = \{\xi_i\}, y = \{\eta_i\}$, with $x \in l_p, y \in l_q$. We substitute

$$\xi = \frac{|\xi_i|}{\left(\sum_{i=1}^{\infty} |\xi_i|^p\right)^{\frac{1}{p}}} \quad \text{and} \quad \eta = \frac{|\eta_i|}{\left(\sum_{i=1}^{\infty} |\eta_i|^q\right)^{\frac{1}{q}}}$$

into inequality (1). There results

$$\frac{|\xi_i|\,|\eta_i|}{\left(\sum_{i=1}^{\infty} |\xi_i|^p\right)^{\frac{1}{p}} \left(\sum_{i=1}^{\infty} |\eta_i|^q\right)^{\frac{1}{q}}} \leqq \frac{|\xi_i|^p}{p\sum_{i=1}^{\infty} |\xi_i|^p} + \frac{|\eta_i|^q}{q\sum_{i=1}^{\infty} |\eta_i|^q}.$$

Summing over i, we arrive at the Hölder inequality for sums

$$\sum_{i=1}^{\infty} |\xi_i \eta_i| \leqq \left(\sum_{i=1}^{\infty} |\xi_i|^p\right)^{\frac{1}{p}} \left(\sum_{i=1}^{\infty} |\eta_i|^q\right)^{\frac{1}{q}}. \tag{3}$$

For $p = q = 2$ it becomes the Buniakovskii inequality for sums.

THE MINKOWSKI INEQUALITY. If $x(t)$ and $y(t)$ belong to the space $L_p(0, 1)$, then

$$\left(\int_0^1 |x(t)+y(t)|^p\,dt\right)^{\frac{1}{p}} \leqq \left(\int_0^1 |x(t)|^p\,dt\right)^{\frac{1}{p}} + \left(\int_0^1 |y(t)|^p\,dt\right)^{\frac{1}{p}}. \tag{4}$$

For the proof, we first remark:

If $z(t) \in L_p(0, 1)$, then $|z(t)|^{p-1} \in L_q(0, 1)$. In fact,

$$(|z(t)|^{p-1})^q = |z(t)|^{(p-1)\frac{p}{p-1}} = |z(t)|^p.$$

Consequently, $(|z(t)|^{p-1})^q$ is an integrable function.

We, furthermore, investigate the integral

$$\int_0^1 |x(t)+y(t)|^p\,dt.$$

Applying the Hölder inequality twice to the functions $|x(t)+y(t)|^{p-1} \in L_q(0, 1)$ and $x(t) \in L_p(0, 1)$, $y(t) \in L_p(0, 1)$, we obtain

$$\begin{aligned}
&\int_0^1 |x(t)+y(t)|^p\,dt \\
&\leqq \int_0^1 |x(t)+y(t)|^{p-1}|x(t)|\,dt + \int_0^1 |x(t)+y(t)|^{p-1}|y(t)|\,dt \\
&\leqq \left(\int_0^1 |x(t)+y(t)|^{(p-1)q}\,dt\right)^{\frac{1}{q}} \left(\int_0^1 |x(t)|^p\,dt\right)^{\frac{1}{p}} \\
&+ \left(\int_0^1 |x(t)+y(t)|^{(p-1)q}\,dt\right)^{\frac{1}{q}} \left(\int_0^1 |y(t)|^p\,dt\right)^{\frac{1}{p}} \\
&= \left(\int_0^1 |x(t)+y(t)|^p\,dt\right)^{\frac{1}{q}} \left[\left(\int_0^1 |x(t)|^p\,dt\right)^{\frac{1}{p}} + \left(\int_0^1 |y(t)|^p\,dt\right)^{\frac{1}{p}}\right].
\end{aligned}$$

We divide both sides by

$$\left(\int_0^1 |x(t)+y(t)|^p\,dt\right)^{\frac{1}{q}}$$

and observe that $1 - \dfrac{1}{q} = \dfrac{1}{p}$. Thus we have obtained the Minkowski inequality (4) for integrals.

Now let $x = \{\xi_i\} \in l_p$ and $y = \{\eta_i\} \in l_p$. We obtain

$$\left(\sum_{i=1}^{\infty} |\xi_i+\eta_i|^p\right)^{\frac{1}{p}} \leqq \left(\sum_{i=1}^{\infty} |\xi_i|^p\right)^{\frac{1}{p}} + \left(\sum_{i=1}^{\infty} |\eta_i|^p\right)^{\frac{1}{p}}. \tag{5}$$

Analogously to the first case, we show that $z = \{\zeta_i\} \in l_p$ implies

$$z' = \{|\zeta_i|^{p-1}\} \in l_p.$$

We investigate
$$\sum_{i=1}^{\infty} |\xi_i + \eta_i|^p.$$

Applying the Holder inequality twice to the sequences $\{|\xi_i + \eta_i|^{p-1}\} \in l^q$ and $\{\xi_i\} \in l_p$, $\{\eta_i\} \in l_p$, we have

$$\sum_{i=1}^{\infty} |\xi_i + \eta_i|^p \leqq \sum_{i=1}^{\infty} |\xi_i + \eta_i|^{p-1} |\xi_i| + \sum_{i=1}^{\infty} |\xi_i + \eta_i|^{p-1} |\eta_i|$$
$$\leqq \left(\sum_{i=1}^{\infty} |\xi_i + \eta_i|^{(p-1)q}\right)^{\frac{1}{q}} \left[\left(\sum_{i=1}^{\infty} |\xi_i|^p\right)^{\frac{1}{p}} + \left(\sum_{i=1}^{\infty} |\eta_i|^p\right)^{\frac{1}{p}}\right]$$
$$= \left(\sum_{i=1}^{\infty} |\xi_i + \eta_i|^p\right)^{\frac{1}{q}} \left[\left(\sum_{i=1}^{\infty} |\xi_i|^p\right)^{\frac{1}{p}} + \left(\sum_{i=1}^{\infty} |\eta_i|^p\right)^{\frac{1}{p}}\right].$$

Dividing both sides by
$$\left(\sum_{i=1}^{\infty} |\xi_i + \eta_i|^p\right)^{\frac{1}{p}}$$
and noting that $1 - \dfrac{1}{q} = \dfrac{1}{p}$, we find the Minkowski inequality (5) for sums.

As a concluding remark, we note that the equality signs in formulas (4) and (5) hold only if almost everywhere in $[0, 1]$,
$$y(t) = k\,x(t) \text{ with } k > 0$$
and
$$\eta_i = k\xi_i \text{ with } k > 0 \text{ for } i = 1, 2, 3, \ldots$$
respectively. We shall prove this for (4) (the proof for (5) is analogous).

First, the equals sign holds in (1) only if $\eta = \xi^a$, as already mentioned. Consequently, the equals sign holds in (2) only if almost everywhere in $[0, 1]$, $|x(t)| = k\,|y(t)|^{p-1}$, because in the deduction of (2) from (1), we can take instead of $|x(t)|$ and $|y(t)|$ proportional magnitudes. It now follows from the deduction of (4) that the equals sign holds only if
$$|x(t) + y(t)| = k_1\,|x(t)|$$
and
$$|x(t) + y(t)| = k_2\,|y(t)|$$
i.e.,
$$|x(t)| = k\,|y(t)|$$
and
$$\operatorname{sign} x(t) = \operatorname{sign} y(t).$$
This means, however, that almost everywhere
$$x(t) = k\,y(t) \text{ with } k > 0,$$
q.e.d.

II. TWO DEFINITIONS OF THE nTH DERIVATIVE OF A FUNCTION OF REAL VARIABLES

There are two definitions for the nth derivative $\alpha^{(n)}(t)$ of a function $\alpha(t)$ at the point t.

1. We introduce the notation

$$\Delta^n_{\Delta t}\alpha(t) = \sum_{k=0}^{n} (-1)^{n-k} \binom{n}{k} \alpha\left(t + \left(k - \frac{n}{2}\right)\Delta t\right)$$

and call $\Delta^n_{\Delta t}\alpha(t)$ a *central difference of nth order* of $\alpha(t)$. Then let

$$\alpha^{(n)}(t) = \lim_{\Delta t \to 0} \frac{1}{(\Delta t)^n} \Delta^n_{\Delta t}\alpha(t)$$

provided this limit exists.

If the right side of this equality tends uniformly to $\alpha^{(n)}(t)$, on the interval $a \leqq t \leqq b$, then $\alpha^{(n)}(t)$ is called *uniform difference derivative of nth order.*

2. We now define the nth derivative of the function $\alpha(t)$ designated by $\alpha^{(n)}(t)_0$. $\alpha^{(n)}(t)_0$ is defined by differentiating the function $\alpha(t)$ successively n times, provided that all preceding derivatives $\alpha'(t)_0, \alpha''(t)_0, \ldots, \alpha^{(n-1)}(t)_0$ exist in a neighborhood of the point t.

Let $\alpha^{(n)}(t)_0$ be defined and continuous on the interval $a \leqq t \leqq b$. Then $\alpha^{(n)}(t)$ also exists and

$$\alpha^{(n)}(t) = \alpha^{(n)}(t)_0.$$

Proof. As we see immediately,

$$\frac{1}{(\Delta t)^n}\Delta^n_{\Delta t}\alpha(t) = \alpha^{(n)}(t + \theta\Delta t)_0, \; -\frac{1}{2} < \theta < \frac{1}{2}.$$

Since the right side tends uniformly to $\alpha^{(n)}(t)_0$ as $\Delta t \to 0$, it follows that

$$\alpha^{(n)}(t) = \lim_{\Delta t \to 0} \frac{1}{\Delta (t)^n} \Delta^n_{\Delta t}\alpha(t) = \alpha^{(n)}(t)_0.$$

We can also prove the converse:

The existence of the continuous uniform difference derivative $\alpha^{(n)}(t)$ on a closed interval implies the existence of the nth derivative $\alpha^{(n)}(t)_0$ and

$$\alpha^{(n)}(t)_0 = \alpha^{(n)}(t).$$

We carry out the proof for $n = 2$. Assume the function $\varphi(t)$ has a continuous second difference derivative $\varphi''(t)$ on the interval $[0, 1]$. We designate by $\psi(t)$ the second iterated integral

$$\psi(t) = \int_0^t \int_0^\tau \varphi''(\tau)d\tau d\tau_1.$$

The second derivative $\psi''(t)_0$ equals the integrand $\varphi''(t)$. Since $\psi''(t)_0 = \varphi''(t)$ is continuous,

$$\psi''(t)_0 = \psi''(t)$$

according to the preceding.

Hence

$$[\psi(t) - \varphi(t)]'' = 0.$$

Now we show that the difference $\psi(t) - \varphi(t)$ can only be linear in t:

$$\varphi(t) = \psi(t) + a + bt.$$

Since the function $\psi(t) + a + bt$ has a second derivative

$$\psi''(t)_0 = \varphi''(t)$$

this implies

$$\varphi''(t)_0 = \varphi''(t).$$

Now let $\alpha''(t) = 0$ on the interval $[0, 1]$. We put

$$\alpha_1(t) = \alpha(t) - \{\alpha(0) + [\alpha(1) - \alpha(0)]t\}.$$

We now have

$$\alpha_1(0) = \alpha_1(1) = 0 \text{ and } \alpha_1''(t) \equiv 0.$$

Let ε be an arbitrary positive number. We consider the function

$$\beta(t) = \alpha_1(t) + \varepsilon t^2.$$

Because $(\varepsilon t^2)'' = 2\varepsilon > 0, \beta''(t) = 2\varepsilon > 0$. For $0 \leqq t \leqq 1$, $\beta(t) \leqq \varepsilon$. If $\beta(t) > \varepsilon$ for some $t \in [0, 1]$, then the maximim of $\beta(t)$ would be greater than ε. Since, however, $\beta(0) = 0$ and $\beta(1) = \varepsilon$, this maximum is assumed for an interior point of $[0, 1]$. At t_0 the second central difference is

$$\Delta^2_{\Delta t}\beta(t_0) = \beta(t_0 + \Delta t) - 2\beta(t_0) + \beta(t_0 - \Delta t) \leqq 0$$

because

$$\beta(t_0) \geqq \beta(t_0 \pm \Delta t).$$

This implies

$$\beta''(t_0) = \lim_{\Delta t \to 0} \Delta^2_{\Delta t}\, \beta(t) \leqq 0$$

and thus we have obtained a contradiction to $\beta''(t_0) > \varepsilon > 0$.

Consequently,

$$\beta(t) \leqq \varepsilon, \ 0 \leqq t \leqq 1.$$

For $0 \leqq t \leqq 1$, also

$$\alpha_1(t) = \beta(t) - \varepsilon t^2 \leqq \beta(t) \leqq \varepsilon.$$

Analogously, one proves the validity of the inequality

$$\alpha_1(t) \geqq -\varepsilon.$$

Hence, we obtain for arbitrary $\varepsilon (\varepsilon > 0)$

$$-\varepsilon \leqq \alpha_1(t) \leqq \varepsilon.$$

This implies

$$\alpha_1(t) \equiv 0$$

and finally

$$\alpha(t) = \alpha(0) + [\alpha(1) - \alpha(0)]\, t = a + bt,$$

q.e.d.

In conclusion, for a function $\varphi(t_1, \ldots, t_n)$ of n real variables, we have

$$\Delta^{(n)}_{t_1, t_2, \ldots, t_n; \Delta t}\, \varphi\,(t_1, t_2, \ldots, t_n) =$$

$$= (\Delta t)^n \left\{\frac{\partial}{\partial t_k}\left[\cdots \frac{\partial}{\partial t_{kn}}\, \varphi(t_1 + \theta_1 \Delta t, \ldots, t_n + \theta_n \Delta t) \ldots\right]\right\}, \qquad \text{(A)}$$

where

$$\Delta^{(n)}_{t_1, t_2, \ldots, t_n; \Delta t}\varphi(t_i) = \sum_{i_1, i_2, \ldots, i_k} (-1)^{n-k} f(t_i'),$$

$t_i' = t_i + \Delta t$ for $i = i_1, i_2, \ldots, i_k$ and is equal to t_i for the remaining i's, and the summation has to be taken over all subsets $(i_1, \ldots, i_k)$, $0 \leqq i_1 \leqq i_2 \leqq \ldots \leqq i_k \leqq n$ of the set $(1, 2, \ldots, n)$. The θ_i are numbers between 0 and 1.

The equality is proved by induction. For $n = 1$, it is the mean value theorem. Assume it is valid for partial differences of $(n - 1)$st order. Then

$$\Delta^{(n)}_{t_1, t_2, \ldots, t_n; \Delta t} \varphi(t_1, t_2, \ldots, t_n) = \Delta^{(n-1)}_{t_1, \ldots, t_{k_1-1}, t_{k_1+1}, \ldots, t_n; \Delta t} \varphi(t_1, \ldots, t_{k_1} + \Delta t, \ldots, t_n)$$

$$- \Delta^{(n-1)}_{t_1, \ldots, t_{k_1-1}, t_{k_1+1}, \ldots, t_n; \Delta t} \varphi(t_1, \ldots, t_{k_1}, \ldots, t_n)$$

$$= \Delta t \frac{\partial}{\partial t_{k_1}} [\Delta^{(n-1)}_{t_1, \ldots, t_{k_1-1}, t_{k_1+1}, \ldots t_n; \Delta t} \varphi(t_1, \ldots, t_{k_1} + \theta_{k_1} \Delta t, \ldots, t_n)] \qquad \text{(B)}$$

(this is obtained by application of the mean value theorem). By induction hypothesis, we obtain

$$\Delta^{(n-1)}_{t_1, \ldots, t_{k_1-1}, t_{k_1+1}, \ldots, t_n; \Delta t} (t_1, \ldots, t_{k_1} + \theta_{k_1} \Delta t, \ldots, t_n)$$

$$= (\Delta t)^{n-1} \frac{\partial}{\partial t_{k_2}} \left\{ \cdots \frac{\partial}{\partial t_{kn}} \right\} \varphi(t_1 + \theta_1 \Delta t, t_2 + \theta_2 \Delta t, \ldots, t_n + \theta_n \Delta t).$$

This and (B) imply (A).

BASIC BIBLIOGRAPHY

for Functional Analysis and Related Subjects

[1] Aleksandrov, P. S. *Introduction to the General Theory of Sets and Functions.* (Russian) Gostekhizdat, 1948.

[2] Akhiezer, H. I. *Lectures on the Theory of Approximations.* (Russian) Gostekhizdat, 1947.

[3] Akhiezer, H. I. and Glazman, I. M. *Theory of Linear Operators in Hilbert Space.* (Russian) Gostekhizdat, 1950.

[4] Akhiezer, H. I. and Krein, M. G. *Some Problems from the Theory of Moments.* (Russian) *DNTVU*, 1938.

[5] Banach, S. *A Course in Functional Analysis.* Kiev, 1948 (Ukranian).

[6] Gel'fand, I. M., Raikov, D. A. and Shilov, G. E. *Commutative Normed Rings.* (Russian) *Uspekhi Math. Nauk* (*UMN*) 1 (1946).

[7] Ditkin, V. A. *Calculus of Operators* (Russian) *UMN*, т. II, vyp. 6 (22), 1947.

[8] Kantorovich, L. V. *Functional Analysis and Applied Mathematics.* (Russian) *UMN*, т. III, vyp. 6 (28) 1948.

[9] Kantorovich, L. V., Vulikh, B. Z. and Pinsker, A. G. *Functional Analysis in Partially Ordered Spaces.* (Russian) Gostekhizdat, 1958.

[10] Krein, M. G. and Rutman, M. A. *Linear Operators that Leave a Cone Invariant in Banach Space.* (Russian) *UMN*, т. III, vyp. 1 (23), 1948.

[11] Lavrent'ev, M. A. and Liusternik, L. A. *Foundations of the Calculus of Variations.* (Russian) т. I, ch II *ONTI*, 1935.

[12] Liusternik, L. A. *Fundamental Concepts in Functional Analysis.* (Russian) *UMN*, vyp. I, 1936.

[13] Liusternik, L. A. *Topology of Functional Spaces and Calculus of Variations in the Large.* (Russian) *Trudy MIAN im. V. A. Steklova*, vyp. XIX, 1947.

[14] Neumark, M. A. 1. *Rings with Involution.* (Russian) *UMN*, т. III, vyp. 5 (27), 1948.
2. *Operator Rings in Hilbert Space.* (Russian) *UMN*, т. IV, vyp. 4 (32), 1949.

[15] Nemytskii, V. V. *The Fixed Point Method in Analysis.* (Russian) *UMN*, vyp. I, 1936.

[16] Petrovskii, I. G. *Lectures on the Theory of Integral Equations.* (Russian) Gostekhizdat, 1951.

[17] Plessner, A. I. *Spectral Theory of Linear Operators, I.* (Russian) *UMN*, vyp. IX, 1941.

[18] Plessner, A. I. and Rokhlin, V. A. *Spectral Theory of Linear Operators, II.* (Russian) *UMN*, т. I, vyp, I (11), 1946.

[19] Pontriagin, L. S. *Continuous Groups.* (Russian) M.–L., 1938.

[20] Radon, I. *On Linear Functional Transformations and Functional Equations.* (Russian) *UMN*, vyp. I, 1936.

[21] Riesz, F. *On Linear Functional Equations.* (Russian) *UMN*, vyp. I, 1936.

[22] Riesz, F. *Linear Operators in Complex Hilbert Space.* (Russian) *UMN*, vyp. IX, 1941.

[23] Riesz, F. *On Functions of Hermitian Operators in Hilbert Space.* (Russian) *UMN*, vyp. IX, 1941.

[24] Riesz, F. *On Some Fundamental Concepts in the General Theory of Linear Functionals.* (Russian) *UMN*, т. I, vyp. 2 (12), 1946.

[25] Sobolev, S. L. *Some Applications of Functional Analysis in Mathematical Physics.* (Russian) Izd. LGU, 1950.

[26] Smirnov, V. I. *A Course in Higher Mathematics, Vol.* 5. (Russian) Gostekhizdat, 1947.

[27] Hadamard, J. *Sur les operations fonctionnelles, C. R.*, 1903.

[28] Akilov, G. P. (Russian) *Nauchnaya sessiya Leningradscogo gos. universitata im. A. A. Zhdanova*, 1950, 14–18.

[29] Artemenko, A. P. *The General Form of a Linear Functional in a Space of Functions of Bounded Variation.* (Russian) *Matem. sborn.* 6 (48), 1939, 215–220.

[30] Bezikobich, A. S. *Almost Periodic Functions.* Cambridge, 1932.

[31] Bohr, H. *Almost Periodic Functions.* (Russian) GTTI., 1934.

[32] Volterra, V. *Sopra le funzioni che dipendono de altri funzioni, Rend. Acad. Lincei*, 6 Vol. VI.

[33] Hilbert, D. *Grundzüge einer allgemeinen Theorie der linearen Integralgleichungen*, 1924.

[34] Grosberg, U. I. *On Linear Functionals in the Space of Functions of Bounded Variation.* (Ukranian) *Uchënye zapiski pedagogicheskogo instituta, Kiev*, т. 2 (1939), 17–23.

[35] Kolmogorov, A. N. *Studia Math.* 5, 1934, 29–33.

[36] Krein, M. G. *Fundamental Properties of Normal Conical Sets in Banach Space.* (Russian) *DAN* 28 (1940), 13–17. *On the Minimal Decomposition of a Linear Functional into Positive Components.* (Russian) *DAN* 28 (1940), 18–22.

[37] Lappo-Danilevskii. *Theory of Functions of Matrices and Systems of Linear Differential Equations.* (Russian) GTTI., 1934.

[38] Levitan, B. *Some Questions from the Theory of Almost Periodic Functions, I.* (Russian) *UMN*, т. II, vyp. 5 (21), 1947.

[39] Liusternik, L. A. *Fundamental Concepts in Functional Analysis.* (Russian) *UMN*, vyp. I, 1936.

[40] Liusternik, L. A. *DAN* (1940), 771–774, 775–778.

[41] Liusternik, L. A. *Matem. sbornik* 41 (1934), 390–401.

[42] Markov, A. A. *On Mean Values and Exterior Densities.* (English) *Math Sbornik* 4 (1938), 165–191.

[43] Natanson, I. P. *Constructive Theory of Functions.* (Russian) Gostekhizdat, 1949.

[44] Nagy, B. *Spektraldarstellung linearer Transformationen des Hilbertschen Raumes*, Berlin, 1942.

[45] Pontriagin, L. S. *Über stetige algebraische Körper*, *Annals of Math.* 33, 1932, 163–174.

[46] Riesz, F. *Sur les operations fonctionnelles linearies*, *C. R.* 1909.

[47] Riesz, F., *C. R*, 1910.

[48] Riesz, F. *Untersuchungen über Systeme integrierbarer Funktionen*, *Math. Ann.*, т. 69, 1910.

[49] Sigalov, A. G. DAN, т. LXX, 5, 1950.

[50] Sukhomlinov, G. A. *On Continuation of Linear Functionals into Complex and Linear Quaternion Space.* (Russian) *Matem. sbornik* т. 3 (45), 1938.

[51] Fréchet, M. *Sur les opérations linéaires*, *J. AM. Math. Soc.* v. V, 1904 and v. VI, 1905.

Fréchet, M. *Sur quelques points du calcul fonctionnel. Rend. del. Cir. Mat. di Palermo*, XXII, 1906.

INDEX